Workshop on Biomedical Natural Language Processing

(BioNLP 2015)

Co-located with ACL-IJCNLP 2015

Beijing, China
30 July 2015

ISBN: 978-1-5108-0943-7

Printed from e-media with permission by:

Curran Associates, Inc.
57 Morehouse Lane
Red Hook, NY 12571

Some format issues inherent in the e-media version may also appear in this print version.

Printed by Curran Associates, Inc. (2015)

Association for Computational Linguistics
209 N. Eighth Street
Stroudsburg, Pennsylvania 18360

Phone: 1-570-476-8006
Fax: 1-570-476-0860

acl@aclweb.org

Additional copies of this publication are available from:

Curran Associates, Inc.
57 Morehouse Lane
Red Hook, NY 12571 USA
Phone: 845-758-0400
Fax: 845-758-2634
Email: curran@proceedings.com
Web: www.proceedings.com

ACL-IJCNLP 2015

BioNLP 2015
Workshop on Biomedical Natural Language Processing

Proceedings of the Workshop

July 30, 2015
Beijing, China

Production and Manufacturing by
Taberg Media Group AB
Box 94, 562 02 Taberg
Sweden

Introduction

BioNLP 2015 received 24 high quality submissions, continuing the fine tradition of the preceding thirteen years of BioNLP. The high quality of the submissions ensured that 12 of those were accepted as full papers / oral presentations and 11 as short papers / poster presentations. The themes in this year's papers and posters show equal interest in clinical text and in biological language processing. The morning session and the keynote presentations focus on the latest developments in biomedical text processing, whereas the afternoon session will present innovations in clinical text processing. This year, researchers continue advancing pathway, event and relation extraction from the literature and information extraction from clinical text, as well as continuing research in languages other than English.

Keynotes

The DARPA Big Mechanism Program

Kevin Knight

DARPA's Big Mechanism Program aims to develop automatic machine-reading technology to distill grounded, causal mechanisms from technical literature, and to assemble those mechanisms into a large, operational model. The first Big Mechanism domain is cancer biology. This talk will describe the goals of the program and the techniques being developed.

Kevin Knight is a Senior Research Scientist and Fellow at the University of Southern California's Information Sciences Institute, and a Professor in the Computer Science Department at USC. He received a Ph.D. in computer science from Carnegie Mellon University and a bachelor's degree from Harvard University. His research interests include natural language processing, statistical modeling, machine translation, language generation, and code breaking.

Machine Reading: Attempting to model and understand biological processes

Christopher Manning
Stanford University

Machine reading calls for programs that read and understand textual descriptions, whereas most current work only attempts to extract atomic facts, often from redundant web-scale corpora. Biological processes are an example of complex phenomena involving a series of events that are connected to one another through various relationships. This work focuses on these processes as a reading comprehension task that requires complex reasoning over a single document. The input is a paragraph describing a biological process, and the goal is to answer questions that require an understanding of the relations between entities and events in the process. To answer questions, we first try to extract from the paragrah a rich structure representing the events of the biological process and relations between them. We represent processes by graphs whose edges describe a set of causal and co-reference event-event relations, and characterize the structural properties of these graphs, so as to be able to better predict them from text descriptions. Then, we map the question to a formal query, which is executed against the extracted structure. We demonstrate that answering questions about Freshman biology via predicted structures substantially improves accuracy over baselines that use shallower representations. This is joint work with Jonathan Berant, Vivek Srikumar, Peter Clark, and other project members.

Christopher Manning is a Professor of Computer Science and Linguistics at Stanford University. His Ph.D. is from Stanford in 1995, and he held faculty positions at Carnegie Mellon University and the University of Sydney before returning to Stanford. He is an ACM Fellow, a AAAI

Fellow an ACL Fellow, and he has coauthored leading textbooks on statistical approaches to natural language processing (Manning and Schuetze 1999) and information retrieval (Manning, Raghavan, and Schuetze, 2008), as well as linguistic monographs on ergativity and complex predicates. His recent work has concentrated on machine learning approaches to various NLP problems, including statistical parsing, named entity recognition, robust textual inference, machine translation, recursive deep learning models for NLP, and large-scale joint inference for NLP.

Overview of BioCreative V Challenge Tasks

Zhiyong Lu

Critical Assessment of Information Extraction in Biology (BioCreative) is a community-wide effort for evaluating text mining and information extraction systems applied to the biological domain. For the past ten years BioCreative challenges have spanned a number of tasks from named entity recognition, to relation extraction, to assisted biocuration. BioCreative V in 2015 is currently underway and consists of five different tracks. In this talk, I will give an overview of each track and show how they are aimed to advance text-mining research and provide practical benefits to real-world applications such as biocuration. Information about BioCreative is available at `www.biocreative.org`

BioCreative 2015 Organizing Committee: `http://biocreative2015.org/organizers`

Zhiyong Lu is Earl Stadtman investigator at NCBI, part of the National Library of Medicine/NIH, where he leads the biomedical text mining research group. His research focuses on developing computational methods for analyzing and making sense of natural language data in biomedical literature and clinical text. Several of his recent research has been successfully adopted in PubMed/PMC and other community resources like SwissProt. Dr. Lu is an Associate Editor for BMC Bioinformatics and serves on the editorial board for the Journal Database. He is also an organizer of the BioCreative challenge. `http://irp.nih.gov/pi/zhiyong-lu`

Acknowledgments

The greatest debt owed by the organizers of a workshop like this is to the authors who graciously continue choosing BioNLP as the venue to share their truly inspired research that resulted in the work submitted for consideration. The next-biggest debt is, without question, to the program committee members (listed elsewhere in this volume) who continue the long-standing tradition of producing three reviews per paper on a tight review schedule and with an admirable level of insight.

Organizers:

Kevin Bretonnel Cohen, University of Colorado School of Medicine
Dina Demner-Fushman, US National Library of Medicine
Sophia Ananiadou, National Centre for Text Mining and University of Manchester, UK
Jun-ichi Tsujii, National Institute of Advanced Industrial Science and Technology, Japan

Program Committee:

Emilia Apostolova, DePaul University, Chicago, USA
Eiji Aramaki, University of Tokyo
Sabine Bergler, Concordia University, Canada
Olivier Bodenreider, National Library of Medicine
Aaron Cohen, Oregon Health and Science University
Kevin Bretonnel Cohen, University of Colorado School of Medicine
Dina Demner-Fushman, US National Library of Medicine
Marcelo Fiszman, National Library of Medicine
Filip Ginter, University of Turku
Cyril Grouin, LIMSI - CNRS, France
Antonio Jimeno Yepes, IBM, Melbourne Area, Australia
Halil Kilicoglu, National Library of Medicine
Jin-Dong Kim, Database Center for Life Science, Japan
Robert Leaman, National Library of Medicine
Zhiyong Lu, National Library of Medicine
Timothy Miller, Children's Hospital Boston
Makoto Miwa, Toyota Technological Institute, Japan
Aurelie Neveol, LIMSI - CNRS, France
Naoaki Okazaki, Tohoku University
Jong Park, KAIST
Thomas Rindflesch, National Library of Medicine
Kirk Roberts, National Library of Medicine
Andrey Rzhetsky, University of Chicago
Yoshimasa Tsuruoka, University of Tokyo, Japan
Karin Verspoor, The University of Melbourne, Australia
John Wilbur, National Library of Medicine
Pierre Zweigenbaum, LIMSI - CNRS, France

Invited Speakers:

Christopher Manning, Stanford University
Kevin Knight, Information Sciences Institute, University of Southern California
Zhiyong Lu, National Library of Medicine

Table of Contents

Conference Program

Thursday, July 30

08:00–08:20 ***Welcome to BioNLP 15***

08:20–10:20 **Reading biomedical literature**

08:20–08:40 *Complex Event Extraction using DRUM*
James Allen, Will de Beaumont, Lucian Galescu and Choh Man Teng

08:40–09:00 *Making the most of limited training data using distant supervision*
Roland Roller and Mark Stevenson

09:00–09:20 *An extended dependency graph for relation extraction in biomedical texts*
Yifan Peng, Samir Gupta, Cathy Wu and Vijay Shanker

09:20–09:40 *Event Extraction in pieces:Tackling the partial event identification problem on unseen corpora*
Chrysoula Zerva and Sophia Ananiadou

09:40–10:00 *Extracting Biological Pathway Models From NLP Event Representations*
Michael Spranger, Sucheendra Palaniappan and Samik Ghosh

10:00–10:20 *Shallow Training is cheap but is it good enough? Experiments with Medical Fact Coding*
Ramesh Nallapati and Radu Florian

10:30–11:00 ***Coffee Break***

11:00–11:45 ***Keynote: "Machine Reading: Attempting to model and understand biological processes" - Christopher Manning***

11:45–12:30 ***Keynote: "The DARPA Big Mechanism Program" - Kevin Knight***

12:30–14:00 ***Lunch Break***

Thursday, July 30 (continued)

14:00–15:00 Poster Session

15:00–15:30 ***Invited Talk: "Overview of BioCreative V Challenge Tasks" - Zhiyong Lu***

15:30–16:00 ***Coffee Break***

16:00–18:00 Clinical text processing

Thursday, July 30 (continued)

18:00 **Closing remarks**

Posters

Lexical Characteristics Analysis of Chinese Clinical Documents
Meizhi Ju, Haomin Li and Huilong Duan

Using word embedding for bio-event extraction
Chen Li, Runqing Song, Maria Liakata, Andreas Vlachos, Stephanie Seneff and Xiangrong Zhang

Measuring the readability of medical research journal abstracts
Samuel J. Severance and K. Bretonnel Cohen

Translating Electronic Health Record Notes from English to Spanish: A Preliminary Study
Weisong Liu and Shu Cai

Automatic Detection of Answers to Research Questions from Medline Abstracts
Abdulaziz Alamri and Mark Stevenson

A preliminary study on automatic identification of patient smoking status in unstructured electronic health records
Jitendra Jonnagaddala, Hong-Jie Dai, Pradeep Ray and Siaw-Teng Liaw

Restoring the intended structure of Hungarian ophthalmology documents
Borbála Siklósi and Attila Novák

Evaluating distributed word representations for capturing semantics of biomedical concepts
MUNEEB TH, Sunil Sahu and Ashish Anand

Investigating Public Health Surveillance using Twitter
Antonio Jimeno Yepes, Andrew MacKinlay and Bo Han

Clinical Abbreviation Disambiguation Using Neural Word Embeddings
yonghui wu, Jun Xu, Yaoyun Zhang and Hua Xu

Representing Clinical Diagnostic Criteria in Quality Data Model Using Natural Language Processing
Na Hong, Dingcheng Li, Yue Yu, Hongfang Liu, Christopher G. Chute and Guoqian Jiang

Complex Event Extraction using DRUM

James Allen[1,2] jallen@ihmc.us
Will de Beaumont[1] wbeaumont@ihmc.us
Lucian Galescu[1] lgalescu@ihmc.us
Choh Man Teng[1] cmteng@ihmc.us

[1]Institute for Human and Machine Cognition, 40 S. Alcaniz Street, Pensacola FL 32502, USA
[2]Department of Computer Science, University of Rochester, Rochester NY 14627, USA

Abstract

Complex mechanisms, such as cell-signaling pathways, consist of many highly interconnected components, yet they are often described in disconnected fragmentary ways. The goal of DRUM (Deep Reader for Understanding Mechanisms) is to develop a system that can read papers and combine results of individual studies into a comprehensive explanatory model. A first step is to automatically extract relevant events and event relationships from the literature. This paper describes initial steps in extending an existing general deep language understanding system, TRIPS, to read biomedical papers. In a preliminary evaluation, our system was the best performing system among the participants, achieving results close to human expert performance. These results suggested that our system is viable for complex event extraction and, ultimately, understanding complex systems and mechanisms.

1. Introduction

Complex mechanisms consist of many highly interconnected components, yet they are often described in disconnected fragmentary ways. Examples include ecosystems, social dynamics and signaling networks in biology. The study of these complex systems is often focused on a small portion of a mechanism at a time. In addition, the huge volume of scientific literature makes it difficult to track the fast developments in the field to achieve a comprehensive understanding of the often distant and convoluted interactions in the system.

The goal of the DRUM (Deep Reader for Understanding Mechanisms) project is to develop a system that can read papers and combine research results of individual studies into a comprehensive explanatory model of a complex mechanism. The system will automatically read scientific papers, extract relevant new model fragments, and compose them into larger models that will expose the interactions and relationships between disparate elements in the mechanism.

A first step towards this goal is to automatically extract relevant events and event relationships from the literature. In this paper we will describe initial steps in extending an existing general deep language understanding system, TRIPS (Allen et al, 2008), to the genre of scientific writing, in particular in the biomedical domain. Events in biomedical research papers are described in a highly specialized and technical language, with complex formulations and nested constructions. We will discuss adaptations made and how the design principles of TRIPS facilitate such adaptations.

We will report on an experimental evaluation of this extended system on extracting events and relationships centered on the Ras signaling pathways from a number of text passages in scientific papers. Our system was the best performing system among those evaluated, achieving results close to human expert performance.

Admittedly this was a small and preliminary evaluation. However, the results suggested our system is viable for complex event extraction. Of note, unlike typical statistical approaches, we did not train on text describing the Ras signaling pathways (or on any other text for that matter). Our results were achieved using a general deep language understanding system, with little domain-specific customization beyond the recognition of named entities and some specialized vocabularies. Most important, our goal does not stop at the surface extraction of events, as is the case for many existing bio-event extraction tasks. With a general deep language understanding system, we are in a good position to develop an *understanding* of the underlying connections in complex models, and the methods developed to achieve that understanding will be readily transferrable to domains other than biology.

2. The TRIPS Architecture

Much recent text processing work has focussed on developing "shallow", statistically driven techniques. TRIPS takes a different approach, using statistical methods as a preprocessing step to provide guidance to a deep parsing system that uses a detailed, hand-built, grammar of English with a rich set of semantic restrictions. Figure 1 shows an overview of the system architecture. In

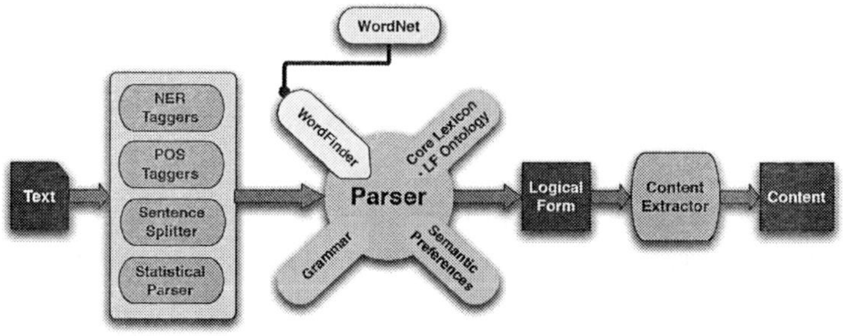

Figure 1. System Architecture.

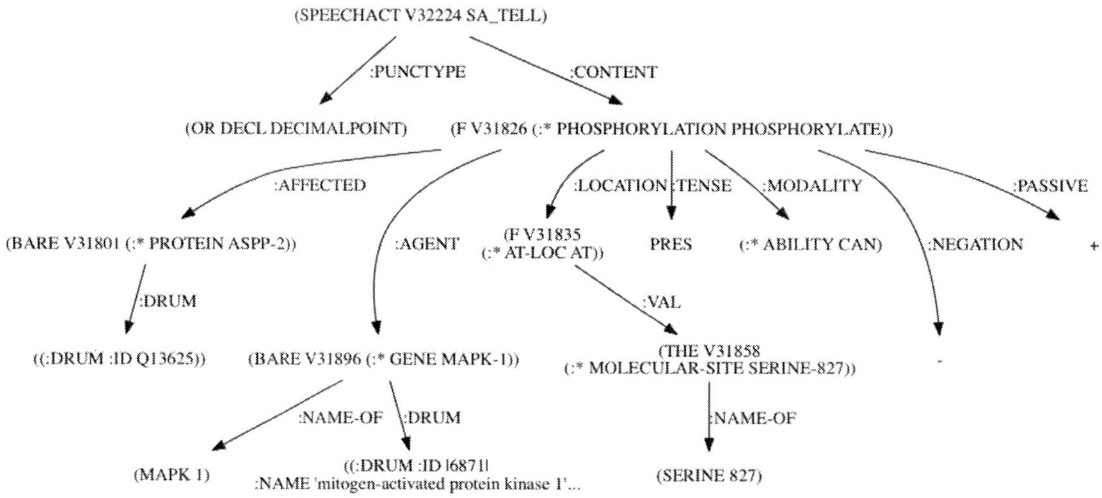

Figure 2. The logical form produced by DRUM of the sentence "*ASPP2 can be phosphorylated at serine 827 by MAPK1.*"

the rest of this section we will describe briefly the main components of the system. The content extractor, customized for biomedical text, will be discussed in more detail in Section 4.

2.1. Parser

The TRIPS grammar is a lexicalized context-free grammar, augmented with feature structures and feature unification. The grammar is motivated by X-bar theory (Jackendoff, 1977), and draws on principles from GPSG (Gazdar et al., 1985), for example head and foot features, and HPSG (Pollard and Sag, 1987, 1994). The search in the parser is controlled by a set of hand-built preferences encoded as weights on the rules, together with domain-general selectional restrictions (encoded in the lexicon and ontology) to eliminate semantically anomalous sense combinations.

The TRIPS parser uses a packed-forest chart representation and builds constituents bottom-up using a best-first search strategy similar to A*, based on rule and lexical weights and the influences of the front end components (described below).

The parser constructs from the input a logical form, which is a semantic representation that captures an unscoped modal logic (Allen, 1995; Manshadi et al., 2008). The logical form includes the surface speech act, semantic types, semantic roles for predicate arguments, and dependency relations. Consider the sentence:

> *ASPP2 can be phosphorylated at serine 827 by MAPK1.*

Figure 2 is a graphical depiction of the logical form of this sentence produced by DRUM. The nodes in the graph represent the word senses and ontology types, together with quantification information, and the edges represent semantic role relations. Of particular interest are two of the core semantic roles: AGENT (here, *MAPK1*), identifying objects that play a causal role in an event; and AFFECTED (here, *ASPP2*), identifying objects that are changed as part of an event. Other roles also provide key information that needs to be extracted. For instance, LOCATION identifies the molecular site (here, *serine 827*) or cellular location (e.g., *cytoplasm*) associated with an event of interest. The logical form also captures tense, modality and aspect information, which is crucial for determining, for example, whether a statement is a stated fact, a conjecture or a possibility (as indicated by the modality).

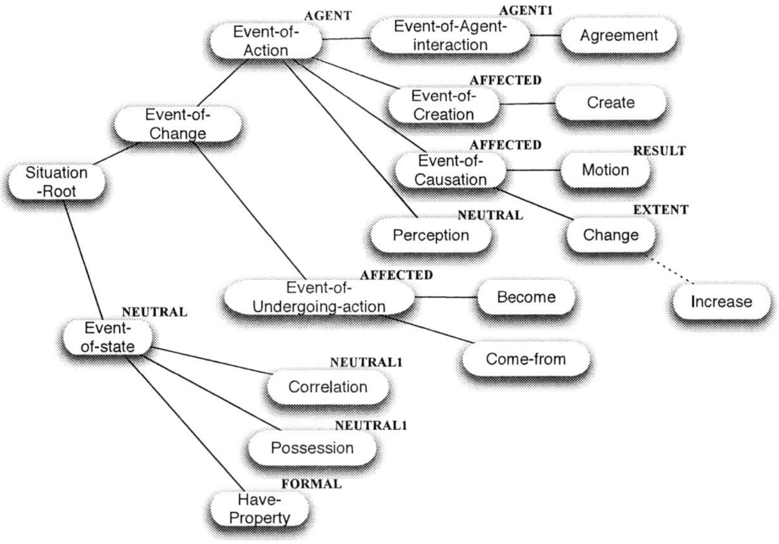

Figure 3. A subset of the TRIPS upper event ontology, showing core roles

2.2. Ontology and Lexicon

The parser draws on a general purpose semantic lexicon and ontology which define a range of word senses and lexical semantic relations. The core semantic lexicon was constructed by hand and contains approximately 7,500 lemmas (generating approximately three times that many words) and 2,000 concepts in the ontology.

The ontology is organized hierarchically and each ontology concept has associated with it possible semantic roles and selectional preferences that further refine the concept. For instance, it can be specified that the AFFECTED role of *phosphorylate* may only take a physical object that is part of a molecule (e.g., a protein or a molecular site). Figure 3 shows a portion of the TRIPS upper ontology for events and their associated core semantic roles.

A TRIPS lexicon entry is composed of two key parts. The first is the ontology type of the word sense, and it receives the roles and restrictions specific to its ontology type together with those inherited from its ancestors in the ontology hierarchy. The second is the grammatical constructions that the word can participate in, in the form of rules that map syntactic patterns to instantiations of objects from the ontology.

2.3. Extending the Lexicon

To attain broad lexical coverage beyond its hand-defined core lexicon, TRIPS uses input from a variety of external resources, some of which will be described in the next sections. Using the built-in subsystem *WordFinder,* TRIPS can augment its lexicon by dynamically building lexical entries with plausible semantic and syntactic structures for virtually any word in WordNet (Fellbaum, 1998), thus extending its coverage to over 100,000 words.

For words not in the core lexicon, WordFinder uses a hand-built mapping between the hypernym information in WordNet (for all the WordNet senses) and the TRIPS ontology. For each identified TRIPS class it gathers all the possible constructions that words of this class in the TRIPS lexicon participate in. It then generates a set of lexical entries for the unknown word by combining each possible ontological class with each possible construction for that class. While this procedure may over-generate, the key is to include the correct constructions among the generated possibilities, since these correct constructions will be the ones realized in parsing sentences (for more information see Allen, 2014).

2.4. Front End Components

To support more robust processing and domain configurability, the core system has the capability to incorporate a variety of statistical and symbolic natural language processing components in the front end, as well as domain-specific components such as specialized named entity recognizers. These include several off-the-shelf natural language tools such as the Shlomo Yona sentencizer[1], the Stanford part-of-speech tagger (Toutanova and Manning, 2000), the Stanford named-entity recognizer (NER) (Finkel et al., 2005) and the Stanford Parser (Klein and Manning, 2003). The output of these and other spe-

[1] http://search.cpan.org/~kimryan/Lingua-EN-Sentence-0.29/

Resource	Entities	References
BRENDA Tissue Ontology	tissues, cell types, cell lines	Gremse et al., 2011
Cell Ontology (CL)	cell types	Diehl et al., 2011
Chemical Entities of Biological Interest (ChEBI)	chemicals, molecule types, cell components	Degtyarenko et al., 2008
Gene Ontology (GO)	molecular functions, biological processes, pathways, cell components, macromolecular complexes	Ashburner et al., 2000
HUGO Gene Nomenclature (HGNC)	genes	Gray et al., 2015
Medical Subject Headings (MeSH®), Supplementary Concept Records (SCR)	drugs and chemicals	Lipscomb, 2000
neXtProt	cell lines, protein families	Gaudet et al., 2015
Pfam	protein families	Finn et al., 2014
Proteomics Standards Initiative for Molecular Interaction (PSI-MI)	molecular interactions, molecule type, macromolecular complexes, genes and proteins, biological roles, units of measurement	Hermjakob et al., 2004
UniProtKB (Swiss-Prot)	proteins	Uniprot Consortium, 2014

Table 4. Sources of domain-specific terminology/concepts and the types of entities incorporated into the TRIPS ontology

cialized preprocessors (e.g., a street address recognizer) is sent to the parser as advice. The parser then decides whether to follow these pieces of advice as it searches for the optimal parse of the sentence.

3. Extensions and Customization for the Biomedical Genre

We describe below several extensions to the general TRIPS system to better handle the text characteristics of the biomedical literature.

3.1. Genre Specialization

The chart produced by the parser is searched using a dynamic programming algorithm to find the least cost sequence of constituents according to a cost table that can be varied by genre. For instance, in dialogue systems speech acts such as CONFIRM (e.g., ok) or GREET (e.g., hello) are expected. For papers in the biomedical domain, such speech acts almost never occur and thus are discounted in favor of TELL statements. Similarly, in dialogue systems utterances are expected to be fairly short and colloquial, whereas in scientific text the sentence structures are expected to be much more formal and involved. The parameters for parsing and the cost table are set accordingly.

In addition, the system can choose to incorporate different front end components. For instance, for the biomedical literature a street address recognizer would not be very useful, but a named entity recognizer for protein names would be most crucial.

Such customizations not only optimize the parser efficiency, but also reduce the potential ambiguities the parser has to deal with, since each additional component offers additional, potentially conflicting, advice the parser has to take into account.

3.2. Lexicon and Ontology Enhancements

The biomedical domain uses specific terminology that is outside the core TRIPS lexicon and ontology. We extended the system's coverage by incorporating domain-specific terminology, with mappings to TRIPS ontology classes. In some cases we introduced new ontology categories to accommodate domain-specific concepts. Table 4 lists the resources used, as well as the types of entities mapped to the TRIPS ontology. Some of these resources organize concepts in ontologies (e.g., using the OBO format (Smith et al., 2007)); for these, we grafted the relevant nodes onto the TRIPS ontology (see Blaylock et al., 2011). For example, most GO biological processes are mapped to the existing TRIPS ontology category `ONT::event-of-change`; however, children of `GO:0007165` (signal transduction) are names/types of signaling pathways, and they are mapped to `ONT::signaling-pathway`—a domain-specific category newly added to the TRIPS ontology. Controlled vocabularies for single entity types (e.g., neXtProt's Cellosaurus) were mapped to single TRIPS ontology types (e.g., `ONT::cell-line`).

In addition, we used the SPECIALIST lexicon (McCray et al., 1994) for obtaining syntactic category information about domain-specific lexical items, which is helpful during parsing; however, since SPECIALIST does not include semantic information, the lexical entries are not mapped into the TRIPS ontology.

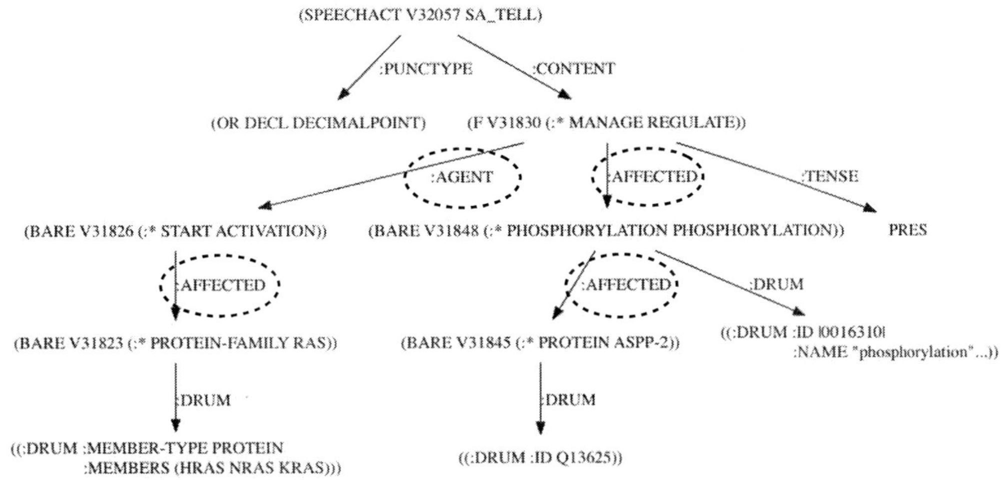

(EVENT ONT::V31830 ONT::**REGULATE** :AGENT ONT::V31826 :AFFECTED ONT::V31848 :TENSE W::PRES :START 0 :END 46)

(EVENT ONT::V31826 ONT::**ACTIVATE** :AFFECTED ONT::V31823 :START 0 :END 15)

(EVENT ONT::V31848 ONT::**PHOSPHORYLATION** :AFFECTED ONT::V31845 :DRUM ((:DRUM :ID GO::|0016310| :NAME "phosphorylation"...)) :START 25 :END 46)

(TERM ONT::V31823 ONT::PROTEIN-FAMILY :NAME W::**RAS** :DRUM ((:DRUM :MEMBER-TYPE ONT::PROTEIN :MEMBERS (HRAS NRAS KRAS))) :START 0 :END 4)

(TERM ONT::V31845 ONT::PROTEIN :NAME W::**ASPP-2** :DRUM ((:DRUM :ID UP::Q13625)) :START 25 :END 31

Figure 5. The logical form of "*RAS activation regulates ASPP2 phosphorylation.*" and the events and terms extracted by DRUM.

3.3. Specialized Constructions

The TRIPS component *WordFinder* can construct lexical entries for words not explicitly found in the core lexicon, using a mapping between WordNet and the TRIPS ontology. This mechanism provides broad coverage of words in general use. However, certain "everyday" words have specialized usage in biology. For instance, "association" is not just a vague relationship but a specific kind of binding between molecules. Some other words are used in idiosyncratic constructions. For instance, "the protein localizes to the nucleus", which means the protein exists in the nucleus, required a novel syntactic template (and semantic characterization). These words pose particular difficulties for our system as our automatically derived general constructions would be inadequate. For such cases we often have to provide hand-tailored lexical entries with appropriate syntactic templates and semantic restrictions to distinguish the everyday and biological senses of the words.

3.4. Handling Nominalizations

Nominalization is prevalent in the biomedical genre (see for instance the example sentence in Figure 5). The TRIPS parser has a general mechanism for handling verb nominalizations. This is enabled by the fact that the ontological information is identical between the verbal and nominal forms of the same event (e.g., *dominate* and *dominance*). The only difference between verbal and nominal forms is the grammatical linking rules involved. For instance, for verbal forms the subject of a certain verb might map to the AGENT role, and the direct object to the AFFECTED role. In nominalizations, the possessive would map to the role identified with the subject of the verbal form, and the object of an *of* prepositional phrase would map to the role identified with the direct object of the verb. While there are a number of different constructions used with nominals, they appear to be generic across the entire set of nominalizations, and a set of a dozen or so generic rules is all that is needed. In addition, virtually all adjunct modifications (e.g., *for three hours*) apply equally well to both verbal and nominal forms using the same adverbial modification rules in the grammar.

4. Event Extraction

From the logical forms produced by the extended TRIPS parser we need to extract the events and event relationships of interest. Because much of the variation expected in sentence constructions is handled by the extended TRIPS system, we are

```
rule-activate (40): ACTIVATE(AGENT, AFFECTED) ← [ONT::start ONT::start-object] (AGENT, AFFECTED)
rule-decrease (20): DECREASE(AGENT, AFFECTED) ← [ONT::decrease] (AGENT, AFFECTED)
```

Figure 6. Specification of the extraction rules for two event types

able to use a relatively compact specification for defining the events and relationships of interest, while coping with fairly complex and nested formulations.

4.1. An Example

Consider the sentence:

> *RAS activation regulates ASPP2 phosphorylation.*

whose logical form is depicted in Figure 5. There are three events in this sentence: the central *regulation* event and two nested events, *activation* and *phosphorylation,* that serve as the arguments to the *regulation* event. The extractions of the three events are also shown in Figure 5, together with the two terms, *RAS* and *ASPP2*, involved in the events. Note that the word "activation" is mapped to the TRIPS ontology type ONT::start. It is this ontology type that triggers the extraction rule for an ACTIVATE event (see Figure 6).

In addition to the AGENT and AFFECTED roles, the :DRUM slot provides DRUM-specific grounding information about the events and entities, mostly derived from bio-resources (see Section 3.2). For example, UP::Q13625 is the UniProt identifier for the protein *ASPP2*.

4.2. Extraction Rule Specification

We capitalized on the TRIPS ontology and parser to develop a compact and easy-to-maintain specification of event extraction rules. Instead of having to write one rule to match each keyword/ phrase that could signify an event, many of these words/phrases have already been systematically mapped to a few types in the TRIPS ontology, using a combination of the TRIPS internal lexicon and the WordFinder component which allows us to attain the coverage of WordNet. For instance, *accumulate, gain, amplify, multiply, boost, double*, among others, all map to the TRIPS ontology type ONT::increase.

In addition, the parser handles various surface structures, and the logical form returned contains normalized semantic roles. For example,

> *RAS activates RAF*
> *RAF is activated by RAS*
> *The activation of RAF by RAS*
> *Activated RAF*
> *RAF activation*

all are parsed into the same basic logical form with the semantic roles AFFECTED: *RAF* and, where applicable, AGENT: *RAS*. Thus, we needed very few (often only one) extraction rule specifications for each event type, covering a wide range of words and syntactic patterns.

Finally, since most events of interest are events of action, the usage patterns of these event words are often essentially identical, modulo the ontology types that signify the events and (less often) the semantic roles that correspond to the event arguments. We generated these rules using a module with standardized rule components, parameterized by only the event-specific ontology types and semantic role mappings. For example, *X activates / decreases / regulates / phosphorylates Y,* though denoting different events, all exhibit the same basic structure with the main semantic roles AGENT and AFFECTED. Complements denoting for example molecular sites and cellular locations for the most part retain the same structure across event types.

Figure 6 shows the stylized specification of two event types, ACTIVATE and DECREASE. The ACTIVATE line is read as follows:

> name of rule: activate
> priority of rule: 40
> name of event to be extracted: ACTIVATE
> semantic role 1: AGENT
> semantic role 2: AFFECTED
> ontology types: ONT::start; ONT::start-object

where the rule priority determines which rule is selected when multiple rules apply, and the ontology types are those in TRIPS that map to the target event type (here, ACTIVATE). The semantic roles may have further constraints on the types that can fill these roles. For instance, only molecular and cellular participants (e.g., proteins, chemicals, nucleus) are of interest in the context of biological events.

Note the similarity between the information for ACTIVATE and DECREASE. The only difference between the two lines is the ontology types that represent the respective event types (ONT::start, ONT::start-object for the former and ONT::decrease for the latter). This compact representation makes it easy to specify, maintain and update the extraction rules.

These rules were developed from general principles rather than based on specific training samples on the Ras signaling pathways. They were subsequently augmented as we learned more about specific biological usages. Although we do base our rules on the biological literature, we emphasize that neither the extraction rules described above nor any of the domain-specific

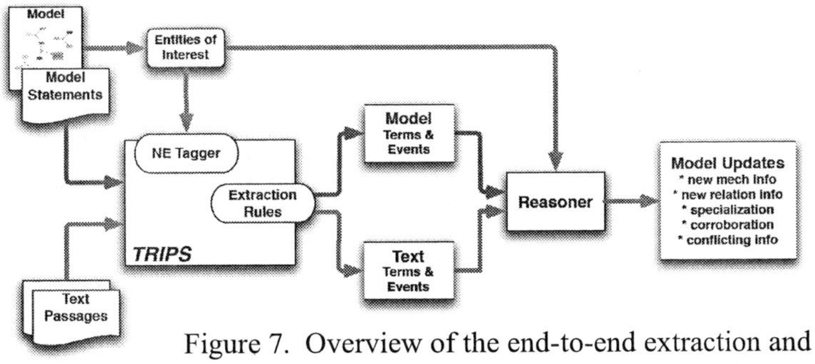

Figure 7. Overview of the end-to-end extraction and reasoning system.

We and others have recently shown that ASPP2 can potentiate RAS signaling by binding directly via the ASPP2 N-terminus [2,6]. Moreover, the RAS-ASPP interaction enhances the transcription function of p53 in cancer cells [2]. Until now, it has been unclear how RAS could affect ASPP2 to enhance p53 function. We show here that ASPP2 is phosphorylated by the RAS/Raf/MAPK pathway and that this phosphorylation leads to its increased translocation to the cytosol/nucleus and increased binding to p53, providing an explanation of how RAS can activate p53 pro-apoptotic functions (Figure 5). Additionally, RAS/Raf/MAPK pathway activation stabilizes ASPP2 protein, although the underlying mechanism remains to be investigated.

Figure 8. Example text passage for evaluation.

enhancements to our system discussed in Section 3 are specific to the language or mechanisms describing the Ras signaling pathways. Thus, we expect our system to have comparable performance on any input describing bio-molecular mechanisms.

5. System Evaluation

We participated in a preliminary evaluation of event extraction, in the context of "reading with a model". A biological model was given in BioPAX (Demir et al., 2010), BEL (Selventa, 2011), and English. Given a set of text passages from scientific papers on the Ras signaling pathways, the goal was to extract from these passages events (and their arguments) that were relevant to the given model and make explicit the links between the extracted events and the model.

BioPAX and BEL do not have the linguistically motivated features and expressivity needed for our approach. To minimize hand coding and to create a uniform system, we created our initial model by reading and processing sentences simplified from the given English model sentences, using the same process as for reading and extracting information from the test passages. The model entities and events such processed were then compared to the entities and events extracted from the text passages. Figure 7 shows an overview of the automated end-to-end extraction and reasoning system.

Two types of events were distinguished here: mechanistic (e.g., X binds to Y) and regulatory/causal relationship (e.g., X increases Y). These were further classified with respect to the given model as: 1) new mechanism and 2) new relationship not in the model; 3) specialization and 4) corroboration of information in the model; and 5) conflict with the model. In addition, each result was to be accompanied by the supporting source text.

The reasoner aligned the extracted entities using their standardized identifiers (e.g., UniProt, HUGO, Gene Ontology). In addition, we derived the relationships between the model and text extractions based on the hierarchical organization of the event types. For instance, a regulation event subsumes a stimulation event, and thus "X regulates Y" corroborates "X stimulates Y" and the latter is a specialization of the former.

6. Results

Several passages, mainly from the results and discussion sections of two scientific papers, were selected as evaluation inputs. An example passage, from (Godin-Heymann et al., 2013), is shown in Figure 8.

The extractions and model comparisons were manually scored by a third party, based on the combined answers provided by two separate teams of biologists (30 events) and the addition of 5 events adopted from system submissions (see below). In "lenient" scoring for precision, incomplete results and results that were correct but irrelevant were excluded, whereas in "strict" scoring these results were counted as incorrect.

Eleven systems of varying degrees of automation participated in the evaluation. We have available only the lenient scores of other teams, as shown in Figure 9. For lenient scoring our system was the best performing system and our performance was close to human performance.

Note that although the human biologists had high precision, there was considerable non-overlap between the answers they provided. This accounted for the approximately 0.50 recall for either of the human teams, using the pooled answers of the two teams as the gold standard.

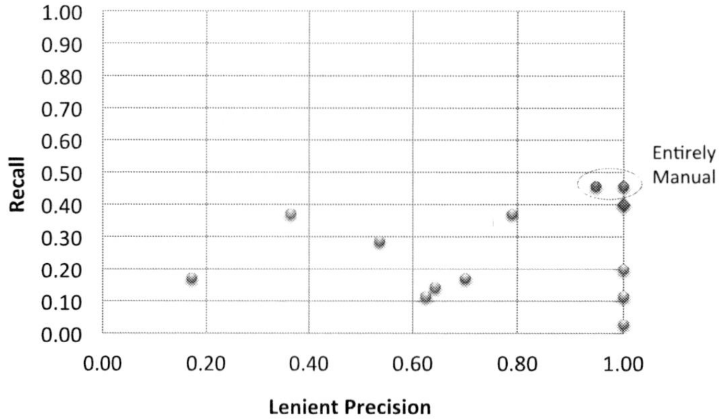

Figure 9. Evaluation results for eleven teams. The diamond ◆ represents the results of our system. The two topmost points are the manual scores of the two teams of human biologists.

Our precision, recall and F1 results for both the lenient and strict scorings are as follows:

	P	R	F1
lenient (strict)	1.00 (0.67)	0.40	0.57

7. Analysis

We believe precision is much more important than recall. A high precision system can generate valuable knowledge nuggets, even if it does not have high throughput, whereas output from a system with high recall but low precision cannot be trusted to be accurate. This is especially the case for such information-rich domains. Because of the huge volume of scientific literature, information is likely to be duplicated in multiple papers, and often also repeated in different forms in the same paper. Therefore, extracting (accurately) even a relatively small portion of the information in these papers could amount to a fair body of knowledge, even if we cannot extract everything from every sentence.

Our system showed promising performance on the evaluation data set. We achieved perfect precision, and recall close to the human experts. The modest recall even for the human experts indicated that this is a fairly difficult domain and there is not a clear-cut way to extract and encode the knowledge represented in these papers. In fact, after considering the submitted results, several additional events extracted by the systems but not by the human experts were incorporated into the gold standard.

We were able to extract some fairly complex, nested, events, similar to the one depicted in Figure 5. The ontology-based extraction and the lexical coverage extended by *WordFinder* allowed us to cope with a variety of expressions. For instance, from "... *ASPP2 can potentiate RAS signaling...*" we were able to map "potentiate" to an INCREASE event even though "potentiate" is not in the TRIPS core lexicon.

Another interesting example is "... *monoubiquitination abrogates GAP-mediated GTP hydrolysis*". This fairly complex sentence illustrates some of the strengths and weaknesses of our system. The system was able to extract two interleaving events:

ev1: REGULATE(AGENT: GAP; AFFECTED: ev2)
ev2: HYDROLYSIS(AFFECTED: GTP)

In the raw processing we also had the following:

ev3: INHIBIT(AGENT: MONOUBIQUITINATION; AFFECTED ev2)

but we failed to identify *what* was being monoubiquitinated and thus were not able to include this extraction in our results. The answer, that *Ras* was being monoubiquitinated, could only be identified with more sophisticated discourse processing.

We identified several main reasons for omissions in our extractions: 1) fragmented parses due to the long and complex sentence structures common in scientific publications; 2) insufficient domain-specific background knowledge, including language patterns specific to biology; 3) need for improved discourse processing and coreference resolution; and 4) lack of inference capabilities and persistent memory of inferences made.

The last point can be illustrated by the sentence "... *the RAS-ASPP interaction enhances the transcription function of p53...*" Here we need to be able to deduce that RAS-ASPP interaction produces a complex of the two, which then participates in further reactions.

As a final example, to be able to make sense of the seemingly simple sentence "*We obtained similar results using K-Ras...*" we need to address all of the above issues. Due to space limitation we will not discuss here the ongoing work towards tackling these challenges.

8. Related Work and Discussion

With the advent of relatively successful text mining strategies (named entity recognition, information extraction and retrieval) for the recognition and normalization of biologically relevant entities, automatic extraction of more complex, relational information from the biomedical literature has become a very active area of research. Shared Tasks (STs) such as the Protein-Protein Interaction (PPI) Task introduced at BioCreative II (Krallinger et al., 2008) and the BioNLP GENIA Event Extraction Task (Kim et al., 2009; Kim et al., 2011; Kim et al., 2013) have spurred a lot of activity in this area, although examples of earlier work certainly exist.

The goal in the PPI task is to extract binary protein-protein interaction pairs from full-text articles. More general biological events (e.g., regulatory events) beyond PPI involve much more varied relationships between entities and, indeed, between events themselves, leading to complex nested structures. The BioNLP STs have evolved to include more complex types of events and arguments. The GENIA ST (in particular 2013 which included coreference) and the Epigenetics and Post-translational Modifications task (EPI) introduced in 2011 (Ohta et al., 2011) are similar to our task. However, there are significant differences, too. We were not provided with gold annotations for entities; all relevant entities (including drugs, cell lines, cell components, sites) had to be extracted, and most of them had to be grounded in a reference database. Protein families were also important, as was the relation between families and the member proteins. Not only were coreferences supposed to be resolved, but, as indicated in Section 5, sometimes complex inferences were required to obtain a target event. In summary, our task was not designed to accommodate specific Information Extraction (IE) techniques; rather, in our evaluation the gold standard was human performance.

We would like to stress that our goal goes beyond IE. The need for deeper semantic approaches has been recognized before (see, e.g., Ananiadou et al., 2010). Still, the field is dominated by ML classifiers (for a list of the top-performing systems in the three BioNLP STs held so far, see Ananiadou et al., 2014). This sometimes results in seemingly paradoxical results, where systems can extract with relatively good performance phosphorylation events, but not ubiquitination events because the training data did not contain enough examples of the latter (Kim et al., 2011).

Indeed, ontological information is rarely used in current systems. GenIE (Cimiano et al., 2005) is an early example of an ambitious ontology-driven system that attempts to identify events based on constructing a full semantic representation of the text (using a semantic lexicon and semantic restrictions), as well as relations between events (using discourse information). The ontology they used, however, was a small, domain-specific one. To our knowledge the system has not been tested on any of the more recent event extraction tasks.

Although semantic (deep) parsing techniques have been rarely used for bio-event extraction, we note the PPI extraction study by Miyao et al. (2009), who found an HPSG-based parser to outperform (particularly in terms of precision) dependency and syntactic parsers, especially when trained on domain-specific corpora. However, they used the predicate-argument structures output by the parser as additional features for a statistical classifier.

In contrast, we do not depend on training with a domain specific corpus (although we have the capability to incorporate modules that do); rather, we extract events directly from the predicate-argument structures represented in the logical form, based on linguistic first principles that can be easily adapted to different domains. The advantage of this approach can be readily seen in this evaluation, in which, with a relatively short (but intensive) ramp up, we were able to outperform all other systems in the extraction of complex events and event relations. Of note, this was despite the fact that our system had lower named entity recognition scores than most others, particularly those with a history of participation in biomedical information extraction shared tasks.

The purpose of this evaluation was not a rigorous ranking of the different participating systems. Rather, we learned key areas we needed to improve. The results of this evaluation suggested that our system is viable for complex event extraction. This is however only the first step in understanding complex models and mechanisms. A general deep language understanding system that can be extended with domain-specific information will allow us to go beyond standard surface extraction tasks and develop the capabilities to truly *understand* big and complex mechanisms.

Acknowledgement

We thank the anonymous reviewers for comments that made this paper better. This work has been partially supported by the DARPA Big Mechanism program under ARO grant W911NF-14-1-0391.

References

Allen, J. F. (1995). Natural Language Understanding. Redwood City, CA, Benjamin Cummings.

Allen, J., M. Swift, et al. (2008). Deep Semantic Analysis of Text. Symposium on Semantics in Systems for Text Processing (STEP), Venice, Italy.

Allen, J. F. (2014). Learning a Lexicon for Broad-coverage Semantic Parsing. ACL Workshop on Semantic Parsing. Baltimore, MD.

Ananiadou, S., S. Pyysalo, J. Tsujii, and D. B. Kell (2010). Event extraction for systems biology by text mining the literature. *Trends in Biotechnology* 28 (7), 381-390.

Ananiadou, S., P. Thompson, R. Nawaz, J. McNaught, and D. B. Kell (2014). Event-based text mining for biology and functional genomics. *Briefings in functional genomics*. doi:10.1093/bfgp/elu015.

Ashburner, M., C. A. Ball, J. A. Blake, D. Botstein, H. Butler, J. M. Cherry, A. P. Davis, K. Dolinski, S. S. Dwight, J. T. Eppig, M. A. Harris, D. P. Hill, L. Issel-Tarver, A. Kasarskis, S. Lewis, J. C. Matese, J. E. Richardson, M. Ringwald, G. M. Rubin, and G. Sherlock (2000). Gene ontology: tool for the unification of biology. *Nature Genetics* 25 (1), 25-29.

Blaylock, N., de Beaumont, W., Allen, J., & Jung, H. (2011). Towards an OWL-based framework for extracting information from clinical texts. In *Proceedings of the 2nd ACM Conference on Bioinformatics, Computational Biology and Biomedicine*, pp. 636-640. ACM.

Cimiano, P., U. Reyle, and J. Šarić (2005). Ontology-driven discourse analysis for information extraction. *Data & Knowledge Engineering* 55 (1), 59-83.

Degtyarenko, K., P. de Matos, M. Ennis, J. Hastings, M. Zbinden, A. McNaught, R. Alcántara, M. Darsow, M. Guedj, and M. Ashburner (2008). ChEBI: a database and ontology for chemical entities of biological interest. *Nucleic Acids Research* 36 (suppl 1), D344-D350.

Demir, E., Cary, M. P., Paley, S., et al. (2010). The BioPAX community standard for pathway data sharing. *Nature Biotechnology*, 28(9):935-942.

Diehl, A. D., A. D. D. Augustine, J. A. Blake, L. G. Cowell, E. S. Gold, T. A. Gondré-Lewis, A. M. M. Masci, T. F. Meehan, P. A. Morel, A. Nijnik, B. Peters, B. Pulendran, R. H. Scheuermann, Q. A. Yao, M. S. Zand, and C. J. Mungall (2011). Hematopoietic cell types: prototype for a revised cell ontology. *Journal of biomedical informatics* 44 (1), 75-79.

Fellbaum, S. (1998, ed.) *WordNet: An Electronic Lexical Database*. Cambridge, MA: MIT Press.

Finkel, J., T. Grenager and C. Manning (2005). Incorporating non-local information into information extraction systems by Gibbs sampling. *Proceedings of the Annual Meeting of the Association for Computational Linguistics*.

Finn, R. D., A. Bateman, J. Clements, P. Coggill, R. Y. Eberhardt, S. R. Eddy, A. Heger, K. Hetherington, L. Holm, J. Mistry, E. L. L. Sonnhammer, J. Tate, and M. Punta (2014). Pfam: the protein families database. *Nucleic Acids Research* 42 (D1), D222-D230.

Gaudet, P., P.-A. Michel, M. Zahn-Zabal, I. Cusin, P. D. Duek, O. Evalet, A. Gateau, A. Gleizes, M. Pereira, D. Teixeira, Y. Zhang, L. Lane, and A. Bairoch (2015). The neXtProt knowledgebase on human proteins: current status. *Nucleic Acids Research* 43 (D1), D764-D770.

Gazdar, G., E. H. Klein, G. K. Pullum; I. A. Sag (1985). *Generalized Phrase Structure Grammar.* Oxford: Blackwell, and Cambridge, MA: Harvard University Press.

Godin-Heymann, N., Y. Wang, E. Slee and X. Lu (2013). Phosphorylation of ASPP2 by RAS/MAPK Pathway Is Critical for Its Full Pro-Apoptotic Function. *PLoS ONE* 8(12): e82022.

Gray, K. A., B. Yates, R. L. Seal, M. W. Wright, and E. A. Bruford (2015). Genenames.org: the HGNC resources in 2015. *Nucleic acids research* 43 (Database issue).

Gremse, M., A. Chang, I. Schomburg, A. Grote, M. Scheer, C. Ebeling, and D. Schomburg (2011). The BRENDA tissue ontology (BTO): the first all-integrating ontology of all organisms for enzyme sources. *Nucleic acids research* 39 (Database issue).

Hermjakob, H., L. Montecchi-Palazzi, G. Bader, J. Wojcik, L. Salwinski, A. Ceol, S. Moore, S. Orchard, U. Sarkans, C. von Mering, B. Roechert, S. Poux, E. Jung, H. Mersch, P. Kersey, M. Lappe, Y. Li, R. Zeng, D. Rana, M. Nikolski, H. Husi, C. Brun, K. Shanker, S. G. N. Grant, C. Sander, P. Bork, W. Zhu, A. Pandey, A. Brazma, B. Jacq, M. Vidal, D. Sherman, P. Legrain, G. Cesareni, I. Xenarios, D. Eisenberg, B. Steipe, C. Hogue, and R. Apweiler (2004). The HUPO PSI's molecular interaction format — a community standard for the representation of protein interaction data. *Nat Biotech* 22 (2), 177-183.

Jackendoff, R. (1977). *X-bar-Syntax: A Study of Phrase Structure.* Linguistic Inquiry Monograph 2. Cambridge, MA: MIT Press.

Kim, J. D., T. Ohta, S. Pyysalo, Y. Kano, and J. Tsujii (2009). Overview of BioNLP'09 shared task on event extraction. In *Proceedings of the Workshop on Current Trends in Biomedical Natural Language Processing* (BioNLP '09), pp. 1-9. ACL.

Kim, J. D., Y. Wang, T. Takagi, and A. Yonezawa (2011). Overview of Genia event task in BioNLP shared task 2011. In *Proceedings of the BioNLP*

Shared Task 2011 Workshop, BioNLP Shared Task '11, pp. 7-15. ACL.

Kim, J.-D., Y. Wang, and Y. Yasunori (2013). The Genia event extraction shared task, 2013 edition - overview. In *Proceedings of the BioNLP Shared Task 2013 Workshop*, Sofia, Bulgaria, pp. 8-15. ACL.

Klein, D. and C. D. Manning (2003). Fast exact inference with a factored model for natural language parsing. In *Advances in Neural Information Processing Systems* 15 (NIPS 2002), Cambridge, MA: MIT Press.

Krallinger, M., F. Leitner, C. Rodriguez-Penagos, and A. Valencia (2008). Overview of the protein-protein interaction annotation extraction task of BioCreative II. *Genome biology* 9 (Suppl 2): S4.

Lipscomb, C. E. (2000). Medical subject headings (MeSH). *Bulletin of the Medical Library Association* 88 (3), 265-266.

Manshadi, M. H., J. F. Allen, et al. (2008). Towards a Universal Underspecified Semantic Representation. 13th Conf. on Formal Grammar. Hamburg, Germany.

McCray, A. T., S. Srinivasan, and A. C. Browne (1994). Lexical methods for managing variation in biomedical terminologies. In *Proceedings of the Annual Symposium on Computer Application in Medical Care*, National Library of Medicine, Bethesda, Maryland., pp. 235-239.

Miyao, Y., K. Sagae, R. Sætre, T. Matsuzaki, and J. Tsujii (2009). Evaluating contributions of natural language parsers to protein–protein interaction extraction. *Bioinformatics* 25 (3), 394-400.

Ohta, T., S. Pyysalo, and J. Tsujii (2011). Overview of the Wpigenetics and Post-translational Modifications (EPI) task of BioNLP Shared Task 2011. In *Proceedings of BioNLP Shared Task 2011 Workshop*, Portland, Oregon, USA, pp. 16-25. ACL.

Pollard, C., and I. A. Sag (1987). *Information-Based Syntax and Semantics. Volume 1. Fundamentals.* CLSI Lecture Notes 13.

Pollard, C., and I. A. Sag (1994). *Head-Driven Phrase Structure Grammar.* Chicago: University of Chicago Press.

Selventa (2011). *Biological Expression Lanaguge V1.0 Language Overview*, Cambridge MA 02140.

Smith, B., M. Ashburner, C. Rosse, J. Bard, W. Bug, W. Ceusters, L. J. Goldberg, K. Eilbeck, A. Ireland, C. J. Mungall, N. Leontis, P. Rocca-Serra, A. Ruttenberg, S.-A. Sansone, R. H. Scheuermann, N. Shah, P. L. Whetzel, and S. Lewis (2007). The OBO foundry: coordinated evolution of ontologies to support biomedical data integration. *Nature Biotechnology* 25 (11), 1251-1255.

Toutanova, K. and C. D. Manning (2000). Enriching the knowledge sources used in a maximum entropy part-of-speech tagger. In *Proceedings of the Joint SIGDAT Conference on Empirical Methods in Natural Language Processing and Very Large Corpora (EMNLP/VLC-2000).*

The UniProt Consortium (2014). UniProt: a hub for protein information. *Nucleic Acids Research* 43 (D1), D204-D212.

Making the most of limited training data using distant supervision

Roland Roller and **Mark Stevenson**
Department of Computer Science
University of Sheffield
Regent Court, 211 Portobello
S1 4DP Sheffield, England
roland.roller,mark.stevenson@sheffield.ac.uk

Abstract

Automatic recognition of relationships between key entities in text is an important problem which has many applications. Supervised machine learning techniques have proved to be the most effective approach to this problem. However, they require labelled training data which may not be available in sufficient quantity (or at all) and is expensive to produce. This paper proposes a technique that can be applied when only limited training data is available. The approach uses a form of distant supervision but does not require an external knowledge base. Instead, it uses information from the training set to acquire new labelled data and combines it with manually labelled data. The approach was tested on an adverse drug data set using a limited amount of manually labelled training data and shown to outperform a supervised approach.

1 Introduction

Relation extraction is a widely explored problem that has been applied to a range of domains (Craven and Kumlien, 1999; Agichtein and Gravano, 2000; Xu et al., 2007) using a variety of techniques (Yangarber, 2003; Bunescu and Mooney, 2006; Neumann and Schmeier, 2012). In the biomedical domain relation extraction has been used to identify a wide range of types of relation, including adverse drug effects (ADE), gene regulations and drug-drug interactions. Community evaluation exercises, such as the BioNLP Shared Task (Kim et al., 2011; Nédellec et al., 2013) or the Drug-Drug Interaction (DDI) challenge (Segura-Bedmar et al., 2013), have shown that supervised learning techniques normally produce better results than other approaches.

Supervised learning techniques rely on labeled training data but these are not available for all relations of interest and are also difficult and time-consuming to create. Other approaches may be more appropriate in situations where training data is limited or unavailable. Minimally supervised approaches, such as seed and bootstrapping techniques (Brin, 1999; Riloff and Jones, 1999; Agichtein and Gravano, 2000), are provided with a small set of seed instances (examples of related information) or patterns and acquire further examples from a large corpus by applying an iterative process. While these approaches do not require labelled training data they often suffer from low precision or semantic drift (Mintz et al., 2009). Distant supervision combines the advantages of minimally supervised and supervised approaches to relation extraction.

Distant supervision makes use of an external knowledge source that provides information about pairs of entities which are related. Sentences containing both entities in a pair are identified from a corpus and used in place of labeled training examples. For example, knowledge that *hair loss* is a drug-related adverse effect of *paroxetine* would allow further positive examples to be identified by searching for other sentences containing the same drug and side-effect. Many knowledge sources only contain positive entity pairs. Therefore negative examples are often generated using a closed-world assumption. Given the known positive entity pairs, negative entity pairs are generated by producing new combinations of entities. Negative example sentences are generated by selecting sentences containing these negative entity pairs.

The example in figure 1 shows the limitations of distant supervision since related entities might express a different relation. This can lead to examples being falsely labelled as positive examples of a relation. Classifiers trained using data generated using distant supervision do not generally

Proceedings of the 2015 Workshop on Biomedical Natural Language Processing (BioNLP 2015), pages 12–20,
Beijing, China, July 30, 2015.

perform as well as those trained using manually labelled data. However, distant supervision allows large data sets to be generated at low cost.

```
There are a few case reports on
[CONDITION:hair loss] associated
with tricyclic antidepressants and
serotonin selective reuptake inhibitors
(SSRIs), but none deal specifically with
[DRUG:paroxetine].
```

Figure 1: Generation of false positives by using automatically labelled data, PMID=10442258

The majority of distant supervision approaches use structured knowledge sources such as Wikipedia (Hoffmann et al., 2010) or Freebase (Mintz et al., 2009; Riedel et al., 2010; Ritter et al., 2013; Augenstein et al., 2014). However there may not be a suitable knowledge base available for a particular relation of interest. This paper addresses the problem of developing relation extraction systems in situations where only a small amount of training data is available.

We introduce a method for relation extraction that can be used when only limited amounts of training data are available. The approach is based on *distant supervision* but, rather than relying on a knowledge base, seed pairs are extracted from Medline articles. Sentences from the Medline Baseline Repository containing these seed pairs are extracted to generate a large distantly labelled training data set. Using this data manually labelled data can be extended and combined to a hybrid *mixture model* which outperforms both the supervised and the distantly supervised models.

This paper makes the following contributions: 1) introduces a method which can be used to train a relational classifier when only a small set of labelled training data is available, 2) provides a method for combining distant supervision with supervised learning methods and 3) presents distant supervision without the need of a knowledge base.

The remainder of the paper is structured as follows. The next section presents the background on relation extraction from biomedical documents. Section 3 introduces the data set which is used for the experiments. The techniques for generating the distantly supervised training data and relational classifier are described in sections 4 and 5. Section 6 describes the experiment and the results. Conclusions are presented in section 7.

2 Related Work

Supervised learning techniques are popular and efficient approaches to detecting relations between entities in natural language. Results using supervised learning methods tend to improve as more training data is available. However the generation of labelled data is cumbersome, expensive and time-consuming. It often requires expert knowledge in restricted domains, such as biomedicine. A new labelled data set is required for each target relation.

In recent years, distant supervision has become very popular. Rather than using manually annotated data, distant supervision uses knowledge about which entity pairs are instances of the target relation to generate automatically labelled data which is used to train a relational classifier. Craven and Kumlien (1999) introduced distant supervision for relation extraction. The authors used the Yeast Protein Database (YPD) as source of knowledge and mapped this information to PubMed articles to generate training examples. The technique has been widely applied particularly outside the medical domain. Many approaches such as (Mintz et al., 2009; Sun et al., 2011; Hoffmann et al., 2011; Krause et al., 2012; Xu et al., 2013) focus on approaches using Freebase as knowledge source to generate automatically labelled data. In recent years distant supervision has also become more popular in the biomedical domain beeing used to detect protein-protein interactions using IntAct (Thomas et al., 2011), protein-residue associations with PDB (Ravikumar et al., 2012) or relationships of the National Drug File-Reference Terminology (NDF-RT) using the UMLS Metathesaurus (Roller and Stevenson, 2014). Liu et al. (2014) focus on the detection of genes in brain regions from literature using the UMLS Semantic Network and Ellendorff et al. (2014) uses the Comparative Toxicogenomics Database (CTD) to detect interactions between genes and chemicals.

The distantly supervised methods of Nguyen and Moschitti (2011) and Pershina et al. (2014) differ slightly from many other approaches. Both combine supervised and distantly supervised models. Nguyen and Moschitti (2011) use a support vector machine and combine the supervised and the distantly supervised classifier with a linear combination. Pershina et al. (2014) instead integrate the manually labelled data directly within their distantly supervised multi-learning approach.

Both approaches show that a combination of a large set of distantly supervised (noisy) data with manually labelled examples can improve the classification results. The combination of noisy data and hand-selected training examples is also used in this paper.

3 Data

The experiments in this work uses the ADE data set (Gurulingappa et al., 2012b) which contains examples of adverse drug effects (ADE). An ADE is a response of a drug which is noxious and unintended, and which occurs at doses normally used in humans for the prophylaxis, diagnosis, therapy of disease, or for the modification of physiological function[1] (Gurulingappa et al., 2012b). ADEs contribute to one of the most common causes of death in industrialised nations and are the fourth leading cause of death in the U.S. (Giacomini et al., 2007). To reduce this risk the side-effects of drugs need to be detected and made publicly available as quickly as possible.

The ADE data set consists of Medline case reports examined by three human annotators. Sentences in these case reports containing adverse effects between *drugs* and *conditions* were extracted and entities annotated to generate the data set. An example relation between a drug and a condition from this data set is shown in figure 2. According to the given sentence the condition *pseudoporphyria* is caused by the two drugs *naproxen* and *oxaprozin*.

```
METHODS: We report two cases of
[CONDITION:pseudoporphyria] caused by
[DRUG:naproxen] and [DRUG:oxaprozin].
```

Figure 2: Example of a drug-related adverse effect taken from PMID=10082597

The ADE corpus only contains examples of positive relations. Negative examples are also required to set-up a meaningful ADE prediction task and to train a supervised ADE classifier. A set of negative examples were generated using the following process.

Named entity recognition is applied to detect drugs and conditions. MetaMap[2] (Aronson and Lang, 2010) was run on the unannotated sentences in the ADE corpus to detect biomedical concepts from the UMLS. MetaMap provides different possible UMLS concept mappings and we select the best (highest ranked) mapping. Each biomedical concept detected by MetaMap now refers to a unique UMLS CUI thereby allowing identical concepts to be merged and assigned semantic types. Using the same approach as Kang et al. (2014), sentences containing concepts with semantic types which belong to the two groups "Chemicals & Drugs" and "Disorders" are extracted and considered as negative examples. Nested relations are not included in our data set.

Training and evaluation sets were then generated. The set of utilised ADE abstracts consists of 1644 publications. 200 abstracts were removed to be used to create training data and the remainder used to form the evaluation set. The training data is created by extracting all positive and negative labelled sentences from the 200 abstracts. In order to provide reliable results we run the same experiment 5 times. Each time we randomly choose a different selection of 200 training and 1444 test abstracts.

4 Automatic Generation of Annotated Training Data

Many of the previous approaches to distant supervision use information about related instances (e.g. drugs and known adverse effects) to automatically generate training data. In the majority of cases this information is obtained from a knowledge base. We employ an alternative approach and make use of information from a small set of abstracts. For example, the sentence shown in figure 2 suggests that there are cases when the drugs *oxaprozin* and *naproxen* cause *pseudoporphyria*. Consequently sentences containing these two *drug-condition* entity pairs (i.e. *oxaprozin-pseudoporphyria* and *naproxen-pseudoporphyria*) are extracted and treated as positive examples.

The data is generated by applying a three stage process (see Figure 3).

1) *Map CUIs to the related entities in the training data set.* We begin by normalising medical concepts. Medical terms can occur in literature with different names, using a different spelling or abbreviations. For instance *Naproxen* can be also described as *Methoxypropiocin*, *MNPA* or *6-Methoxy-alpha-methyl-2-naphthaleneacetic Acid*. UMLS maps these different names to the same

[1] World Health Organization (WHO) glossary of terms used in Pharmacovigilance.

[2] http://metamap.nlm.nih.gov/, MetaMap version 13 with UMLS 2013AA

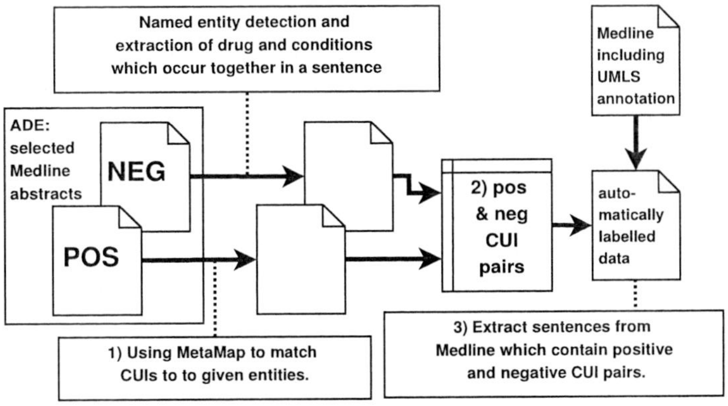

Figure 3: Automatic generation of training data for ADE relations

CUI, *C0027396*. We run MetaMap with the same configuration on the sentences containing positive examples. In many cases it is possible to assign a MetaMap annotation to the existing related entities.

We only assign a CUI to an entity if MetaMap identifies a CUI that can be mapped to the entity in its full length (not only a substring). Negative training examples already include CUI information for each entity (see section 3).

2) *Extract a set of positive and negative seed instance pairs.* In the next step, we extract all CUI pairs from the positive ADE examples and add them to a set of positive instance pairs P. We also extract CUI pairs of negative ADE examples and add them to a negative instance pair set N. Each CUI pair which occurs in both sets (P and N), is removed from N. Considering the 200 training abstracts of the first setup (of five) it is possible to extract 310 different positive CUIs pairs and 869 negative CUI pairs. 12 CUI pairs occur in both sets. Therefore the number of different CUI pairs in N is reduced to 857.

3) *Extract sentences containing positive and negative seed instances from abstracts.* The distantly labelled training data is generated using the Medline Baseline Repository (MBR)[3], a large collection of biomedical abstracts annotated using MetaMap[4]. We use 3,000,000 abstracts published between 1997-2003. Then sentences from this subset containing positive and negative CUI pairs are extracted and labelled as positive and negative examples.

[3]http://mbr.nlm.nih.gov/

[4]MetaMap annotations use UMLS release 2011AB, http://mbr.nlm.nih.gov/Download/MetaMapped_Medline/2012/

Regarding the the 200 training abstracts of the first setup a total of 7868 sentence containing positive instance pairs and 14,4315 sentence containing negative instance pairs were identified and extracted. Although 310 different positive and 857 different negative CUI pairs were extracted from the 200 abstracts (see above), only 290 different positive and 441 different negative CUI pair combinations were detected within the portion of MBR used for this experiment. It is also interesting to note that only 13 positive CUI pairs occur more than 100 times within the 7868 positive examples. The most frequent positive CUI pairs are listed in table 1. 213 of the positive CUI pairs occur fewer than 10 times.

The automatically generated data has a strong bias. To generate an automatically labelled training data with a similar bias as the test set we reduce the amount of negative examples to the same ratio as the manually labelled examples.

5 Relation Extraction

We use the Java Simple Relation Extraction[5] (jSRE) (Giuliano et al., 2006) which is based on LibSVM (Chang and Lin, 2011). jSRE includes an implementation of the shallow linguistic kernel which provides reliable classification results and has been used also for other experiments on the ADE data set (Gurulingappa et al., 2012a; Kang et al., 2014).

The shallow linguistic kernel is a combination of the *global context kernel* and the *local context kernel*. The global context kernel considers n-grams of the words (and other information such as stemmed words and part of speech tags) between

[5]https://hlt.fbk.eu/technologies/jsre

frequency	drug	condition
#1352	C0019134='Heparin'	C0272285='Heparin-induced thrombocytopenia'
#1199	C0026549='Morphine'	C0030193='Pain'
#980	C0023175='Lead'	C0020538='Hypertensive disease'
#396	C0031507='Phenytoin'	C0036572='Seizure'

Table 1: Most frequent positive CUI pairs found in the automatically labelled data set

the two entities. The local context kernel considers only a limited amount of information around each entity.

Sentences from the training and test data are parsed using the Charniak-Johnson Parser (Charniak and Johnson, 2005) to generate part of speech tags. Next, words are reduced to their stem using the Porter Stemmer (Porter, 1997).

We use three different methods within the experiments: supervised relation extraction, distantly supervised relation extraction and a relation extraction using a mixture-model. The supervised model uses a set of abstracts (1-200) from the training data as input. The distantly supervised model takes the automatically generated data based on the MetaMap annotated Medline Baseline Repository as input. The mixture-model merges the automatically generated and manually labelled training data to form a combined training set.

6 Experiment

In this experiment we examine different sizes of manually labelled training data. Starting with a single abstract for training we slowly increase the number of seed abstracts to 200. In parallel we generate for each training set a different distantly labelled data set using the given ADE seed facts of the training data. The more information the manually labelled data contains, the more different seeds can be extracted which increases the size of the distantly labelled data. Thereafter we combine in each step both data sets to a mixture-model.

In order to provide reliable results we repeat this experiment five times (five evaluation rounds) with a different selection of abstracts for training and test. In each evaluation round the abstracts utilised for training are chosen randomly. The remaining abstracts are used for evaluation. During a specific evaluation round (increasing training data) the test set remains unchanged. The results of the experiments are presented in table 2 and figure 4. The results represent the mean of all five different evaluation rounds.

The results show that the performance for all models improves as the amount of data increases. Performance of the supervised classifier increases sharply as the number of abstracts is increased from 1 to 10 abstracts. Increasing the size of the training data to 50 abstracts produces a further improvement of approximately 30%. These results demonstrate that even small amounts of training data are sufficient to provide reasonable results on the ADE data set.

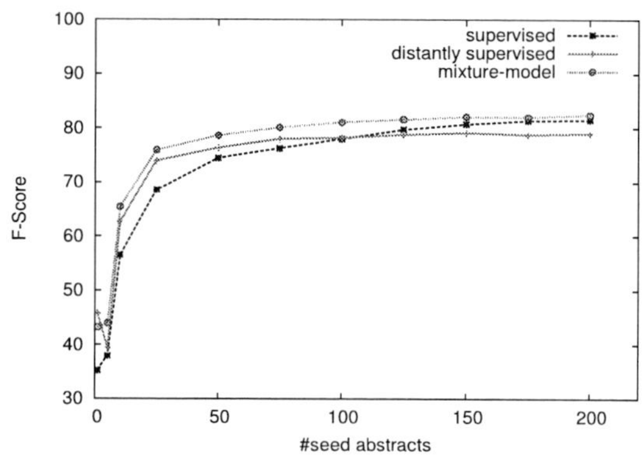

Figure 4: Effect of varying number of seed abstracts

Performance of the distantly supervised classifier shows a similar pattern. Increasing the number of seed abstracts results in a larger distantly labelled training data set which improves classification results. The distantly supervised classifier outperforms the supervised one when there are fewer than 100 seed abstracts. The reason for this is the supervised classifier does not have access to a sufficient volume of training data while the distant supervision is able to generate more. As the number of seed abstracts increases the situation is reversed with the supervised classifier outperforming the distantly supervised one. When more than 100 abstracts are available the supervised classifier has the advantage of having access to enough accurately labelled examples to train a relation ex-

#SA	supervised model			distant supervision			mixture model		
	prec.	**rec.**	**F1**	**prec.**	**rec.**	**F1**	**prec.**	**rec.**	**F1**
1	52.75	42.33	35.18	42.20	53.95	**45.84**	43.54	50.40	43.20
5	68.48	32.53	37.92	76.78	37.47	39.45	78.40	38.76	**43.98**
10	66.85	51.17	56.53	71.90	61.33	62.66	73.90	61.97	**65.43**
25	68.01	69.88	68.57	69.01	81.48	73.99	71.88	81.58	**75.96**
50	72.29	76.88	74.48	69.27	86.68	76.39	72.62	86.46	**78.66**
75	73.77	79.18	76.27	68.35	91.10	78.01	73.43	88.30	**80.13**
100	75.41	80.85	78.00	67.79	92.56	78.24	73.80	89.99	**81.09**
125	75.79	84.16	79.75	69.11	91.65	78.77	74.91	89.77	**81.64**
150	76.89	85.06	80.77	70.15	90.99	79.19	75.81	89.65	**82.13**
175	77.14	86.15	81.39	68.50	93.03	78.84	74.45	91.40	**82.04**
200	77.32	86.28	81.54	68.77	92.98	79.02	75.01	91.63	**82.47**

Table 2: Effect of varying size of training data set

traction system. The distantly supervised classifier still has access to more data but it is not as accurate.

#SA	manual lab.		distantly lab.		seeds	
	pos	**neg**	**pos**	**neg**	**pos**	**neg**
10	67	121	510	891	15	54
25	180	232	1048	1404	38	103
50	388	485	2026	2580	81	213
75	590	756	2643	3398	123	330
100	804	1024	3818	4851	172	448
150	1200	1447	5663	6863	248	636
200	1632	1900	8289	9607	336	834

Table 3: ADE training data size (mean across five runs)

The mixture model produces the best results of all approaches when 5 or more abstracts are used. This result is interesting since the manually labelled data is simply extended using a simple form of distant supervision that is straightforward to apply. The mixture model tends to achieve higher precision but lower recall than the distantly supervised approach, possibly because the training data used by the mixture model is more accurate and contains fewer "false positive" examples. On the other hand the precision and recall of the mixture model are often higher than the supervised model. The increase in recall is presumably caused by having access to additional training data and the precision scores suggest that the classifier is not harmed by some of these containing noisy labels.

The difference in performance between the supervised and the mixture-models gets smaller as the number of seed abstracts increases.

Table 3 shows the mean size of the different sets of training data. The amount of distantly labelled data is much larger than the manually labelled data at each classification step. Larger amounts of manually labelled data increase the number of ADE seed instances that can be extracted which leads to more distantly supervised examples.

7 Discussion and Conclusion

This paper introduced a new distantly supervised method for relation extraction that was applied to the identification of ADE relations from biomedical documents. The approach is able to use information from an existing training data set to automatically acquire new training data. Using this data, a relational classifier can be trained to detect and extract similar information in natural language. The classifier is able to provide comparable results to a supervised classifier using a small gold standard as input. Furthermore we presented a mixture model using manually labelled and distantly labelled data which is able to outperform a classifier using only (a small set of) gold standard data. This result is notable since distantly supervised data tends to be much noisier than manually labelled data and therefore produce less accurate classifiers.

Distant supervision is a well explored technique for relation extraction that has proven to be effective. Our proposed methods differs slightly in the way seed instances are generated. Rather than using a knowledge base we directly extract positive and negative seed pairs from an existing data set and use them for distant supervision.

We plan to extend the work described in this paper in various ways. Firstly we would like to experiment with alternative classifiers such as applying dependency features and stacking or merging to combine different kernel models. We would also like to explore different techniques for combining the supervised and the distantly supervised model.

Acknowledgements

The authors are grateful to the Engineering and Physical Sciences Research Council for supporting the work described in this paper (EP/J008427/1).

References

Eugene Agichtein and Luis Gravano. 2000. Snowball: extracting relations from large plain-text collections. In *Proceedings of the fifth ACM conference on Digital libraries*, DL '00, pages 85–94, New York, NY, USA. ACM.

A. Aronson and F. Lang. 2010. An overview of MetaMap: historical perspective and recent advances. *Journal of the American Medical Association*, 17(3):229–236.

Isabelle Augenstein, Diana Maynard, and Fabio Ciravegna. 2014. Relation extraction from the web using distant supervision. In *Proceedings of the 19th International Conference on Knowledge Engineering and Knowledge Management (EKAW 2014)*, Linköping, Sweden, November.

Sergey Brin. 1999. Extracting patterns and relations from the world wide web. In *Selected papers from the International Workshop on The World Wide Web and Databases*, WebDB '98, pages 172–183, London, UK, UK. Springer-Verlag.

Razvan Bunescu and Raymond J. Mooney. 2006. Subsequence kernels for relation extraction. In *Submitted to the Ninth Conference on Natural Language Learning (CoNLL-2005)*, Ann Arbor, MI, July.

Chih-Chung Chang and Chih-Jen Lin. 2011. LIBSVM: A library for support vector machines. *ACM Transactions on Intelligent Systems and Technology*, 2:27:1–27:27.

Eugene Charniak and Mark Johnson. 2005. Coarse-to-fine n-best parsing and maxent discriminative reranking. In *Proceedings of the 43rd Annual Meeting on Association for Computational Linguistics*, ACL '05, pages 173–180, Stroudsburg, PA, USA. Association for Computational Linguistics.

Mark Craven and Johan Kumlien. 1999. Constructing biological knowledge bases by extracting information from text sources. In *In Proceedings of the Seventh International Conference on Intelligent Systems for Molecular Biology (ISMB)*, pages 77–86. AAAI Press.

Tilia Ellendorff, Fabio Rinaldi, and Simon Clematide. 2014. Using large biomedical databases as gold annotations for automatic relation extraction. In *Proceedings of the Ninth International Conference on Language Resources and Evaluation (LREC'14)*, Reykjavik, Iceland.

Kathleen M. Giacomini, Ronald M. Krauss, Dan M. Roden, Michel Eichelbaum, Michael R. Hayden, and Yusuke Nakamura. 2007. When good drugs go bad. In *Nature, 466(7139)*, pages 975–977.

Claudio Giuliano, Alberto Lavelli, and Lorenza Romano. 2006. Exploiting shallow linguistic information for relation extraction from biomedical literature. In *Proceedings of the 11th Conference of the European Chapter of the Association for Computational Linguistics (EACL 2006)*, Trento, Italy.

Harsha Gurulingappa, Abdul MateenRajput, and Luca Toldo. 2012a. Extraction of potential adverse drug events from medical case reports. *Journal of Biomedical Semantics*, 3(1):15.

Harsha Gurulingappa, Abdul Mateen Rajput, Angus Roberts, Juliane Fluck, Martin Hofmann-Apitius, and Luca Toldo. 2012b. Development of a benchmark corpus to support the automatic extraction of drug-related adverse effects from medical case reports. *Journal of Biomedical Informatics*, 45(5):885–892. Text Mining and Natural Language Processing in Pharmacogenomics.

Raphael Hoffmann, Congle Zhang, and Daniel S. Weld. 2010. Learning 5000 relational extractors. In *Proceedings of the 48th Annual Meeting of the Association for Computational Linguistics*, ACL '10, pages 286–295, Stroudsburg, PA, USA. Association for Computational Linguistics.

Raphael Hoffmann, Congle Zhang, Xiao Ling, Luke Zettlemoyer, and Daniel S. Weld. 2011. Knowledge-based weak supervision for information extraction of overlapping relations. In *Proceedings of the 49th Annual Meeting of the Association for Computational Linguistics*, ACL '11, pages 541–550.

Ning Kang, Bharat Singh, Chinh Bui, Zubair Afzal, Erik van Mulligen, and Jan Kors. 2014. Knowledge-based extraction of adverse drug events from biomedical text. *BMC Bioinformatics*, 15(1):64.

Jin-Dong Kim, Sampo Pyysalo, Tomoko Ohta, Robert Bossy, Ngan Nguyen, and Jun'ichi Tsujii. 2011. Overview of bionlp shared task 2011. In *Proceedings of BioNLP Shared Task 2011 Workshop*, pages 1–6, Portland, Oregon, USA, June. Association for Computational Linguistics.

Sebastian Krause, Hong Li, Hans Uszkoreit, and Feiyu Xu. 2012. Large-scale learning of relation-extraction rules with distant supervision from the web. In *Proceedings of the 11th International Conference on The Semantic Web - Volume Part I*, ISWC'12, pages 263–278, Berlin, Heidelberg. Springer-Verlag.

Mengwen Liu, Yuan Ling, Yuan An, Xiaohua Hu, Alan Yagoda, and Rick Misra. 2014. Relation extraction from biomedical literature with minimal supervision and grouping strategy. In *Proceedings of IEEE Conference on Bioinformatics and Biomedicine (BIBM14)*, pages 444–449.

Mike Mintz, Steven Bills, Rion Snow, and Dan Jurafsky. 2009. Distant supervision for relation extraction without labeled data. In *Proceedings of the Joint Conference of the 47th Annual Meeting of the ACL and the 4th International Joint Conference on Natural Language Processing of the AFNLP: Volume 2 - Volume 2*, ACL '09, pages 1003–1011, Stroudsburg, PA, USA. Association for Computational Linguistics.

Claire Nédellec, Robert Bossy, Jin-Dong Kim, Jung-Jae Kim, Tomoko Ohta, Sampo Pyysalo, and Pierre Zweigenbaum. 2013. Overview of bionlp shared task 2013. In *Proceedings of the BioNLP Shared Task 2013 Workshop*, pages 1–7, Sofia, Bulgaria, August. Association for Computational Linguistics.

Günter Neumann and Sven Schmeier. 2012. Exploratory search on the mobile web. In *In 4th International Conference on Agents and Artificial Intelligence (ICAART 2012)*, pages 110–119, Vilamoura, Algarve, Portugal.

Truc-Vien T. Nguyen and Alessandro Moschitti. 2011. Joint distant and direct supervision for relation extraction. In *Proceedings of The 5th International Joint Conference on Natural Language Processing (IJCNLP'11)*, pages 732–740. Association for Computational Linguistics.

Maria Pershina, Bonan Min, Wei Xu, and Ralph Grishman. 2014. Infusion of labeled data into distant supervision for relation extraction. In *Proceedings of the 52nd Annual Meeting of the Association for Computational Linguistics (Volume 2: Short Papers)*, pages 732–738, Baltimore, Maryland, June. Association for Computational Linguistics.

M. F. Porter. 1997. Readings in information retrieval. chapter An Algorithm for Suffix Stripping, pages 313–316. Morgan Kaufmann Publishers Inc., San Francisco, CA, USA.

KE Ravikumar, Haibin Liu, Judith Cohn, Michael Wall, and Karin Verspoor. 2012. Literature mining of protein-residue associations with graph rules learned through distant supervision. *Journal of Biomedical Semantics*, 3(Suppl 3):S2.

Sebastian Riedel, Limin Yao, and Andrew McCallum. 2010. Modeling relations and their mentions without labeled text. In *Proceedings of the European Conference on Machine Learning and Knowledge Discovery in Databases (ECML PKDD '10)*.

Ellen Riloff and Rosie Jones. 1999. Learning dictionaries for information extraction by multi-level bootstrapping. In *Proceedings of the Sixteenth National Conference on Artificial Intelligence and the Eleventh Innovative Applications of Artificial Intelligence Conference Innovative Applications of Artificial Intelligence*, AAAI '99/IAAI '99, pages 474–479, Menlo Park, CA, USA. American Association for Artificial Intelligence.

Alan Ritter, Luke Zettlemoyer, Mausam, and Oren Etzioni. 2013. Modeling missing data in distant supervision for information extraction. In *Association for Computational Linguistics Vol. 1 (TACL)*.

Roland Roller and Mark Stevenson. 2014. Applying umls for distantly supervised relation detection. In *Proceedings of the Fifth International Workshop on Health Text Mining and Information Analysis (Louhi 2014)*, Gothenburg, Sweden.

Isabel Segura-Bedmar, Paloma Martínez, and María Herrero Zazo. 2013. Semeval-2013 task 9 : Extraction of drug-drug interactions from biomedical texts (ddiextraction 2013). In *Second Joint Conference on Lexical and Computational Semantics (*SEM), Volume 2: Proceedings of the Seventh International Workshop on Semantic Evaluation (SemEval 2013)*, pages 341–350, Atlanta, Georgia, USA, June. Association for Computational Linguistics.

Ang Sun, Ralph Grishman, and Satoshi Sekine. 2011. Semi-supervised relation extraction with large-scale word clustering. In *Proceedings of the 49th Annual Meeting of the Association for Computational Linguistics: Human Language Technologies*, pages 521–529, Portland, Oregon, USA, June. Association for Computational Linguistics.

Philippe Thomas, Illés Solt, Roman Klinger, and Ulf Leser. 2011. Learning protein protein interaction extraction using distant supervision. In *Proceedings of Robust Unsupervised and Semi-Supervised Methods in Natural Language Processing*, pages 34–41.

Feiyu Xu, Hans Uszkoreit, and Hong Li. 2007. A seed-driven bottom-up machine learning framework for extracting relations of various complexity. In *Proceedings of the 45th Annual Meeting of the Association of Computational Linguistics*, pages 584–591, Prague, Czech Republic, June. Association for Computational Linguistics.

Wei Xu, Raphael Hoffmann, Le Zhao, and Ralph Grishman. 2013. Filling knowledge base gaps for distant supervision of relation extraction. In *Proceedings of the 51st Annual Meeting of the Association for Computational Linguistics (Volume 2: Short Papers)*, pages 665–670, Sofia, Bulgaria, August. Association for Computational Linguistics.

Roman Yangarber. 2003. Counter-training in discovery of semantic patterns. In *Proceedings of the 41st Annual Meeting on Association for Computational Linguistics - Volume 1*, ACL '03, pages 343–350, Stroudsburg, PA, USA. Association for Computational Linguistics.

An extended dependency graph for relation extraction in biomedical texts

Yifan Peng[1] **Samir Gupta**[1] **Cathy H. Wu**[1,2] **K. Vijay-Shanker**[1]
[1]Department of Computer Information & Sciences
[2]Center for Bioinformatics and Computational Biology
University of Delaware
Newark, DE 19716
{yfpeng,sgupta,wuc,vijay}@udel.edu

Abstract

Kernel-based methods are widely used for relation extraction task and obtain good results by leveraging lexical and syntactic information. However, in biomedical domain these methods are limited by the size of dataset and have difficulty in coping with variations in text. To address this problem, we propose Extended Dependency Graph (EDG) by incorporating a few simple linguistic ideas and include information beyond syntax. We believe the use of EDG will enable machine learning methods to generalize more easily. Experiments confirm that EDG provides up to 10% f-value improvement over dependency graph using mainstream kernel methods over five corpora. We conducted additional experiments to provide a more detailed analysis of the contributions of individual modules in EDG construction.

1 Introduction

With growing amount of biomedical information available in textual form, there has been considerable interest in applying NLP techniques and machine-learning (ML) methods to biomedical literature. Some of these projects involve extracting relations such as protein-protein interaction (Krallinger et al., 2008).

In biomedical domain, most relation extraction work is currently applied on the abstracts of articles. These abstracts by nature are dense with information and often use constructions such as appositives and relative clauses. The abundance of textual variations can thus be problematic for ML systems, especially with small training corpora.

One solution to this issue is to find a suitable level of abstraction in the text representation so that ML methods become easier to generalize. Use of syntax and parse information provides one such abstraction. Using syntactic dependency information has become prevalent in biomedical relation extraction. It has been suggested dependency links are close to the semantic relationship needed for the next stage of interpretation (Covington, 2001).

There have been significant advances in the development of advanced machine learning and kernel methods and the use of sophisticated parameter tuning in the biomedical domain. In this work, we focus on the representation of the text used in learning rather than the machine learning technique, with the hope that advances in both directions will be improve the performance of the relation extraction systems. In this paper we propose Extended Dependency Graph (EDG), which includes information about text that goes beyond syntax. We will define EDG and discuss how we construct it from a given sentence by using some simple linguistic notions.

The hypothesis we test here is that EDG allows ML techniques to generalize more easily. To determine the effect of EDG, we conducted experiments on protein-protein interaction (PPI) extraction. For this purpose, we used two kernels: a simple kernel based on edit distance (Erkan et al., 2007) and a more elaborate kernel that is one of the top performing kernels on the PPI task (Airola et al., 2008). We compared the performance of both kernels using dependency graph and EDG on 5 corpora. Our results suggest EDG provides up to 10% f-value improvement over dependency graph. On 3 out of 5 corpora the results are better than the overall best system in the study of (Tikk et al., 2010), as well as an ensemble method that builds on them (Miwa et al., 2009a). We also evaluate the contributions of the individual components included in EDG.

Proceedings of the 2015 Workshop on Biomedical Natural Language Processing (BioNLP 2015), pages 21–30,
Beijing, China, July 30, 2015.

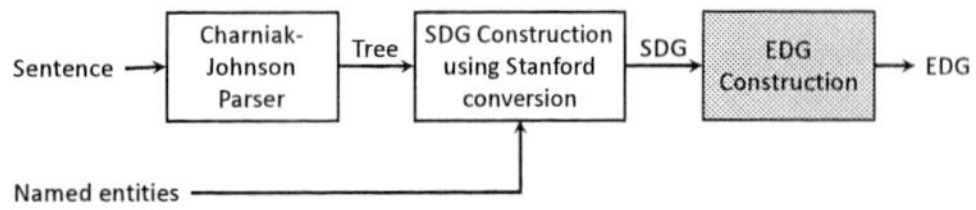

Figure 1: Framework.

2 Related work

Many kernel-based relation extraction systems have employed lexical and syntactic information (Bunescu and Mooney, 2005; Zhou et al., 2007; Ning and Qi, 2011). There has been a growth in the use of more complex kernels and sophisticated parameter tuning methods to improve the results (Zhang et al., 2006; Choi and Myaeng, 2010). In PPI task, machine learning methods using rich feature vectors (Miwa et al., 2009b), edit distance kernel (Erkan et al., 2007), dependency tree kernel (Chowdhury et al., 2011), all-path graph kernel (Airola et al., 2008), or their combination and variations (Miwa et al., 2009a; Zhang et al., 2012) have been proposed.

Our focus is on improving the representation of information in natural texts, rather than on developing new kernels. There have been several attempts to leverage syntax and shallow semantic argument structure (Miwa et al., 2010; Van Landeghem et al., 2010; Van Landeghem et al., 2012; Liu et al., 2013; Oepen et al., 2014; Peng et al., 2014; Nguyen et al., 2015). Though the focus of these works was not to utilize the information with machine learning methods, they offer insight on utility of information beyond syntax. We develop the EDG approach for relation extraction based on these ideas.

3 Method

Figure 1 illustrates the overall architecture with the core component highlighted: EDG construction. The input is a sentence with named entities marked. We use Charniak-Johnson parser and Stanford conversion tool to get the basic syntactic dependency graph (SDG). Our approach focuses on how to leverage simple linguistic principles and information beyond syntax to construct EDG from SDG.

3.1 Extended dependency graph (EDG)

In this paper, we use EDG to represent the structure of the sentence. Like in the case of many dependency graph representations used in relation extraction, the vertices in a EDG are labelled with information such as the text, part-of-speech, and the word lemma. If an entity mention spans multiple tokens in a sentence, we merge their corresponding vertices (called contracting vertices) into one vertex.

EDG has two types of dependencies. The syntactic dependencies that are obtained from collapsed dependencies output by applying Stanford dependencies converter on a syntactic parsing tree (De Marneffe and Manning, 2008). The other type of dependencies are the numbered arguments based on the guidelines of PropBank (Bonial et al., 2012). Because we are currently focusing on binary relation extraction, we use only *arg0* and *arg1* (probably better stated as not-*arg0*) in EDG. Figure 2 shows EDGs of three text fragments with syntactic edges appearing above the words and numbered argument edges appearing below. From a relation extraction perspective, the syntactic dependencies in Figure 2 are less relevant but their numbered arguments between two entity mentions are same.

There are two motivations for using numbered arguments. One is to "provide consistent argument labels across different syntactic realizations of the same verb" (Bonial et al., 2012) with the intention of making generalizations easier downstream. The other is to add/propagate new *arg0* and *arg1* using reasoning that goes beyond syntax.

Following these two motivations, we will first discuss how to capture *arg0* and *arg1* using different syntactic dependencies obtained from Stanford dependencies. Then we will describe relations such as *is-a*, *member-collection*, and *part-whole* and how to propagate *arg0* and *arg1* using them.

3.2 Syntax based *arg0* and *arg1*

We follow approaches of SemRep (Rinaldi et al., 2006) and PASMED (Nguyen et al., 2015) to obtain the basic edges *arg0* and *arg1* from the syntactic dependencies. For example, EDG will include an *arg0* from a verb to the noun if the syntactic dependency is *nsubj* or *agent* and include an *arg1* if the dependency is *nsubjpass* or *dobj*.

In addition, we consider situation where verbs in gerund form are used as noun modifiers. Figure 3 shows a compound noun phrase. We know that there is a PPI between "retinoblastoma" and "protein", because we can rewrite the phrase into

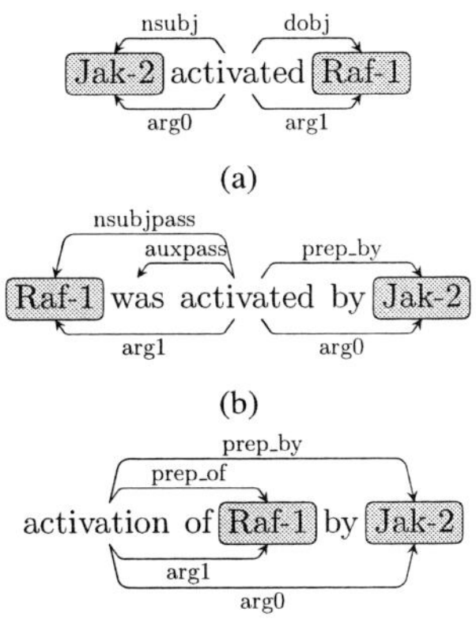

(a)

(b)

(c)

Figure 2: Sample EDGs with an active (a), passive (b), or normalized (c) verb.

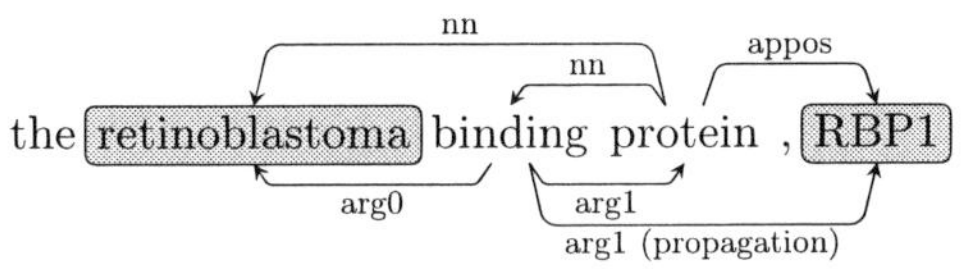

Figure 3: A sample compound noun phrase.

"retinoblastoma binds to protein, RBP1". Therefore, we add *arg1* from "binding" to "protein" in Figure 3. This operation will introduce cyclicity because the gerund is included in the noun phrase headed by "protein". We posit that these edges are useful when found in combination with other construction, such as appositive. We will discuss how to propagate *arg1* from the gerund "binding" to "RBP1" later.

Next we consider two cases of argument elision.

Elided argument relation Here we consider cases when the argument of a predicate is not explicit but implicit. Figure 4 shows a sentence where *arg0(interaction, Presenilin 1)* can be inferred. The SDG includes a *prep_via* from the first verb "suppresses" to the nominalized verb "interaction", to indicate the PP attachment to the verb. In this case, we add an edge *arg0* from the nominalized verb to the *arg0*-argument of the first verb. In constructing EDG, we also consider *prep_through* as well as *prep_by* when a gerund verb, rather than a nominalized verb, follows it.

Reduced relative clauses Relative clause is a clause that modifies a noun phrase. There are two types of relative clauses that frequently appear in biomedical text. Full relative clauses are introduced by relative pronouns, such as "which"

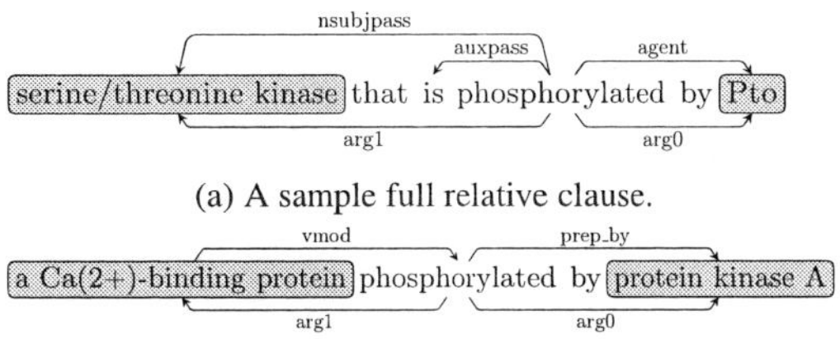

(a) A sample full relative clause.

(b) A sample reduced relative clause.

Figure 5: Sample relative clauses.

and "that". Reduced relative clauses start with a gerund or past participle and have no overt subject.

The PropBank annotation guidelines (Bonial et al., 2012) posit a numbered argument link from the relative clause verb to the trace in the parse tree which also indicates the referent noun phrase. For full relative clauses, we follow the normal procedure for verbs (Figure 5a). For reduced relative clauses, since we use the dependency structure that includes no traces, we use the edge *vmod* in the SDG from the head of the noun phrase to the reduced relative clause's verb (Figure 5b). The direction of this edge indicates that the relative clause is syntactically included in the larger noun phrase. For the *arg* edge, we reverse the direction of *vmod* and create an edge from the relative clause's verb, as shown in Figure 5b. When compared to Figure 5a, the *arg* construction unifies the treatment for full relative clauses.

Notice that although in both cases, the *arg1* is not an incident on named entities, it might still lead to the named entity through the propagation of edges as discussed in the next subsection.

3.3 Going Beyond Syntax

Here we consider the propagation of *arg* using information that goes beyond syntax.

Co-reference If an edge *arg* from a vertex *v* reaches a pronominal node, we add a new edge *arg* from *v* to any named entity the pronoun co-refers to. To detect the coreference we use the implementation of the technique described in (Qiu et al., 2004). For the acronyms with long-form and short-form, we treat them in the same way as coreference. We add extra edge *arg* when there is an *arg* incident on the long-form. We use the acronym detector of (Schwartz and Hearst, 2003) to add acronyms missed in SDG. Interestingly, SDG uses *appos* for both acronym and appositive.

Appositive Reconsider the fragment "the

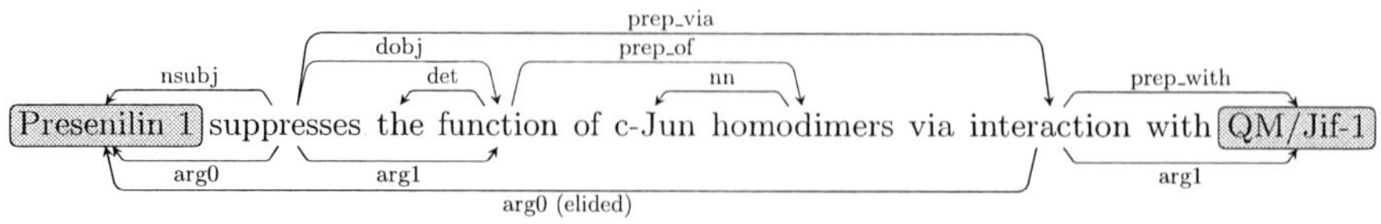

Figure 4: A sample elided argument relation.

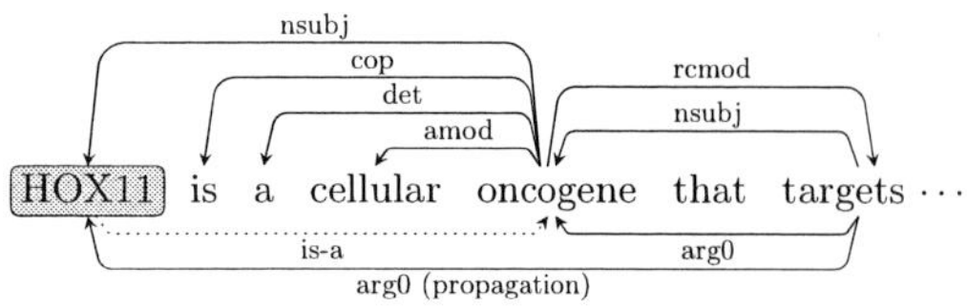

Figure 6: A sample *is-a* relation.

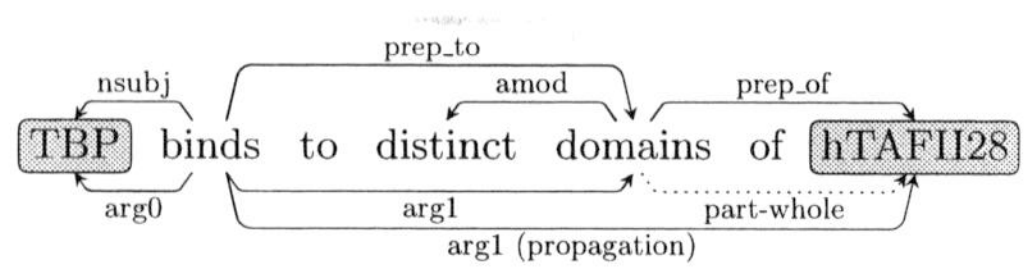

Figure 8: A sample *part-whole* relation.

retinoblastoma binding protein, RBP1" in Figure 3. Using the construction discussed thus far, the *arg1* will reach "protein". Further, SDG uses an edge *appos* from "protein" to "RBP1" for appositional modifier. We integrate *arg1* and *appos* to construct another edge *arg1* from "binding" to the actual named entity "RBP1".

Is-A In addition to appositive, we consider other forms of *is-a* relation mentioned textually, but cannot be directly found from the syntactic dependences. For example, in Figure 6, there is no edge in SDG to explicitly capture the *is-a* relation. It is worth noting that the edge *nsubj* itself does not indicate the *is-a* relation, but together with two other edges *cop* and *det*, we can figure it out. Hence we add a new edge from "oncogene" to "HOX11" to reflect this relation in EDG (dotted edge). Afterwards, we propagate *arg0* from "targets" to "HOX11".

Besides the pattern shown in Figure 6, we also identify "known as", "designated as", "considered as", "identified as" and "act as" as patterns that signal *is-a* relations. These patterns contain and extend rules in (Snow et al., 2005; Hearst, 1992).

Member-collection links a generic reference (called *collection*) to a group of entity mentions (called *members*). Like in Figure 7, typical key words that can identify *member-collection* relations are "including" and "such as". We consider the cases where mention group follows the keywords and the generic reference precedes these words. After the detection, we propagate *arg* from the collection to its members.

Part-whole links an entity part to its mention, typically denoting construction of larger entities out of smaller ones. Just like "breaking the glass of the window" can be stated as "breaking the window", in biomedical tasks an action on a larger unit can often be inferred from a mention of the action applied on its part. That is, in Figure 8, after we detect a *part-whole* relation, an edge *arg1* incident on the part is propagated to the object that contains it.

In this paper, we focus on three types of patterns to recognize *part-whole* relations. The first is the preposition phrase such as "domain of *e*". Here "domain" indicates the part and *e* indicates the larger entity mention the "domain" belongs to. Other keywords indicating parts include "fragment", "portion", and "region". The second structural elements is a compound nominal like "*e* domain". The third group exploits keywords such as "contain", "consist", and "compose". For each *part-whole* relation, we propagate edges from the part to its entity mention.

4 Experiments

We evaluated our method on protein-protein interaction (PPI) extraction task, where the system identifies whether a given protein pair in a sentence has PPI relationship or not. We used SDG or EDG as input representation of the sentences, which includes the named protein entities.

4.1 Kernels

We tested the effect of using EDG on two kernels that have been employed for PPI extraction.

Edit distance kernel is based on the edit distance among the shortest paths between entities in the dependency graph and is based on the minimal number of operations (deletion, insertion, substitution at word level) needed to transform one path (p_1) into the other (p_2). Following (Erkan et al.,

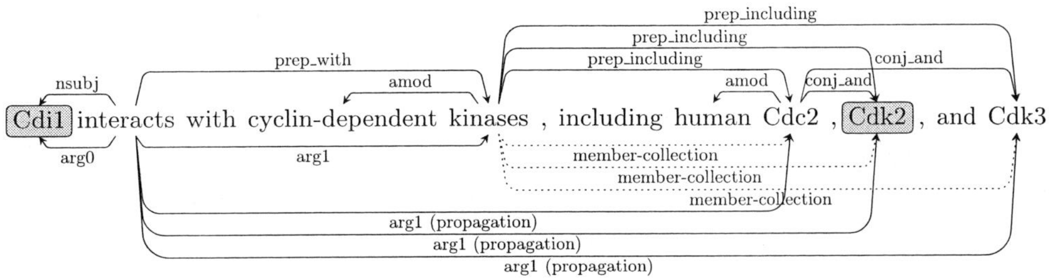

Figure 7: A sample *member-collection* relation.

2007), this number is normalized by the length of the longer path and converted into a similarity measure.

$$sim_e(p_1, p_2) = e^{-\gamma \text{editdist}(p_1, p_2)} \quad (1)$$

When comparing two shortest paths, we considered the word lemma and the edge labels . We also renamed the candidate pair in the sentence as "E1" and "E2" and the remaining proteins provided in the annotation as "EX". For example, the following are the shortest paths of Figure 2a, 3, and 8.

(a) E1 ← *arg0* ← activate → *arg1* → E2
(b) E1 ← *arg0* ← bind → *arg1* → E2
(c) E1 ← *arg0* ← bind → *arg1* → E2

Therefore, the edit distance between (a) and (b) is 1 because the predicate verbs are different. The distance between (b) and (c) is 0. It shows the generalizability of using EDG.

All-paths graph kernel is a practical instantiation of a graph kernel framework (Gärtner et al., 2003). It counts weighted shared paths of all possible lengths in a graph (Airola et al., 2008). All-paths graph kernel uses two graph representations: (1) a dependency graph where all edges on the shortest paths between the candidate pair receive a weight of 0.9 and other edges receive a weight of 0.3; and (2) a linear graph where each word node is connected by an edge to its succeeding word node with weight 0.9.

We used word (not lemma) and edge labels to compute the all-paths graph kernel. Similar to the case with the edit distance kernel, we replaced the protein names in a sentence with "E1", "E2" and "EX". We use the APG software (`http://mars.cs.utu.fi/PPICorpora/GraphKernel.html`) to train and test the kernel. The software uses sparse regularized least squares method instead of SVM.

Table 1: Basic statistics of the corpora.

Corpus	Sentences	# Positives	# Negatives
AIMed	1,955	1,000	4,834
BioInfer	1,100	2,534	7,132
HPRD50	145	163	270
IEPA	486	335	482
LLL	77	164	166

4.2 Experimental setup

We evaluated our method on five PPI corpora that have been used in the community: AIMed (Bunescu et al., 2005), BioInfer (Pyysalo et al., 2007), HPRD50 (Fundel et al., 2007), IEPA (Ding et al., 2002), and LLL (Nédellec, 2005). These corpora have different sizes (Table 1) and vary slightly in their definition of PPI (Pyysalo et al., 2008).

(Tikk et al., 2010) conducted a comparison of a variety of PPI extraction systems on these corpora (`http://mars.cs.utu.fi/PPICorpora`). We used the same experimental setup to evaluate our methods: self-interactions were excluded from the corpora and 10-fold document-level cross-validation is used for evaluation.

For our experiments, we used the Charniak-Johnson parser (Charniak and Johnson, 2005) and the Stanford conversion tool with "Collapsed" setting to obtain SDG (De Marneffe and Manning, 2008). The edit distance kernel was trained with LIBSVM (Chang and Lin, 2011). The APG kernel was trained with APG software.

Both these kernels have several parameters, whose settings can influence the performance. In this paper, we did not perform exhaustive systematic parameter search and optimization. We believe such parameter tuning techniques might lead to further improvements.

For the edit kernel, we set γ to 4.5, which was

the value used in the original application of edit kernel on these corpora (Erkan et al., 2007). We set c in SVM to 10, which was the average best value used in (Tikk et al., 2010). For the APG kernel, we used the default settings of implementation of (Airola et al., 2008) which uses a grid parameter search for each iteration of the 10-fold cross validation. The parameter search selects the best setting based on a random set of 1,000 samples from the training sets (9 folds). If there are less than 1,000 samples, the software used the whole training set. Note that the test sets (the remaining fold) were not used for the parameter tuning.

4.3 Results

Performance, as measured by precision, recall, and F-value, is shown in Table 2. To provide context, we also include the results published in (Tikk et al., 2010) and (Miwa et al., 2009a). The first reports the results of the APG kernel (Airola et al., 2008) that was found to be a leading performer on these 5 corpora in the study reported in (Tikk et al., 2010). The second set of results is those of an ensemble method that combines different systems.

Although we are using the same corpora in the study of (Tikk et al., 2010), and the same implementation of the APG kernel, the results in Row 1 and Row 6 in the table are not the same. The differences are possibly due to the fact that different parsers were used and how parameters were chosen. However, we want to emphasize that all our own measurements (e.g., in Rows 3-5 or Rows 6-8) are directly comparable to each other because the same parameter settings were used for each corpus.

The first part of Table 2 shows results using the edit distance kernel with original dependency graph (Row 3), and with the complete EDG (Row 4). We also experimented with different configurations of EDG by dropping one of the extra edge types added in EDG. The results obtained by the best configuration are reported in Row 5. On three of the corpora, the best results are obtained by using the full EDG. However, better results were obtained on HPRD50, when the *member-collection* relations were not included and on LLL, when the *is-a* relations were not included. In the next subsection we will address why these relations were not included.

Overall, comparing Rows 3 and 4, we obtain F-value improvements using EDG over using SDG on 4 corpora (except LLL), with around 10% gains on AIMed and HPRD50 and noticeable gain in recall. For 3 of the corpora (AIMed, HPRD50 and IEPA), there is an increase in both precision and recall. For BioInfer, the gain in precision slightly exceeds the loss in recall whereas in LLL the gain in precision is slightly lower than the loss in recall. When Row 5 is used for comparison, we obtain an improvement in F-value for all 5 corpora with improvement in recision and recall in 4 corpora (BioInfer being the exception). We now see over 18% F-value improvement on HPRD50.

Despite weak performance of the edit kernel using the baseline SDG, the performance of this kernel with full EDG is close to or exceeds the results of the leading PPI systems using kernel methods (Rows 1 and 2) on 4 corpora and exceeds them on these 4 corpora when results of Row 5 is considered.

The second part of Table 2 (Rows 6–8) shows results using the APG kernel. The EDG (Best) in Row 8 is achieved on AIMed, BioInfer and LLL by dropping the *is-a* relation and on HPRD50 by not including the *member-collection* relations. We see F-value gains on 4 corpora through the use of EDG.

Comparing the results on the edit distance and APG kernels, we find that the more complex APG kernel (the best one overall in (Tikk et al., 2010) study) gets generally better results than Edit kernel using the baseline SDG. However, the use of EDG not only closes the gap between the kernels but in fact, edit kernel with EDG obtains higher F-value than APG with SDG or EDG in 4 of the 5 corpora.

To provide the comparision with non-kernel methods, we also include the results published in (Miwa et al., 2009b), which is the state-of-the-art system on the five corpora. This paper develops several systems that use a rich feature vector, combining analysis from different parsers and the values obtained from multiple kernels including the APG's score. L2-SVM and SVM-CW are among the leading SVM-based systems proposed in this paper.

Row 9 shows the results of L2-SVM on these corpora. We observe that both edit kernel and APG kernel with EDG (Best) gets improvements on two of the corpora. Row 10 shows the results of SVM modified for corpora weighting (SVM-CW). Using one of the corpora as the target corpus, SVM-CW weights the remaining corpora (called

Table 2: Evaluation results. Performance is reported in terms of Recall/Precision/F-value.

Kernel	AIMed	BioInfer	HPRD50	IEPA	LLL
1 (Tikk et al., 2010)	53.6/59.9/56.2	61.3/60.2/60.7	69.8/68.2/67.8	82.6/66.6/**73.1**	98.0/68.0/78.4
2 (Miwa et al., 2009a)	68.8/55.0/60.8	71.1/65.7/**68.1**	76.1/68.5/70.9	78.6/67.5/71.7	86.0/77.6/80.1
Edit kernel					
3 SDG	40.0/61.4/48.4	64.7/49.5/56.1	55.8/68.4/61.5	69.6/74.7/72.0	89.6/71.7/79.7
4 EDG	57.3/65.3/**61.1**	57.6/59.9/58.7	66.9/75.7/71.0	69.9/76.2/72.9	85.4/74.1/79.3
5 EDG (Best)	–	–	76.7/83.3/**79.9**	–	92.1/78.2/**84.6**
All-paths graph kernel					
6 SDG	69.0/48.0/56.6	73.5/58.8/65.3	69.3/60.1/64.4	77.9/65.4/71.1	87.8/69.9/77.8
7 EDG	66.0/52.3/58.3	72.1/56.1/63.1	71.2/62.7/66.7	75.2/65.3/69.9	82.9/69.4/75.6
8 EDG (Best)	71.3/51.1/59.5	69.2/58.7/63.5	76.1/62.6/68.7	76.1/68.2/71.9	87.2/75.3/80.8
Feature vector (Miwa et al., 2009b)					
9 L2-SVM	63.2	66.2	67.2	73.0	80.3
10 SVM-CW	64.0	66.7	72.7	75.2	85.9

the source corpora) with "goodness" for training on the target corpus, adjusting the effect of their compatibility and incompatibility (Miwa et al., 2009b). Thus, their results are not directly comparable with our results. However we obtain improvements using edit kernel with EDG (Best) on HPRD50.

4.4 Contribution of individual relation

Table 3 compares the effects of different techniques in EDG on five corpora using the edit distance kernel. We first evaluated SDG obtained from the Stanford conversion tool with "CCProcessed" setting (Row 2) for processing conjunctions, and next added only syntax based *arg0* and *arg1* (Row 3). After that, we added in succession referential links (including coreference, appositive, and *is-a*), *member-collection*, and *part-whole* detection in the EDG construction step by step (Row 4–6). Overall, using "CCProcessed" increases the F-values on all five corpora. EDG constructed using syntax based *arg* achieves additional increases on 4 out of 5 corpora (exception was IEPA). Every subsequent step generally provides more improvements on F-values. However, we observed that on HPRD50, *member-collection* decreased F-value. Therefore we tried to switch off this part in the EDG construction but included the rest of the relations and achieved a higher F-value of 79.9% on this corpus (Row 7). This corresponds to the same result we displayed in Row 5 (EDG Best) in Table 2. On the LLL corpus, as components were successively added, we noticed a drop in F-value when referential linking was added. So similarly by turning off *is-a* detection and including all other EDG edges enabled us to obtain the EDG best F-value of 84.6% on LLL.

We also identified that *is-a* decreased F-values on IEPA, however no further improvement could be made by switching it off. We plan to further analyze this result in the future.

Additionally, due to the gap in the performance between our system and (Miwa et al., 2009a) on BioInfer, we analyzed the error cases and noticed several cases similar to the following example. The candidate pair of named enitites are marked in bold.

- This process involves other **actin-binding proteins**, such as **cofilin** and coronin.

Using techniques as shown in Figure 3, we can create *arg0 (binding, actin)* and *arg1 (binding, proteins)* in EDG and also detect *member-collection* relation between "actin-binding proteins" and "cofilin". With propogation, an interaction between "actin" and "cofilin" can be predicted. However, this relation is annotated as a negative, but instead the annotation in BioInfer includes a positive relation between "actin-binding proteins" and "cofilin". Because of similar examples in BioInfer, the *member-collection* and *is-a* and propagation failed to improve the results in BioInfer.

5 Conclusion

In this paper, we strive to find a level of abstraction that is more suitable for tasks such as relation extraction. For this purpose, we introduced techniques to create a new dependency graph representation (EDG) that goes beyond syntactic dependencies. We evaluated the efficacy of EDG

Table 3: Contributions of different part in SDG and EDG using edit kernel. Performance is reported in terms of Recall/Precision/F-value.

Kernel	AIMed	BioInfer	HPRD50	IEPA	LLL
1 SDG (Collapsed)	40.0/61.4/48.4	64.7/49.5/56.1	55.8/68.4/61.5	69.6/74.7/72.0	89.6/71.7/79.7
2 SDG (CCProcessed)	46.4/58.9/51.9	56.2/57.1/56.6	58.9/67.6/63.0	70.2/74.8/72.4	89.6/73.5/80.8
3 EDG (syntax based *arg*)	48.1/61.2/53.9	56.3/58.5/57.4	66.9/73.2/69.9	69.3/74.4/71.7	89.0/74.1/80.9
4 EDG (above, coref, app, isa)	52.2/58.6/55.2	56.7/58.3/57.5	65.6/77.0/70.9	69.0/74.0/71.4	87.2/72.2/79.0
5 EDG (above, mem-coll)	53.2/59.2/56.0	57.1/58.6/57.8	64.4/77.8/70.5	69.6/76.4/72.8	85.4/74.5/79.6
6 EDG (above, part-whole)	57.3/65.3/61.1	57.6/59.9/58.7	66.9/75.7/71.0	69.9/76.2/72.9	85.4/74.1/79.3
7 EDG (Best)	57.3/65.3/61.1	57.6/59.9/58.7	76.7/83.3/79.9	69.9/76.2/72.9	92.1/78.2/84.6

with the edit distance and APG kernels and applied them on 5 different PPI-related datasets. We obtained improvements in F-value by using EDG. We find that despite the simplicity of the edit kernel and its weak performance with the baseline graph, results comparable to state-of-the-art systems using kernel methods are obtained on different corpora with the inclusion of EDG.

While the use of EDG has led to gain in recall as well as precision mostly, the recall drops with BioInfer dataset. We would like to analyze this result further in the future. One of our main motivations for developing EDG is to develop methods to learn with small datasets and whether the abstraction captured in EDG allows for easier generalization. The testing of learning with small datasets and use in context of active learning will be investigated in the future.

We plan to test the use of EDG on other relation extraction tasks in the biomedical domain. We also plan to investigate richer features and their combinations in conjunction with the use of EDG.

Acknowledgments

Research reported in this manuscript is supported by the National Science Foundation under Grant No. DBI-1062520.

References

Antti Airola, Sampo Pyysalo, Jari Björne, Tapio Pahikkala, Filip Ginter, and Tapio Salakoski. 2008. All-paths graph kernel for protein-protein interaction extraction with evaluation of cross-corpus learning. *BMC bioinformatics*, 9(Suppl 11):S2.

Claire Bonial, Jena Hwang, Julia Bonn, Kathryn Conger, Olga Babko-Malaya, and Martha Palmer, 2012. *English PropBank Annotation Guidelines*. Center for Computational Language and Education Research Institute of Cognitive Science University of Colorado at Boulder.

Razvan C. Bunescu and Raymond J. Mooney. 2005. A shortest path dependency kernel for relation extraction. In *HLT-EMNLP*, pages 724–731, Stroudsburg, PA, USA.

Razvan C. Bunescu, Ruifang Ge, Rohit J. Kate, Edward M. Marcotte, Raymond J. Mooney, Arun K. Ramani, and Yuk Wah Wong. 2005. Comparative experiments on learning information extractors for proteins and their interactions. *Artificial Intelligence in Medicine*, 33(2):139–155.

Chih-Chung Chang and Chih-Jen Lin. 2011. LIBSVM: A library for support vector machines. *ACM Transactions on Intelligent Systems and Technology*, 2(27):1–27.

Eugene Charniak and Mark Johnson. 2005. Coarse-to-fine n-best parsing and maxent discriminative reranking. In *Proceedings of the 43rd Annual Meeting on Association for Computational Linguistics*, pages 173–180. Association for Computational Linguistics.

Sung-Pil Choi and Sung-Hyon Myaeng. 2010. Simplicity is better: revisiting single kernel ppi extraction. In *Proceedings of the 23rd International Conference on Computational Linguistics*, pages 206–214. Association for Computational Linguistics.

Faisal Md. Chowdhury, Alberto Lavelli, and Alessandro Moschitti. 2011. A study on dependency tree kernels for automatic extraction of protein-protein interaction. In *Proceedings of BioNLP 2011 Workshop*, pages 124–133, Portland, Oregon, USA, June. Association for Computational Linguistics.

Michael A. Covington. 2001. A fundamental algorithm for dependency parsing. In *Proceedings of the 39th Annual ACM Southeast Conference*, pages 95–102.

Marie-Catherine De Marneffe and Christopher D Manning. 2008. The stanford typed dependencies representation. In *Coling 2008: Proceedings of the workshop on Cross-Framework and Cross-Domain Parser Evaluation*, pages 1–8. Association for Computational Linguistics.

Jing Ding, Daniel Berleant, Dan Nettleton, and E Wurtele. 2002. Mining MEDLINE: Abstracts, sentences, or phrases. In *Proceedings of the Pacific Symposium on Biocomputing*, volume 7, pages 326–337.

Günes Erkan, Arzucan Özgür, and Dragomir R Radev. 2007. Semi-supervised classification for extracting protein interaction sentences using dependency parsing. In *EMNLP-CoNLL*, volume 7, pages 228–237.

Katrin Fundel, Robert Küffner, and Ralf Zimmer. 2007. Relex - relation extraction using dependency parse trees. *Bioinformatics*, 23(3):365–371.

Thomas Gärtner, Peter Flach, and Stefan Wrobel. 2003. On graph kernels: Hardness results and efficient alternatives. In *Conference on Learning Theory*, pages 129–143.

Marti A. Hearst. 1992. Automatic acquisition of hyponyms from large text corpora. In *COLING*, pages 539–545, Stroudsburg, PA, USA. Association for Computational Linguistics.

Martin Krallinger, Florian Leitner, Carlos Rodriguez-Penagos, Alfonso Valencia, et al. 2008. Overview of the protein-protein interaction annotation extraction task of biocreative ii. *Genome biology*, 9(Suppl 2):S4.

Haibin Liu, Karin Verspoor, Donald C Comeau, Andrew MacKinlay, and W John Wilbur. 2013. Generalizing an approximate subgraph matching-based system to extract events in molecular biology and cancer genetics. In *Proceedings of the BioNLP Shared Task 2013 Workshop*, pages 76–85.

Makoto Miwa, Rune Sætre, Yusuke Miyao, and Jun'ichi Tsujii. 2009a. Protein-protein interaction extraction by leveraging multiple kernels and parsers. *International Journal of Medical Informatics*, 78(12):e39–46.

Makoto Miwa, Rune Sætre, Yusuke Miyao, and Jun'ichi Tsujii. 2009b. A rich feature vector for protein-protein interaction extraction from multiple corpora. In *Proceedings of the 2009 Conference on Empirical Methods in Natural Language Processing: Volume 1-Volume 1*, pages 121–130. Association for Computational Linguistics.

Makoto Miwa, Rune Sætre, Yusuke Miyao, and Jun'ichi Tsujii. 2010. Entity-focused sentence simplification for relation extraction. In *Proceedings of the 23rd International Conference on Computational Linguistics*, COLING '10, pages 788–796, Stroudsburg, PA, USA. Association for Computational Linguistics.

Claire Nédellec. 2005. Learning language in logic - genic interaction extraction challenge. In *Proceedings of the 4th Learning Language in Logic Workshop*, volume 7, pages 31–37.

Nhung TH Nguyen, Makoto Miwa, Yoshimasa Tsuruoka, Takashi Chikayama, and Satoshi Tojo. 2015. Wide-coverage relation extraction from medline using deep syntax. *BMC bioinformatics*, 16(1):107.

Xia Ning and Yanjun Qi. 2011. Semi-supervised convolution graph kernels for relation extraction. In *SDM*, pages 510–521. SIAM.

Stephan Oepen, Marco Kuhlmann, Yusuke Miyao, Daniel Zeman, Dan Flickinger, Jan Hajic, Angelina Ivanova, and Yi Zhang. 2014. Semeval 2014 Task 8: Broad-coverage semantic dependency parsing. In *Proceedings of the 8th International Workshop on Semantic Evaluation (SemEval 2014)*, pages 63–72.

Yifan Peng, Manabu Torii, Cathy H. Wu, and K. Vijay-Shanker. 2014. An NLP framework for fast development of pattern-based biomedical relation extraction systems. *BMC bioinformatics*, 15:285.

Sampo Pyysalo, Filip Ginter, Juho Heimonen, Jari Björne, Jorma Boberg, Jouni Järvinen, and Tapio Salakoski. 2007. BioInfer: A corpus for information extraction in the biomedical domain. *BMC bioinformatics*, 8(1):50.

Sampo Pyysalo, Antti Airola, Juho Heimonen, Jari Björne, Filip Ginter, and Tapio Salakoski. 2008. Comparative analysis of five protein-protein interaction corpora. *BMC bioinformatics*, 9(Suppl 3):S6.

Long Qiu, Min yen Kan, and Tat seng Chua. 2004. A public reference implementation of the rap anaphora resolution algorithm. In *Proceedings of the Fourth International Conference on Language Resources and Evaluation (LREC 2004)*, pages 291–294.

Fabio Rinaldi, Gerold Schneider, Kaarel Kaljurand, Michael Hess, and Martin Romacker. 2006. An environment for relation mining over richly annotated corpora: the case of genia. *BMC bioinformatics*, 7(Suppl 3):S3.

A.S. Schwartz and M.A. Hearst. 2003. A simple algorithm for identifying abbreviation definitions in biomedical text. In *Proceedings of the Pacific Symposium on Biocomputing*, volume 8, pages 451–462.

Rion Snow, Daniel Jurafsky, and Andrew Y. Ng. 2005. Learning syntactic patterns for automatic hypernym discovery. In *Advances in Neural Information Processing Systems 17*, pages 1297–1304, Cambridge, MA. MIT Press.

Domonkos Tikk, Philippe Thomas, Peter Palaga, Jörg Hakenberg, and Ulf Leser. 2010. A comprehensive benchmark of kernel methods to extract protein-protein interactions from literature. *PLoS computational biology*, 6(7):e1000837.

Sofie Van Landeghem, Sampo Pyysalo, Tomoko Ohta, and Yves Van de Peer. 2010. Integration of static relations to enhance event extraction from text. In *Proceedings of the 2010 Workshop on Biomedical Natural Language Processing*, BioNLP '10, pages

144–152, Stroudsburg, PA, USA. Association for Computational Linguistics.

Sofie Van Landeghem, Jari Björne, Thomas Abeel, Bernard De Baets, Tapio Salakoski, and Yves Van de Peer. 2012. Semantically linking molecular entities in literature through entity relationships. *BMC bioinformatics*, 13(Suppl 11):S6.

Min Zhang, Jie Zhang, Jian Su, and Guodong Zhou. 2006. A composite kernel to extract relations between entities with both flat and structured features. In *Proceedings of the 21st International Conference on Computational Linguistics and the 44th annual meeting of the Association for Computational Linguistics*, pages 825–832, Stroudsburg, PA, USA.

Yijia Zhang, Hongfei Lin, Zhihao Yang, Jian Wang, and Yanpeng Li. 2012. Hash subgraph pairwise kernel for protein-protein interaction extraction. *IEEE/ACM Transactions on Computational Biology and Bioinformatics (TCBB)*, 9(4):1190–1202.

Guodong Zhou, Min Zhang, Dong Hong, and Ji Qiaoming Zhu. 2007. Tree kernel-based relation extraction with context-sensitive structured parse tree information. In *EMNLP-CoNLL*, pages 728–736.

Event Extraction in pieces: Tackling the partial event identification problem on unseen corpora

Chrysoula Zerva
National Center of Text Mining,
School of Computer Science,
University of Manchester, UK
c.zerva@cs.man.ac.uk

Sophia Ananiadou
National Center of Text Mining,
School of Computer Science,
University of Manchester, UK
sophia.ananiadou@manchester.ac.uk

Abstract

Biomedical event extraction systems have the potential to provide a reliable means of enhancing knowledge resources and mining the scientific literature. However, to achieve this goal, it is necessary that current event extraction models are improved, such that they can be applied confidently to unseen data with a minimal rate of error. Motivated by this requirement, this work targets a particular type of error, namely partial events, where an event is missing one or more arguments. Specifically, we attempt to improve the performance of a state-of-the-art event extraction tool, EventMine, when applied to a new cancer pathway curation corpus. We propose a post-processing ranking approach based on relaxed constraints, in order to reconsider the candidate arguments for each event trigger, and suggest possible new arguments. The proposed methodology, applicable to the output of any event extraction system, achieves an improvement in argument recall of 2%-4% when applied to EventMine output, and thus constitutes a promising direction for further developments.

1 Introduction

In text mining, events are currently the most complex information unit that can be extracted from raw text, in terms of their ability to capture n-ary dynamic relations between entities and/or other events as indicated in Figure 1. Their dynamic properties mean that events constitute the closest equivalent to human-extracted information. The structured information representation of events can be used to enrich current knowledge sources such as ontologies and databases in an automated manner. This can be particularly useful for researchers in the biomedical domain, who use complicated models to represent molecular reactions, pathways etc. In order to improve these models, biologists currently need to sift through a continuously growing mountain of literature (Ananiadou et al., 2014). Thus, an automated means to extract knowledge, using event extraction technology, and to exploit this knowledge to augment existing models, would be an immense asset within biomedical research.

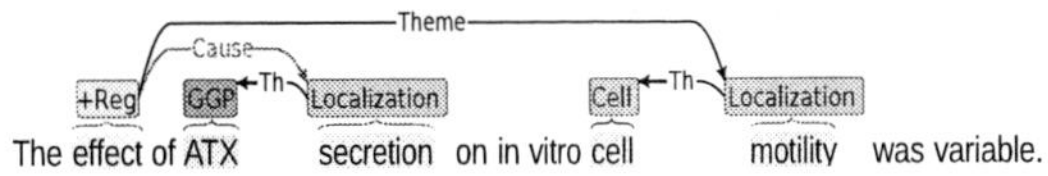

Figure 1: Event Extraction example: Localisation events nested as arguments to Positive Regulation (+Reg) event [1]

Motivated by the above, the Big Mechanism project (Cohen, 2014) aims to augment cancer pathway models automatically with events extracted from biomedical literature. To this end, event extraction systems need to be able not only to extract high quality events that cover a wide range of biomedical event types but also to robustly do so even when applied to unseen data. Indeed, the expectation is that event extraction systems will be successful in carrying out this task even when parameters such as text type or domain, are altered, without the need to retain the system.

However, the structure of current event extraction systems can hinder the ability to achieve the above goal. Since event extraction has so far been treated as a supervised learning task, the performance of systems is heavily dependent on the annotation, context and domain of the training data, and may drop significantly when one of the initial

[1] Sentence taken BioNLP 2013 CG corpus (Nédellec et al., 2013). Annotation visualised with BRAT annotation tool (Stenetorp et al., 2012).

Proceedings of the 2015 Workshop on Biomedical Natural Language Processing (BioNLP 2015), pages 31–41,
Beijing, China, July 30, 2015. ©2015 Association for Computational Linguistics

specifications changes, even within the same domain. Especially in pipelined architectures, which consist of sequential classification tasks, additional annotation constraints are learned from the training corpus at each stage in the pipeline. While these additional constraints improve the model's precision, they render it less adaptable to deviating event structures. One of the consequences of this is the failure to retrieve some of the information that should be associated with the event, leading to so-called partial event identification, where some of the event arguments are missing. Although such errors may not be of vital importance in all domains, they can be extremely detrimental when attempting to link an event to a biomedical model, since they can lead to erroneous or useless assertions.

The work described here focusses on resolving the problem by applying a generic constraint relaxation post-processing strategy to the output of an event extraction system (EventMine (Miwa and Ananiadou, 2013)), with the aim of reducing the number of recognised events that have missing arguments. Motivated by an analysis of the Big Mechanism testing corpus described in Section 3, we relax the annotation constraints related to argument roles and subsequently reconsider all the entities within a sentence that are valid argument candidates, by exploiting syntactical dependencies. We employ the confidence values obtained from an *Adaboost* (Freund and Schapire, 1997) classifier to rank candidate arguments for each event trigger, and to determine which of them constitute valid additions to the event. Using this approach, we are able to improve the *recall* on partial events identified by EventMine by at least 2% and, importantly, we gain fruitful insights into factors that could further improve performance.

2 Background: The Event Extraction Task in Biomedicine

In this section we provide an overview of the event extraction procedure, focussing on biomedical events. Our emphasis is on the details of pipelined event extraction, since this is the approach employed by the EventMine system, which we use to perform event extraction. Finally, we review the main approaches for adapting event extraction to new or unseen data.

2.1 Event Structure

In text mining, events refer to units representing dynamic, n-ary relations between named entities. In the biomedical domain, this definition can be narrowed to units representing molecular interactions stated within textual documents (Björne and Salakoski, 2011).

The typical structure of events (as defined and used in BioNLP shared tasks, e.g., (Kim et al., 2009), (Nédellec et al., 2013)) includes an obligatory *predicate/trigger*, i.e., a word sequence in text that characterises the event type. Potentially, an event may also have one or more *arguments*, i.e., entities in text that are semantically linked to the trigger. Considering the trigger and the arguments as nodes, the links between them can be considered as directed edges (from the trigger to the argument), which represent the role that the argument plays with respect to the trigger. As events are dynamic elements, the same entity can participate in different events, and may assume different roles in each event. Also, since events are solid information units, they can themselves act as arguments to other events, leading to the extraction of complex/nested events (Björne et al., 2010). These characteristics can be observed in the example of Fig 1 presented in Section 1.

2.2 Event Mining Architecture

In order to extract structures of the complexity illustrated in Figure 1, current state-of-the-art systems break event extraction down into multiple classification tasks that have to be solved in order to produce the final structured event representation. The learning process to carry out these tasks can be undertaken either sequentially in a pipelined manner, as in EventMine (Miwa and Ananiadou, 2013) and TEES (Björne and Salakoski, 2013), or as joint learning task, as for FAUST (Riedel and McCallum, 2011). EventMine, the system employed in this work, utilises the pipelined approach, and consists of the following modules:

- *Event trigger classifier:* Identifies spans of text that act as triggers and annotates them with the corresponding type (event label).
- *Argument detector:* Links each trigger with at most one argument and annotates the edge (link) with the corresponding argument role type.
- *Multiple argument detector:* Adds additional arguments to the pairs of the previous step, final-

ising each event structure.

- *Modification detector:* Identifies event modifications (negation and speculation)

All the above are formulated as multi-class tasks that are learned in a supervised way, using one-versus-rest SVM implementation of LibLinear (Fan et al., 2008). EventMine is able to perform with state-of-the-art accuracy, achieving F-score of 52% on the CG and 53% on PC task of the latest BioNLP shared task (Nédellec et al., 2013), rendering it a suitable tool for this study.

2.3 Adaptation and generalisation approaches

One of the problems usually encountered with supervised models, such as those used by EventMine, is that they are specifically tailored to features of the corpus on which they have been trained. As a result, their functionality is restricted to the trigger, argument and role types that they have been trained to identify and extract. For example, some corpora focus only on protein-protein interactions, while others include chemical reactions, anatomical entities and or a combination of the above. Intuitively, in order to capture events that encompass all the above types, either one would have to re-annotate a corpus with all the required types of interest, or use some combination of either the corpora or the models trained on them.

Since corpus annotation is an expensive and time consuming task, various computational approaches to combining information have been proposed. A particularly straightforward approach is to combine the models in a stacking manner as in (Wolpert, 1992), where a method inspired from cross-validation is used to train different models on subsets of the different corpora, and then use the validation set to learn how to combine their outputs to obtain the desired result. More recently, a range of domain adaptation techniques have been proposed that try to adapt to a new corpus by either selectively training on the instances and/or features that are expected to maximize performance (Chen et al., 2011; Xia et al., 2013), or by attempting to tailor feature distributions to the one of the new corpus with various methods such as kernel based ones (Daumé III, 2009; Kulis et al., 2011) or transfer component analysis (Pan et al., 2011). Finally, (Miwa et al., 2013) suggests the use of a filtering model, which considers the overlap of the available corpora and filters redundant and contradicting labelling across different corpora and then merges the corpora in order to train a single model on their combination. The filtering, as Miwa explains, is heuristically achieved by limiting the generation of negative instances in each corpus to only those cases in which the corresponding surface expression matches at least one positive instance of an annotated type in any corpus that shares that type. The method, referred to as wide coverage, when implemented in EventMine outperforms other stacking and domain adaptation methods as shown in (Miwa et al., 2013). Accordingly, it was the chosen approach for this work.

3 Corpora and Annotation Considerations

3.1 Training Corpora

For training the wide coverage method was applied on the combination of the training sets of the following corpora, treated as described in (Miwa et al., 2013) : Genia09 of BioNLP '09 (Kim et al., 2009), Genia11, EPI & ID of BioNLP '11 (Kim et al., 2011), DNA-methylation (Ohta et al., 2011), ePTM (Pyysalo et al., 2011), mTOR (Caron et al., 2010), and MLEE (Pyysalo et al., 2012).

3.2 Testing Corpora

The corpus that provided the motivation for this work, henceforth referred to as BM, is a small annotated set of six passages extracted from full-text biomedical research papers in PubMed [2]. It concerns cancer pathway curation and was manually annotated with biomedical named entities and events by expert biologists participating in the Big Mechanism project (Cohen, 2014). In total it consists of 155 event and 247 named entity annotations. The range of the entities and events annotated render it a valid candidate for the application of the wide coverage approach described in Section 2.3, because the entities span across *Chemical, Protein* and *Cell* instances, while the event types cover pathways, various protein interactions (*Binding, Regulation*, etc) and other cancer related events. Since there is no single related training corpus with similar annotations, a model that can learn labels from different corpora is necessary to facilitate recognition of all of the above event and entity types.

[2] PubMed ids: PMC2872605, PMC3058384

The annotation scheme in BM corpus differs from the uniform annotation scheme used in the training corpora in the following ways:

- The entity labels are different to those used in the training corpora (see Fig 3 and section 4.1)
- No distinction is made between different event types in the test corpus annotations
- A simplified edge type annotation was followed in the BM corpus, discriminating only between simple arguments and arguments indicating the site (cellular location) where the event took place. As opposed to the BioNLP schema the edge annotations in BM contain less semantic information (there is no discrimination between roles such as Instrument, Participant, Cause etc)

Figure 2 illustrates an event annotated according to both BioNLP guidelines and to the BM corpus guidelines. The simplifications of the BM corpus

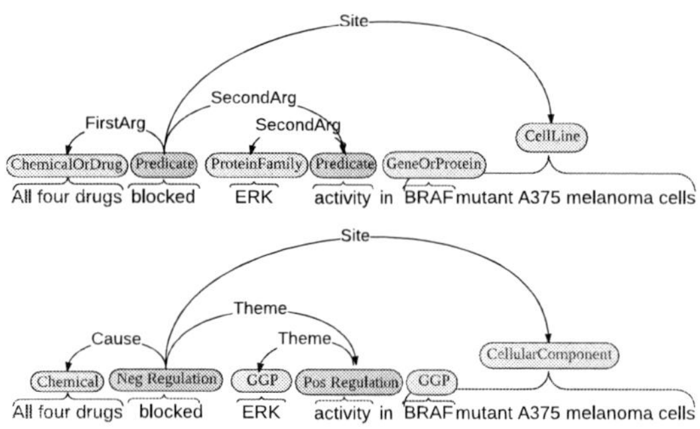

Figure 2: Example of event annotation in the BM corpus (top) versus BioNLP (bottom) : we can observe the different labels used for entities, events and edges

annotation scheme, compared to the scheme used in the training corpora, motivated our approach to relaxing the constraints used to link arguments to event triggers, as described in the previous section.

Due to the small size of BM corpus presented above, our experiments were repeated on the MLEE corpus (Pyysalo et al., 2012), using the development set as a test case. The MLEE corpus was chosen because as BM it displays a wide range of entities and events that spanned across different levels of biological organisation (molecular to organ) instead of focusing on protein reactions. The experiments were repeated twice, once including the MLEE training corpus in the training data, and once keeping it only for test purposes, in order to provide a comparison between application on seen and unseen data (see Section 6). For the purposes of this study, the edge annotations in MLEE are simplified to *Arg1* and *Arg2*, such that it is in line with the annotation scheme of the BM corpus, allowing for the relaxed constraint approach to be applied and for a better comparison of the results.

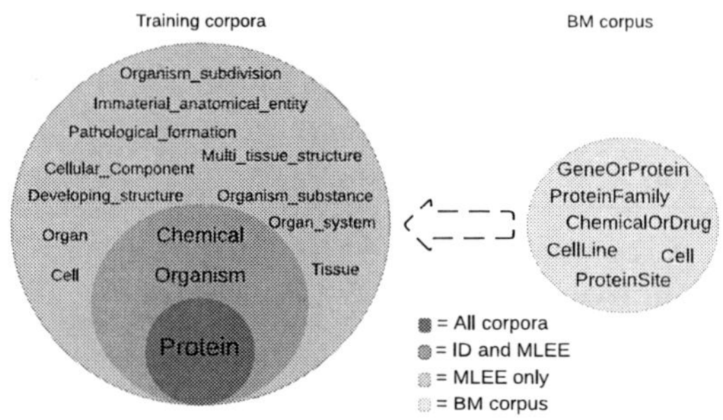

Figure 3: Initial named entity labels for the corpora used in the experiments. BM corpus labels were adapted to the training corpora as explained in section 4.1

4 Methodology

4.1 Adapting the entity labels without supervision

Since there is no widely accepted standard in the community in terms of the annotation labels for named entities, a common issue when processing multiple corpora, is overlapping annotations. In other words, different labels may be used to describe the same entity type, and that exactly is the case for the BM corpus when compared to the training corpora.[3] Hence, when testing on unseen data, it is necessary to map the labels of the test corpus to the ones that the model is trained to recognise, in order to obtain optimal results. For example, proteins were annotated as *GeneOrProtein* in the BM corpus and as *Protein* in the training corpora. For the filtering and unification of annotations instead of manually identifying the overlapping annotations, a heuristic automated label filtering method was implemented, in order to map the labels of the target/test corpus (TL) to those of the source/training one (SL). To that end, label similarity was calculated based on the following heuristic formula:

[3]The training corpora also contained conflicting / overlapping annotations initially that were priorly resolved in a similar manner

$$TL_i \rightarrow SL_j,$$

$$SL_j \rightarrow \underset{k}{\operatorname{argmax}}\left(\frac{\#(AnnE\_TL_i \cap AnnE\_SL_k)}{\#AnnE\_SL_k}\right) \quad (1)$$

where $AnnE\_TL_i$ corresponds to an annotated text span under the label TL_i in the target corpus, while $AnnE\_SL_j$ to an annotated text span under the label SL_j. The aforementioned text spans, can be single or multi-word tokens. Using this method, each label from the test corpus was assigned to the most similar label in the training corpus. For BM the labels were adapted as following:

BM corpus		Training Corpora
GeneOrProtein	→	*Protein*
ProteinFamily	→	*Protein*
ProteinSite	→	*Protein* [4]
ChemicalOrDrug	→	*Chemical*
CellLine	→	*Cellular_component*
SubcellularLocation	→	*Protein*

Table 1: Mapping between BM and trining corpora named entity annotations

It should be mentioned that in the case of the BM corpus, while there were obviously synonymous labels, the actual overlap found by the above technique was less than 10% for all labels, nonetheless still valid. In general this technique allows corpora to be added or removed without the need to fully revise the corpus. The same method could be used for different annotation types, such as event or edge type annotation.

4.2 Re-evaluating argument candidates of correct partial events

We hypothesise that, owing to the complexity of event patterns sought by the model, it sometimes fails to identify the complete set of arguments for an event, even if those arguments are correctly identified in the text as entities. This leads to the identification of partial events, such as the one presented in Figure 4. In order to identify the missing arguments, we aim to reduce the complexity of learned patterns, while complying with the annotation of the BM corpus.

Thus, we apply a relaxed post-processing step to the event extraction results, such that constraints regarding the learned roles of arguments are no longer imposed. Approaches that relax rule or pattern constraints have previously been shown to

[4]No overlap found, manual decision.

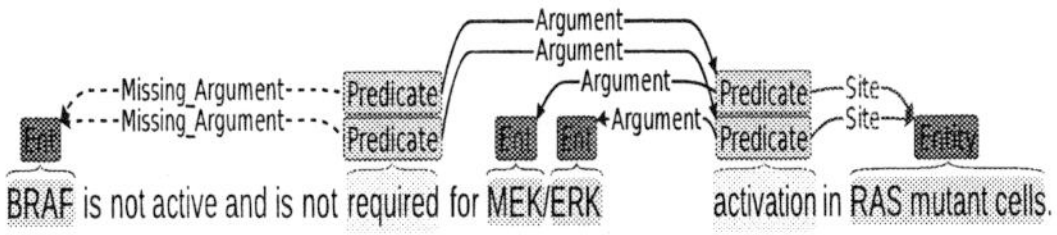

Figure 4: Example of partial event from the test corpus: *BRAF* should also be linked to the event but is missed by the EventMine model

constitute an efficient method of achieving generalisation and allowing models to be better adapted to other natural language processing tasks such as named entity recognition in (Tatar and Cicekli, 2011) and (Zhou and Su, 2003) or information extraction in (Ciravegnia and Lavelli, 2004). In our case, the relaxed constraints permit a re-evaluation of the possible relations between event triggers and recognised entities in a given sentence.

We implemented a ranking approach such that for each identified event trigger, all the entities in the same sentence that are not already linked to it by the EventMine model are ranked according to their likelihood of being related to the trigger.

The entity ranking is based on syntactic dependencies between word tokens for each sentence. The underlying assumption is that for an entity to be linked to an event trigger as an argument, there has to be some syntactic relation between the two terms. Syntactic analysis is undertaken by the Enju syntactic parser (Miyao et al., 2008), which has a model trained on biomedical corpora. Since each dependency can be seen as a link between two words, we can consider dependency relations as structured dependency graphs. Dependency graphs have been used before in event extraction (Buyko et al., 2009; Liu et al., 2013) with Liu's approach, on subgraph matching of directed dependency graphs, achieving high precision but low recall. Aiming for high recall, we take a different approach; in the undirected graphs, we expect the path between a trigger word and its related arguments to be shorter than the path between the same trigger word and other, non related entities in the same sentence. We thus consider the shortest dependency path length as the main feature for ranking. For example from the Enju output for the partial event shown in Figure 5, we can see that the shortest dependency path length between the entity BRAF and the event trigger is equal to 1 (direct link). To facilitate full exploitation of the dependency graph, we considered the following:

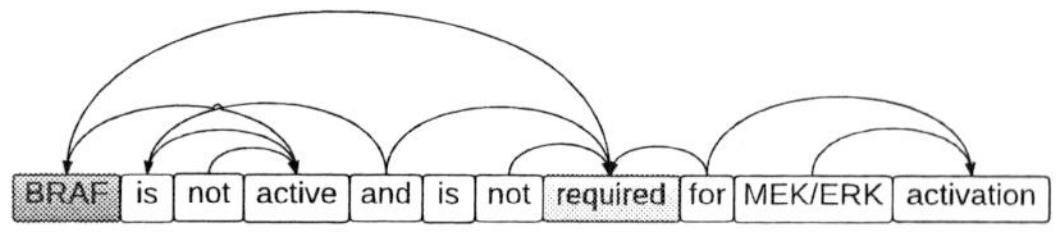

Figure 5: Dependency link representation example for a Biomedical sentence as analysed by Enju

- **Dependency type:** Enju provides the dependency type (prepositional, coordination, noun modifier, etc) for each dependency link/edge. This type can be exploited either to assign different weights to each dependency edge of the argument-trigger path, or to consider different path patterns. Such manipulation has been employed in rule-based event extraction in (Kilicoglu and Bergler, 2009), achieving good accuracy but low recall. Also, extracting specific path patterns renders the approach dependent on a particular parser, thus limiting the independence and adaptability of its application. Since the focus of this study was on recall and adaptability, the dependency type information was ignored, except for the case that follows.
- **Flattening coordination:** We decided that preprocessing was necessary to resolve coordination dependencies, such that a given entity would have the same distance to a trigger, regardless of the existence of a coordination argument dependency. Accordingly, in the calculation of the shortest paths, all coordination dependencies (labelled as *coord_arg* by Enju) are flattened as shown in Figure 6.

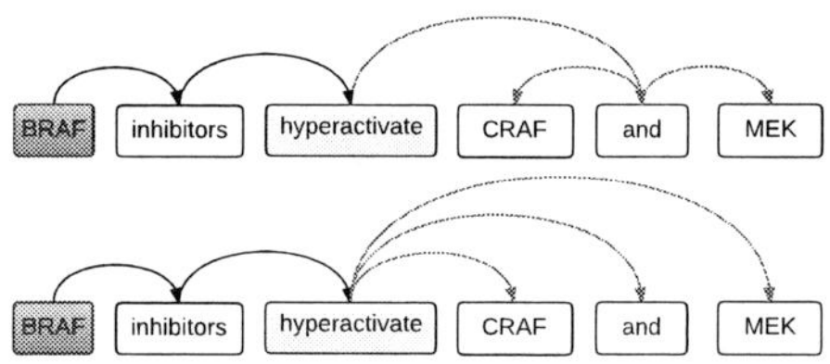

Figure 6: Flattening coordination dependencies

- **Nested Events:** To handle nested events, extracted events were also considered as entities, using the trigger as the representative text span, used to calculate the distance to the trigger of the top-level event and the rest of the features.

In order to obtain the rankings, we firstly consider our problem as one of *binary classification*, where the task is to classify each entity with respect to each trigger, as a valid (positive case) or non-valid (negative case) argument. Then, in training a classifier on the above binary classification task, we can employ the prediction confidence of the classifier model in order to rank the entities with respect to the event. The top ranked entity is selected and added to the event. In order to train a strong classifier model, a greater number of attributes that indicate the relation of an entity to a trigger were considered and implemented as additional features for the classifier. The main feature classes of the final feature set are listed below:

- Shortest dependency path (numeric)
- Entity Type (nominal)
- Participation in other events (binary)
- PoS (Part of Speech) (nominal)
- Context PoS (surrounding tokens) (nominal)
- Relative position to the event trigger (before/after) (nominal)
- Dependency on a prepositional token - type of prepositional token (binary-nominal)
- Event type (nominal)
- Token distance to trigger (numeric).

For the binary classification task, after comparison of an SVM, a logistic regression and an Adaboost classifier (implemented with random tree models), the AdaBoost classifier was chosen as it outperformed the rest by at least 10% (10 fold cross validation F-score on training set: 0.93). [5]

We also tried to avoid the addition of spurious arguments to events. Our initial experiments revealed that a considerable number of events require either a single argument or no arguments. For some event types such as *Gene_Expression*, such cases constituted more than 80% of the events. In order to avoid the addition of spurious arguments, and inspired by (Rahman and Ng, 2009) , an artificial *"null"* named entity instance was created for each event, and assigned to the events in the training set that did not require a second (or even a first) argument. Thus, the classifier would consider and rank the null entity along with the rest for each event.

Finally, to account for entities that are indirectly linked to events, i.e. those which occur as arguments of nested events, for each trigger, entities

[5] It should be noted that, while Adaboost appears to be most efficient for the purposes of our study, our classification task is only binary, and it is not straightforward to assume that it would outperform SVM in the rest of the EventMine pipeline, without additional testing

belonging to its parent event or its nested events were excluded from ranking. Furthermore, entities that were already assigned to a different event having the same trigger, as in Figure 7, were considered mutually exclusive.

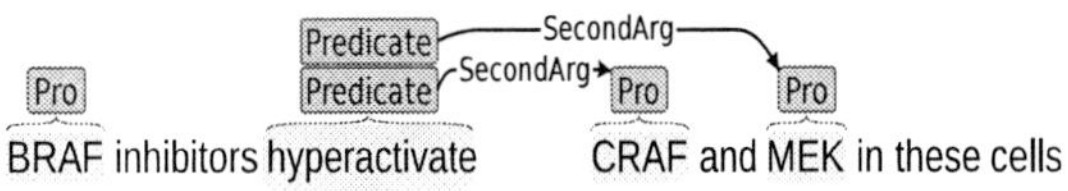

Figure 7: CRAF and MEK are mutually exclusive

5 Evaluation

In order to evaluate the performance of our method we compare the identified arguments with the ones annotated in the gold corpus.

For each event annotated by EventMine, we define argument recall and precision as :

$$Recall = \frac{arg_{EM} \cap arg_{gold}}{arg_{gold}} \quad (2)$$

$$Precision = \frac{arg_{EM} \cap arg_{gold}}{arg_{EM}} \quad (3)$$

where arg_{EM} is the set of arguments EventMine identifies for this event and arg_{gold} the corresponding set of arguments identified in the gold standard. [6]

6 Results and Discussion

6.1 Experimental Results

We applied and evaluated the ranking methodology to the corpora described in Section 3 and the results are shown in Tables 2 and 3.

	Precision		Recall		Fscore		Percent.
	EM	+R	EM	+R	EM	+R	in corpus
Phosphorylation	0.93	0.82	0.86	0.93	0.89	0.87	10
Planned_process	0.5	0.5	0.5	0.5	0.50	0.50	2
Negative_regulation	0.91	0.86	0.83	0.83	0.87	0.84	16
Localization	1	1	0.67	0.67	0.80	0.80	1
Regulation	0.88	0.88	0.88	0.88	0.88	0.88	3
Gene_expression	0.8	0.8	0.75	0.75	0.77	0.77	7
Binding	0.47	0.47	0.43	0.47	0.45	0.47	2
Positive_regulation	0.66	0.59	0.61	0.63	0.63	0.61	44
Total	0.67	0.63	0.62	0.64	0.64	0.63	

Table 2: Results on BM corpus before (EM) and after Re-ranking (+R)

For both corpora, our ranking method leads to an increase in recall compared to the default EventMine application. However, our method also results in a decrease in precision. In order to appreciate the impact of missing arguments on unseen data we repeat the experiment using the MLEE training corpus during training for EventMine. In this case (see Table 4), the recall is higher and the improvement from the post-processing step not significant, suggesting that the post-processing methodology is advantageous mostly when EventMine is applied to new domains.

[6] For events that are not matched in the gold standard both values are zero.

	Precision		Recall		Fscore		Percent.
	EM	+R	EM	+R	EM	+R	in corpus
Protein_catabolism	0.5	0.5	0.5	0.5	0.50	0.50	2
Phosphorylation	0.69	0.59	0.69	0.69	0.69	0.64	5
Dissociation	0.78	0.44	1	1	0.88	0.61	1
Transcription	0.5	0.5	0.5	0.5	0.50	0.50	2
Negative_regulation	0.5	0.48	0.38	0.5	0.43	0.49	9
Regulation	0.53	0.43	0.4	0.47	0.46	0.45	4
Gene_expression	0.88	0.88	0.86	0.86	0.87	0.87	27
Localization	0.63	0.61	0.69	0.76	0.66	0.68	6
Positive_regulation	0.72	0.65	0.63	0.67	0.67	0.66	32
Binding	0.66	0.68	0.56	0.64	0.61	0.66	11
Total	0.7	0.67	0.64	0.68	0.67	0.67	

Table 3: Results on MLEE corpus before (EM) and after Re-ranking (+R)

	Precision		Recall		Fscore		Percent.
	EM	+R	EM	+R	EM	+R	in corpus
Protein_catabolism	0.6	0.6	0.6	0.6	0.60	0.60	1
Death	0.71	0.71	0.71	0.71	0.71	0.71	1
Transcription	0.5	0.5	0.5	0.5	0.50	0.50	1
Localization	0.76	0.67	0.77	0.77	0.76	0.72	5
Development	0.55	0.55	0.55	0.55	0.55	0.55	4
Regulation	0.49	0.45	0.45	0.46	0.47	0.45	7
Breakdown	0.78	0.78	0.78	0.78	0.78	0.78	1
Positive_regulation	0.68	0.64	0.63	0.64	0.65	0.64	22
Growth	0.88	0.88	0.88	0.88	0.88	0.88	3
Phosphorylation	0.69	0.56	0.69	0.69	0.69	0.62	2
Blood_vessel_development	0.96	0.96	0.75	0.75	0.84	0.84	16
Dissociation	0.67	0.42	1	1	0.80	0.59	1
Cell_proliferation	0.92	0.92	0.92	0.92	0.92	0.92	1
Pathway	0.54	0.46	0.46	0.48	0.50	0.47	1
Planned_process	0.81	0.77	0.7	0.71	0.75	0.74	10
Negative_regulation	0.76	0.7	0.66	0.67	0.71	0.68	10
Gene_expression	0.88	0.88	0.88	0.88	0.88	0.88	9
Binding	0.73	0.71	0.65	0.68	0.69	0.69	4
Tissue_remodeling	1	1	1	1	1.00	1.00	1
Total	0.76	0.73	0.68	0.69	0.72	0.71	

Table 4: Results on MLEE corpus before (EM) and after Re-Ranking (+R) (MLEE training set added to the training corpora of EventMine)

Moreover, we can observe in Tables 3 and 4 that the event types recognised are not 100% overlapping. Indeed, since in the first case EventMine is not trained on the MLEE corpus, the set of event types that it is trained to recognise only partially overlaps with the event types annotated in MLEE. As such, in a large number of cases, even though the event trigger is correctly extracted, it is attributed an event type other than the one annotated in the gold standard. For example some of the *BreakDown* events (in Table 4) tend to be recognised as *Negative Regulation* when the model is not trained on the MLEE events (Table 3). We thus wanted to examine the impact of erroneous event type identification on the linking and precision of argument linking, given that EventMine models learn different annotation constraints for each event type. Table 5 compares the performance achieved by our method when the event type assigned by EventMine matches the label in the gold standard, with the overall performance. It can be observed that when the labels do match the performance increases significantly. Thus, it seems that part of error in linking arguments to an event derives from an erroneous recognition of the type of the argument, that is often linked to events that the model is not trained to recognise properly.

	Overall Results		Same label in GS		
	Precision	Recall	Precision	Recall	Percentage
Protein_catabolism	0.5	0.5	1	1	0.67
Phosphorylation	0.69	0.69	1	1	1
Dissociation	0.78	1	1	1	0.33
Transcription	0.5	0.5	1	1	1
Negative_regulation	0.5	0.38	1	0.75	0.93
Localization	0.53	0.4	0.89	0.67	0.93
Gene_expression	0.88	0.86	1	1	0.91
Regulation	0.63	0.69	0.79	0.87	1
Binding	0.72	0.63	0.93	0.81	1
Positive_regulation	0.66	0.56	0.87	0.75	0.94

Table 5: Performance of EventMine for matching type annotations versus overall results

6.2 Analysis of Results and Performance Considerations

The results shown in the previous section are promising in terms of recall. However, there is still considerable room for improvement, especially in terms of decreasing the added noise, so as to minimise the drop in precision. Below we present the most important observations regarding our results and we analyse the errors produced.

- **Correct identification of the partial event but erroneous identification of the missing argument:** Of the noisy events, 60% constituted cases that were correctly identified as partial events, but where the ranking algorithm failed to identify the correct entity to link to the trigger. This was a common pattern in cases where the argument was an event, but the ranking system actually selected one of that event's arguments instead of the whole event, as illustrated in Figure 8. It is important to note that in some of such cases, the event trigger was not annotated by EventMine in the first place. Thus, it was impossible for our method to capture it. This emphasises the strong dependency of our method on EventMine's performance. A possible solution to this problem, which will be considered as future work, is to reformulate the problem as a joint learning task, in which one classifier would focus on ranking single named entity candidates and the other on ranking event candidates, and they would be combined in the test corpus in order to choose the most likely solution. Such an approach would, however, have increased complexity, and its results remain to be tested. [7]

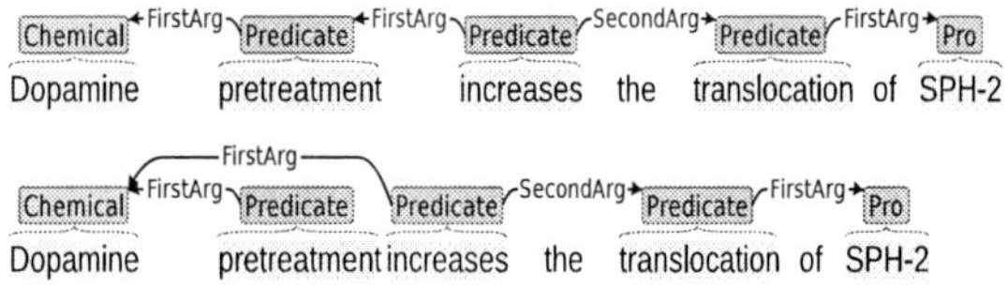

Figure 8: Linking the nested event argument instead of the trigger: Compare correct annotation (top) with produced one (bottom)

- **Entities related to the event in a complementary manner:** In a considerable number of erroneous cases, the ranking system identified arguments that were not annotated in the gold corpus, but which nevertheless were related to the trigger. Two distinctive patterns emerged, as illustrated in Figure 9.
  1. Aliases of the original argument, used in the same sentence (usually a superclass)
  2. Text spans with multiple annotations that are linked multiple times to the event as separate entities

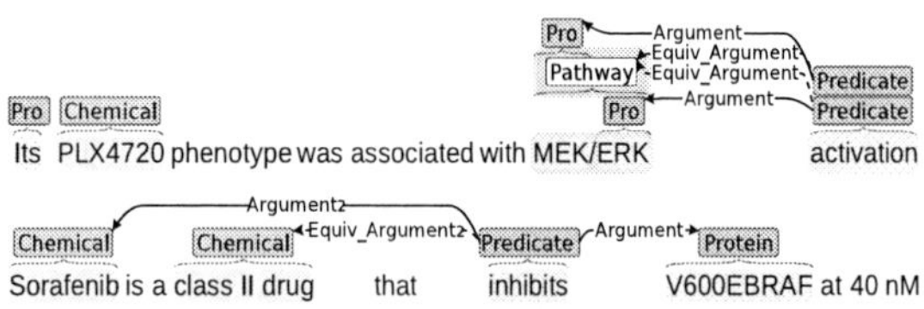

Figure 9: Multi annot. (top): Pathway entity erroneously considered a valid additional argument Alias (bottom): Sorafenib and its superclass both considered valid argument candidates that are not mutually exclusive

- **Overfitting to "null" instances:** As can be deduced from the result tables (2, 3 and 4), there was a considerable percentage of partial events whose missing arguments were still not fully identified by our method. In those cases, the classifier ranked the "null" instance mentioned

[7] However the performance will still depend on the recall of the event extraction system.

in Section 4 as the best option. Investigation revealed that for partial events, the correct missing argument was ranked second after the "null" instance in more than 50% of cases (null instance suggestions accounted for 80%-70% of the total suggestions). A possible solution to further increase the recall would be to drop the "null" instance implementation, and use a confidence threshold instead. However, such a method would be more *ad hoc*, having severe implications on the generalisability of the model.

As a final note, it should be mentioned that in some cases, our method made suggestions that could correct events containing errors (i.e., correct trigger but wrong argument). While these cases were not considered in the scope of this work, it would be interesting to investigate how our method could be adapted/expanded to suggest argument corrections as well as additions.

7 Conclusions and Future Work

Our novel approach to improving event extraction results has successfully shown that identifying and ranking additional arguments by relaxing annotation constraints can aid in improving the argument recall and reducing partial (and sometimes even erroneous) event extraction. Of particular note is the demonstration that our approach has the greatest impact when applied to unseen data. As such, we consider that our results are extremely promising, even though there is still a large margin for further improvements and experimentation.

An important feature of our approach is that the methodology employed is generic enough to be applied to output of any other event extraction architecture (particularly pipelined ones) or any other biomedical corpus without significant modification. Future testing on different corpora and annotation schemes will help to reinforce the robustness and generalisability of our method.

However, this study has already revolved various promising areas for further investigation, in terms of both increasing recall and reducing noisy additions. Of particular interest would be to see whether employing methods with multiple classifiers (co-training, joint-learning or ensemble methods) would improve the performance and reduce the noise. Such an approach could target either classifiers trained on different argument types (named entities or entire events) or even classifiers specialising in particular event types. However, this would constitute a whole new area of research and experimentation.

A further aspect, only minimally considered in this work, is the influence of the training instances and labels on the performance. On the MLEE corpus, it was observed that for events whose automatically assigned event type did not match the gold standard, argument recall and precision also deteriorated. Hence, we can deduce that improving the accuracy of event type assignment would have a positive impact on event extraction performance. The same conclusion could hold also for the named entity labels; as mentioned in Section 4, the BM corpus was initially annotated with a different NE label-set that was automatically (without supervision) aligned with the training corpus annotations in order for the trained model to be applied to it. However, instead of adapting the testing corpus annotations, it would be worthwhile to provide efficient unsupervised methods for adapting the labels in the training corpus to those in the testing corpus. Such an approach could boost the precision without compromising recall by reducing the impact of training on instances (events) that are not related to the ones in the test set. To that end, it would be interesting to combine the wide coverage approach (Miwa et al., 2013) with domain adaptation approaches such as the ones mentioned in Section 2.3 or simply instance reweighting ones such as (Jiang and Zhai, 2007).

The above considerations will be vital in facilitating the incorporation of constraint relaxation as an integral part of the EventMine architecture, rather that as a post-processing step. This will help to enhance EventMine's properties of generalisability and adaptability, and thus allow it it to achieve more robust performance. However, the challenge will be to consider the constraint relaxation and adaptation problem globally, rather than only for argument role annotation constraints.

Acknowledgments

This work was supported by the DARPA funded *Big Mechanism* Project, as well as by the EPSRC funded *Centre for Doctoral Training in Computer Science* scholarship. We would like to thank Dr. Riza Theresa Batista-Navarro and Dr. Ioannis Korkontzelos for the useful discussions and feedback at critical points. Finally, we would like to thank our referees for their constructive input.

References

Sophia Ananiadou, Paul Thompson, Raheel Nawaz, John McNaught, and Douglas B Kell. 2014. Event-based text mining for biology and functional genomics. *Briefings in functional genomics*, page elu015.

Jari Björne and Tapio Salakoski. 2011. Generalizing biomedical event extraction. In *Proceedings of the BioNLP Shared Task 2011 Workshop*, BioNLP Shared Task '11, pages 183–191, Stroudsburg, PA, USA. Association for Computational Linguistics.

Jari Björne and Tapio Salakoski. 2013. Tees 2.1: Automated annotation scheme learning in the bionlp 2013 shared task. In *Proceedings of the BioNLP Shared Task 2013 Workshop*, pages 16–25.

Jari Björne, Filip Ginter, Sampo Pyysalo, Jun'ichi Tsujii, and Tapio Salakoski. 2010. Complex event extraction at pubmed scale. *Bioinformatics*, 26(12):i382–i390.

Ekaterina Buyko, Erik Faessler, Joachim Wermter, and Udo Hahn. 2009. Event extraction from trimmed dependency graphs. In *Proceedings of the Workshop on Current Trends in Biomedical Natural Language Processing: Shared Task*, BioNLP '09, pages 19–27, Stroudsburg, PA, USA. Association for Computational Linguistics.

Etienne Caron, Samik Ghosh, Yukiko Matsuoka, Dariel Ashton-Beaucage, Marc Therrien, Sébastien Lemieux, Claude Perreault, Philippe P Roux, and Hiroaki Kitano. 2010. A comprehensive map of the mtor signaling network. *Molecular systems biology*, 6(1).

Minmin Chen, Kilian Q Weinberger, and John Blitzer. 2011. Co-training for domain adaptation. In *Advances in neural information processing systems*, pages 2456–2464.

F. Ciravegnia and A. Lavelli. 2004. Learning-pinocchio: adaptive information extraction for real world applications. *Natural Language Engineering*, 10:145–165, 6.

Paul R. Cohen. 2014. Darpa's big mechanism program. `http://www.darpa.mil/Our_Work/I2O/Programs/Big_Mechanism.aspx`.

Hal Daumé III. 2009. Frustratingly easy domain adaptation. *arXiv preprint arXiv:0907.1815*.

Rong-En Fan, Kai-Wei Chang, Cho-Jui Hsieh, Xiang-Rui Wang, and Chih-Jen Lin. 2008. Liblinear: A library for large linear classification. *The Journal of Machine Learning Research*, 9:1871–1874.

Yoav Freund and Robert E Schapire. 1997. A decision-theoretic generalization of on-line learning and an application to boosting. *Journal of computer and system sciences*, 55(1):119–139.

Jing Jiang and ChengXiang Zhai. 2007. Instance weighting for domain adaptation in nlp. In *ACL*, volume 7, pages 264–271.

Halil Kilicoglu and Sabine Bergler. 2009. Syntactic dependency based heuristics for biological event extraction. In *Proceedings of the Workshop on Current Trends in Biomedical Natural Language Processing: Shared Task*, pages 119–127. Association for Computational Linguistics.

Jin-Dong Kim, Tomoko Ohta, Sampo Pyysalo, Yoshinobu Kano, and Jun'ichi Tsujii. 2009. Overview of bionlp'09 shared task on event extraction. In *Proceedings of the Workshop on Current Trends in Biomedical Natural Language Processing: Shared Task*, pages 1–9. Association for Computational Linguistics.

Jin-Dong Kim, Sampo Pyysalo, Tomoko Ohta, Robert Bossy, Ngan Nguyen, and Jun'ichi Tsujii. 2011. Overview of bionlp shared task 2011. In *Proceedings of the BioNLP Shared Task 2011 Workshop*, pages 1–6. Association for Computational Linguistics.

B. Kulis, K. Saenko, and T. Darrell. 2011. What you saw is not what you get: Domain adaptation using asymmetric kernel transforms. In *Computer Vision and Pattern Recognition (CVPR), 2011 IEEE Conference on*, pages 1785–1792, June.

Haibin Liu, Lawrence Hunter, Vlado Kešelj, and Karin Verspoor. 2013. Approximate subgraph matching-based literature mining for biomedical events and relations. *PloS one*, 8(4):e60954.

Makoto Miwa and Sophia Ananiadou. 2013. Nactem eventmine for bionlp 2013 cg and pc tasks. In *Proceedings of BioNLP Shared Task 2013 Workshop*, pages 94–98.

Makoto Miwa, Sampo Pyysalo, Tomoko Ohta, and Sophia Ananiadou. 2013. Wide coverage biomedical event extraction using multiple partially overlapping corpora. *BMC bioinformatics*, 14(1):175.

Yusuke Miyao, Rune Sætre, Kenji Sagae, Takuya Matsuzaki, and Jun'ichi Tsujii. 2008. Task-oriented evaluation of syntactic parsers and their representations. In *ACL*, volume 8, pages 46–54.

Claire Nédellec, Robert Bossy, Jin-Dong Kim, Jung-Jae Kim, Tomoko Ohta, Sampo Pyysalo, and Pierre Zweigenbaum. 2013. Overview of bionlp shared task 2013. In *Proceedings of the BioNLP Shared Task 2013 Workshop*, pages 1–7.

Tomoko Ohta, Sampo Pyysalo, Makoto Miwa, and Jun'ichi Tsujii. 2011. Event extraction for dna methylation. *J. Biomedical Semantics*, 2(S-5):S2.

Sinno Jialin Pan, I.W. Tsang, J.T. Kwok, and Qiang Yang. 2011. Domain adaptation via transfer component analysis. *Neural Networks, IEEE Transactions on*, 22(2):199–210, Feb.

Sampo Pyysalo, Tomoko Ohta, Makoto Miwa, and Junichi Tsujii. 2011. Towards exhaustive event extraction for protein modifications. In *Proceedings of BioNLP 2011 Workshop*, pages 114–123.

Sampo Pyysalo, Tomoko Ohta, Makoto Miwa, Han-Cheol Cho, Jun'ichi Tsujii, and Sophia Ananiadou. 2012. Event extraction across multiple levels of biological organization. *Bioinformatics*, 28(18):i575–i581.

Altaf Rahman and Vincent Ng. 2009. Supervised models for coreference resolution. In *Proceedings of the 2009 Conference on Empirical Methods in Natural Language Processing: Volume 2-Volume 2*, pages 968–977. Association for Computational Linguistics.

Sebastian Riedel and Andrew McCallum. 2011. Fast and robust joint models for biomedical event extraction. In *Proceedings of the Conference on Empirical Methods in Natural Language Processing*, pages 1–12. Association for Computational Linguistics.

Pontus Stenetorp, Sampo Pyysalo, Goran Topić, Tomoko Ohta, Sophia Ananiadou, and Jun'ichi Tsujii. 2012. Brat: a web-based tool for nlp-assisted text annotation. In *Proceedings of the Demonstrations at the 13th Conference of the European Chapter of the Association for Computational Linguistics*, pages 102–107. Association for Computational Linguistics.

Serhan Tatar and Ilyas Cicekli. 2011. Automatic rule learning exploiting morphological features for named entity recognition in turkish. *Journal of Information Science*, 37(2):137–151.

David H Wolpert. 1992. Stacked generalization. *Neural networks*, 5(2):241–259.

Rui Xia, Chengqing Zong, Xuelei Hu, and E. Cambria. 2013. Feature ensemble plus sample selection: Domain adaptation for sentiment classification. *Intelligent Systems, IEEE*, 28(3):10–18, May.

Guodong Zhou and Jian Su. 2003. System for recognising and classifying named entities, December 31. US Patent App. 10/585,235.

Extracting Biological Pathway Models From NLP Event Representations

Michael Spranger *
Sony Computer Science Laboratories Inc.
Tokyo, Japan
michael.spranger@gmail.com

Sucheendra K. Palaniappan *
INRIA,
Campus de Beaulieu,
Rennes, France
sucheendra.palaniappan@inria.fr

Samik Ghosh
The Systems Biology Institute,
Minato-ku,
Tokyo, Japan
ghosh@sbi.jp

Abstract

This paper describes an an open-source software system for the automatic conversion of NLP event representations to system biology structured data interchange formats such as SBML and BioPAX. It is part of a larger effort to make results of the NLP community available for system biology pathway modelers.

1 Introduction

Biological pathways represent important insights into the flow of information within a cell by encoding the sequence of interactions among various biological players (such as genes, proteins etc.) in response to certain stimuli (or spontaneous at times) which leads to a change in the state of the cell. Studying and analyzing these pathways is crucial to understanding biological systems.

Traditionally, pathways are represented as maps which are constructed and curated by expert curators who manually read numerous biomedical documents, comprehend and assimilate the knowledge into maps. This process is aided by a variety of graphical tools such as CellDesigner (Funahashi et al., 2008).

Such manual pathway curation comes with a number of problems. Most importantly: 1) the amount of time and therefore cost for detailed pathway maps is high. 2) As new research findings are published these pathway need to be updated or augmented. Often, the speed at which molecular research is progressing, means it is hard to keep pathways in sync. 3) Many times the interpretation of details is left to the judgment of the curator, which leads to considerable variability of pathways.

Considering these limitations, there has been an increased emphasis on using Natural Language Processing (NLP) techniques for automated pathway curation. The BioNLP Shared Task - Pathway Curation (BioNLPST-PC) competition (Nédellec et al., 2013; Ohta et al., 2013) was focused on this specific problem. From the NLP perspective the extraction of biological knowledge is posed as an event detection problem with standard NLP event detection algorithms used to extract the biological information from text (Ananiadou et al., 2010).

Although there has been a lot of work on the problem of automatic pathway extraction from text, to our knowledge there has been little effort to make the extracted information available in standard pathway formats. The majority of pathway data is represented, stored and exchanged using standard formats such as SBML (Hucka et al., 2003) and BioPAX (Demir et al., 2010). Contrary to these formats existing NLP extraction systems often use a data format called the "standoff format", to represent their results. While the standoff format is often described as easily convertible into SBML and BioPAX, no actual software seems to exist to automate this conversion. This paper tries to fill this gap by describing a software system for the conversion of NLP event representations to the system biology structured data interchange formats SBML and BioPAX. We also provide open sourced software tools `st2sbml` and `st2biopax` to convert from stand-off to SBML/BioPAX format. The software tools and additional information about the contents of this paper can be found on our supplementary webpage[1].

2 NLP Event Representations

Existing NLP systems often use an event representation format comprised of a set of annotation rules and file formats to represent pathway events and entities (Kim et al., 2011). For the purpose of

* These two authors contributed equally to this paper and the software system.

[1] `https://github.com/sbnlp/standoff-conversion`

Proceedings of the 2015 Workshop on Biomedical Natural Language Processing (BioNLP 2015), pages 42–51,
Beijing, China, July 30, 2015.

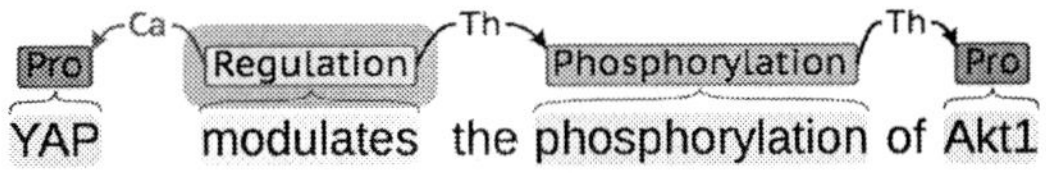

Figure 1: Graphical representation of the event representation of Example 1

this paper we base ourselves on the standoff representation (ST) proposed for the BioNLP Shared Task 2011, 2013 (Nédellec et al., 2013).

Annotations in ST link spans of texts through character offsets to `entities` (e.g. Proteins, Genes etc.) and `events` (Positive Regulation etc.). Events and entities are represented line by line with links between them.

The following is an example sentence and a possible event representations.

(1) YAP modulates the phosphorylation of Akt1.

```
T1 Protein 0 3 YAP
T2 Protein 37 41 Akt1
T3 Regulation 4 13 modulates
T4 Phosphorylation 17 32 phosphorylation
E1 Phosphorylation:T4 Theme:T2
E2 Regulation:T3 Theme:E1 Cause:T1
```

Each annotation starts with a unique annotation-ID. The annotations-IDs encodes the annotation type in the first letter (T - text bound annotation, E - event annotation). This is followed by the annotation-type. For instance, the text bound annotation T1 is of type protein, whereas T3 is of type Regulation. Text bound annotations also encode the start and end position as well as the text they annotate. Text bound annotation T1 for instance ranges from character 0 to character 3 of the annotated text and the actual text is "YAP".

Event annotations build on top of text bound annotation. The annotations-ID for an event is followed by an event-type and the reference to the text bound annotation. For instance, E1 is a Phosphorylation event and the corresponding text is T4 "phosphorylation". Additionally, event annotations encode roles. T2 is the theme of E1, which in this case means that "Akt1" is undergoing a phosphorylation. Events can also be used as theme. For example the theme of E2 is E1, which means that the phosphorylation is regulated by "YAP". Different roles are possible depending on the type of the event.

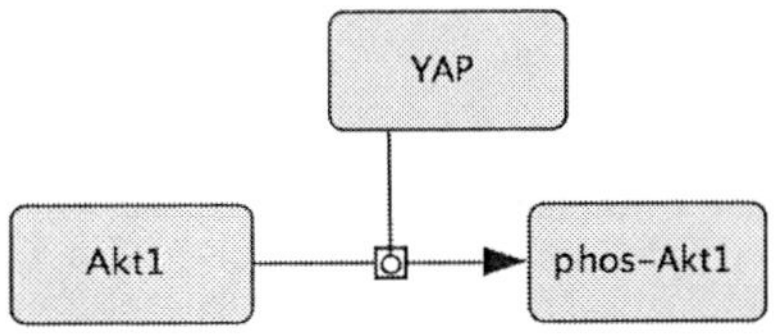

Figure 2: Example 1 converted into SBML (viewed with CellDesigner)

3 From Event Representations to SBML

Systems Biology Markup Language, or short SBML (Hucka et al., 2003), is a XML-based markup language to describe, store and communicate biological models. It is among the most widely used formats with numerous software support. SBML essentially encodes models using biological players called `sbml:species`[2]. `sbml:species` can participate in interactions, called `sbml:reaction`. Species participate in interaction as `sbml:reactant`, `sbml:product` and `sbml:modifier`. The basic idea being that some quantity of reactant is consumed to produce a product. Reactions are influenced by modifiers.

SBML supports mathematical representations of the underlying dynamics of the reactions and is essentially used to simulate models. Due to this, there is no SBML vocabulary to specify different types of reactions (such as transcription, phosphorylation etc.) or species (such as protein, DNA etc.). Alternatively, species and reactions can be annotated and uniquely specified using MIRIAM resources and annotations (Novere et al., 2005). We use controlled vocabulary from the Systems Biology Ontology (SBO) and the Gene Ontology (GO). This information is also useful to convert SBML files to other formats such as SBGN (Le Novere et al., 2009) using tools such as VANTED (Junker et al., 2006).

Figure 2 shows Example 1 converted into an SBML model using the mapping algorithm described in the following paragraphs.

3.1 Mapping Algorithm

The conversion of standoff formatted information to an SBML model consists of five steps.

[2]We will refer to SBML vocabulary using the prefix "sbml".

Standoff Entity	SBO term	SBO name
Complex	SBO:0000253	non-covalent complex
Gene_or_gene_product	SBO:0000245	macromolecule
Dna	SBO:0000251	deoxyribonucleic acid
DnaRegion	SBO:0000251	deoxyribonucleic acid
Drug	SBO:0000247	simple chemical
Ion	SBO:0000327	non-macromolecular ion
Protein	SBO:0000252	polypeptide chain
Rna	SBO:0000250	ribonucleic acid
RnaRegion	SBO:0000250	ribonucleic acid
Gene	SBO:0000354	informational molecule segment
Small Molecule	SBO:0000247	simple chemical
Simple_molecule	SBO:0000247	simple chemical

Table 1: Mapping of Annotation-type to SBO term.

Step 1: Initialize the Model Firstly, read the event annotation files and create a memory internal representation of triggers and events. We initialize an empty SBML model with a single `sbml:compartment` named "default".

Step 2: Create sbml:species For each entity in the standoff format, a `sbml:species` is added to the SBML model. This only applies to standoff entities that can be mapped to an SBO term. Then the following is done 1) map the annotation-ID of the trigger to the id in the `sbml:species`, 2) create a meta id by appending "metaid_0000" and annotation-ID; meta id facilitates that annotations to this species can uniquely refer to it 3) add the annotation-text as the name of the `sbml:species`, 4) map the annotation-type to an SBO term and add to the `sbml:species` (see Table 1)

For instance, the standoff line

```
T2 Protein 37 41 Akt1
```

will be mapped to

```
<species sboTerm="SBO:0000252"
 id="T2" name="Akt1"
 metaid="metaid_0000T2"
 compartment="default"/>
```

On the other hand, a line such as

```
T39  Entity 641 648 nucleus
```

will not be used to create a species in the SBML model, because "Entity" cannot be mapped to an SBO term. Here, "nucleus" actually refers to a compartment which is not directly deducible from the entity definition in the standoff format. To deal with such cases, we need to take into account their role in Events something that is described in the next few paragraphs.

Standoff Event	SBO/GO term	SBO/GO name
Conversion	SBO:0000182	conversion
Acetylation	SBO:0000215	acetylation
Deacetylation	GO:0006476	Protein Deacetylation
Methylation	SBO:0000214	Methylation
Demethylation	GO:0006482	Protein Demethylation
Phosphorylation	SBO:0000216	phosphorylation
Dephosphorylation	SBO:0000330	Methylation
Ubiquitination	SBO:0000224	Ubiquitination
Deubiquitination	GO:0016579	Protein Deubiquitination
Degradation	SBO:0000179	degradation
Catabolism	GO:0009056	Catabolic Process
Catalysis	SBO:0000172	Catalysis
Protein_catabolism	GO:0009056	Catabolic Process
Association	SBO:0000177	non-covalent binding
Binding	SBO:0000177	non-covalent binding
Dissociation	SBO:0000180	dissociation
Regulation	GO:0065007	biological regulation
Positive_regulation	GO:0048518	positive regulation
Activation	SBO:0000412	biological activity
Negative_regulation	GO:0048519	negative regulation
Inactivation	SBO:0000412	biological activity
Gene_expression	GO:0010467	Genetic Production
Transcription	SBO:0000183	Transcription
Translation	SBO:0000184	Translation
Localization	GO:0051179	Localization
Transport	SBO:0000185	Transport Reaction
Pathway	SBO:0000375	Process

Table 2: Mapping of annotation-type to SBO/GO term.

Step 3: Create sbml:reaction Most events are added to the SBML model as `sbml:reaction`. For instance, the text trigger and event annotation corresponding to E1 in Example 1 result in the following SBML description

```
<reaction metaid="metaid_0000E1"
          sboTerm="SBO:0000216"
          id="E1"
          name="Phosphorylation"
          reversible="false">
 <annotation> ... </annotation>
</reaction>
```

The SBO/GO term is assigned according to the mapping depicted in Table 2. The reaction id is based on the event id (E10). The metaid of the form "metaid_0000 + id" is also added and the `sbml:reaction` name is the event-type. Lastly, all reactions are constructed as non reversible.

In a second step sbml:reactant, sbml:product and sbml:modifier are added to SBML reactions based on the roles of events.

Theme is the entity that undergoes the effects of the event. It is mapped to the sbml:reactant of the SBML reaction. For this a reactant reference is created and the species corresponding to the entity is linked to that reference via the id of the species (annotation-id of the entity).

Product can be specified for Binding, Dissociation[3] and Conversion events. Product is mapped to sbml:product of the corresponding reaction. The entities appearing in the product role are used for creating a product reference with the same entity.

Cause is an entity/event causing the event. Cause is eventually mapped to entities which are then mapped to the reaction as sbml:modifier (via modifier reference).

Information in Site (which describes the site on the Theme entity that is modified in the event) is added to the "Notes" section of the SBML reaction as there seems to be no direct way to represent this information in SBML. Notes are human-readable annotations that can be added to SBML reactions.

Step 4: Handle Localization and Transport Events Localization and Transport events are handled differently from other events. They occur with additional roles besides Theme.

AtLoc describes the location/compartment at which the entity/species is located not an actual reaction. Hence, localization events with AtLoc roles do not end up as reactions in SBML. Instead, first we check if a sbml:compartment described by the AtLoc role exists, else a new sbml:compartment is created (see the nucleus example discussed earlier). Next, the compartment of the theme entity of the event is set to the corresponding sbml:compartment.

[3]In data used for evaluation we also encountered Dissociation events with Participant and Complex roles. They are mapped to sbml:product and sbml:reactant respectively.

FromLoc/ToLoc Transport and Localization events can also include FromLoc and ToLoc roles which describes the transport of the theme entity/species from some location/compartment to another. Consequently, we create a reaction where the Theme entity/species starts out in the compartment described by FromLoc (sbml:reactant) and ends up in the compartment described by the ToLoc (sbml:product) role. If the FromLoc/ToLoc sbml:compartment does not exist when creating the sbml:reaction, a new sbml:compartment is created corresponding to FromLoc/ToLoc.

Step 5: Handle Gene Expression Events We model Gene expression events (e.g. Transcription and Translation) as reactions in SBML. However, this class of reactions does not have the sbml:reactant role. For Transcription events (process in which a gene sequence is copied to produce RNA) if the type of Theme is RNA, it gets mapped to sbml:product. If the type of Theme is DNA, then it gets mapped to the sbml:modifier of the Transcription sbml:reaction.

Translation events are handled in a similar manner.

Step 6: Handle Regulation Events In principle regulation events such as Positive/Negative Regulation, Activation and Inactivation can be handled as described in Step 3 when the Theme and Cause are species. If Theme and Cause are species then they are added to a regulation reaction as reactant and modifier respectively.

However, the standoff format definition also allows regulation events where Theme and Cause are themselves events[4]. For example, the following standoff lines describe a Positive regulation of a Phosphorylation event.

```
T14 Protein 776 782 eIF-4E
T15 Protein 852 859 insulin
T43 Phosphorylation 820 835
phosphorylation
T44 Positive_regulation 839 848
    increased
E21 Phosphorylation:T43 Theme:T14
E22 Positive_regulation:T44 Cause:T15
    Theme:E21
```

[4]In some of the data used to test our conversion we also encountered Catalysis events which had event themes. They are handled exactly as Positive Regulation events.

If the Theme is an event, then we do not create a reaction but simply add the Cause entity as a modifier to the reaction corresponding to the Theme event of the regulation. For the example above this means that the Phosphorylation reaction E21 is positively regulated (modified) by insulin (T15).

In reality though things are a bit more complicated since the Theme event might itself not exist as a reaction. For instance, there could be an event description as follows:

```
E23     Positive_regulation:T35 Cause:T21
Theme:E13
E13     Positive_regulation:T36 Theme:E21
```

Here, the event E23 has Theme E13, which itself is a Positive regulation with Theme E21. However, E13 itself does not correspond to a reaction. In this case the algorithm recursively tracks down the Theme event across multiple event annotations until it finds an event that exists in the SBML model as a reaction (In this case E21 is identified as the Theme for E23).

In case the Cause is an event, the product of the Cause event is used as a modifier. If the reaction corresponding to the Cause event does not have a product yet, then a corresponding product species is first created and added to the model.

Step 7: Optional Cleanup and Annotation Operations As a last step optional cleanup/enhancement operations can be performed. They can be used to ensure consistency of the resulting SBML model.

Add UniProt information We use the annotation -text to retrieve information about species from UniProt. The UniProt ID is added as controlled vocabulary to the corresponding SBML species. Other information is added as XML annotation and XHTML notes. This includes information about alternate names, gene names, gene ids where available and appropriate.

Remove unused species Not all entities end up as products, reactants or modifiers of an SBML reaction. In many cases, the named entity recognizer might recognize some entity but no links to events is established. However, the entities might have been added to the model (see Step 1). Entities not partaking in any reaction can be removed automatically.

Complete reactions Our software supports automatic adding of products and reactants for reactions that were not explicitly annotated in that way. For instance, all phosphorylation events can extended with corresponding sbml:product species. The completion takes into account that certain reactions such as Gene expression reactions do not have reactants.

Here is an example of what we mean. For a Phosphorylation reaction, the first pass of the algorithm maps the Theme to sbml:reactant and no sbml:product is added. For example, E1 (in Example 1) would have Akt1 as a sbml:reactant. To complete this reaction a new sbml:species with name phoAkt1 is created representing the phosphorylated form of Akt. phoAkt1 is added as the sbml:product to the reaction E1 (See Figure 2).

Remove reactions without reactants, products In other defunct cases the standoff file might include events that cannot be translated into reactions with reactants and/or products. For example, we encountered in real data that a reaction might only have a modifier (Cause). Such reactions are automatically removed if requested by the user.

3.2 Implementation

We used python and the python version of libSBML to develop the conversion algorithm. libSBML was used for generating and accessing the SBML model content. We used a custom implementation of a Standoff parser which translates the line-wise description of standoff triggers and events in a1/a2 and ann files into a memory structure of triggers (id, type, text) and events (id, type, roles). These structures are the basis for generating and completing the SBML model. The conversion is fast. It scales linearly with the number of entities, events and roles.

3.3 Discussion

The conversion of standoff format files to SBML is quite straightforward with a few exceptions where events cannot be mapped directly to an SBML reaction as is the case with Localization events that have an AtLoc role. Moreover, not all entities end up as sbml:species. Cellular components

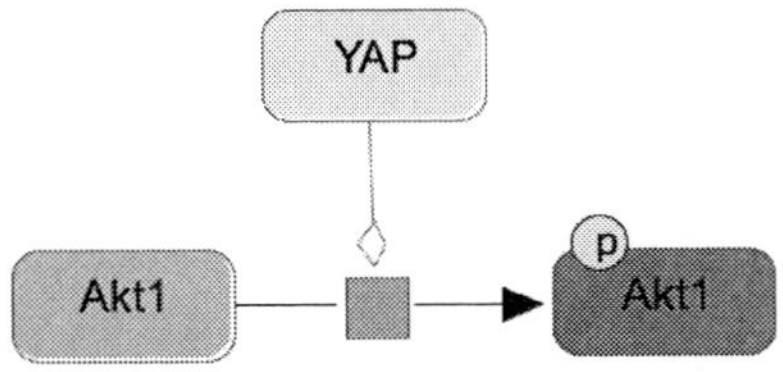

Figure 3: Example 1 converted into BioPAX (viewed with the ChiBE editor)

used in Localization and Transport events, for instance, end up as compartments. Another example are Regulation events that have events as `Theme`. In all of these cases, events in the standoff do not have a direct correspondent in the sbml model.

The algorithm is open to extension. For instance, in order to integrate a new event with `Theme`, `Cause`, `Product`, `Site` roles only a new SBO mapping needs to be defined.

SBML has graphical editing tool support through, for example, CellDesigner. Although CellDesigner uses SBML as its base format, there are a lot of tool specific custom XML annotations that convey a more fine grained view of `sbml:species` and `sbml:reactions` for visualization purposes. Our focus in this paper is the conversion to pure SBML format without tool-based customizations.

4 From Event Representations to BioPAX

Biological Pathway Exchange (BioPAX) is another widely used pathway data format based on RDF/OWL. It is used for storage, analysis, integration and exchange of pathway models (Demir et al., 2013) . BioPAX, unlike SBML is more fine grained in its explicit handling of different types of biological players (`bp:PhysicalEntity`[5]), and their interactions (`bp:Interaction`).

The BioPAX conversion algorithm is similar in structure to the SBML conversion. Figure 3 shows Example 1 converted into a BioPAX model using the mapping algorithm described in the following paragraphs.

Step 1: Initialize the model Read the event files, parse and create a memory internal representation of triggers and events. Create an empty model BioPAX model.

[5]We will henceforth use the the prefix bp to refer to BioPAX vocabulary

Step 2: Create bp:PhysicalEntity Each entity in the standoff format is mapped to the corresponding `bp:PhysicalEntity` class in BioPAX. `bp:PhysicalEntity` is a superclass of molecules such as proteins, DNA, RNA, Small molecules, Complex etc. Depending on the granularity of the description of the entity, the element is initialized as a sub class of `bp:PhysicalEntity`. The mapping is described in Table 3. The created `bp:PhysicalEntity` is assigned a unique-id which is the same as the annotation-ID of the entity. The name of the `bp:PhysicalEntity` is assigned the annotation-text. For instance, the protein T8, described in the previous section will be encoded as:

```
<bp:Protein rdf:about="T8">
 <bp:name rdf:datatype =
"http://www.w3.org/2001/XMLSchema#string">
IkappaBs</bp:name>
</bp:Protein>
```

Standoff Entity	BioPAX class
Cellular_component	prefix.CellularLocationVocabulary
Complex	prefix.Complex
DNA	prefix.Dna
Drug	prefix.PhysicalEntity
Entity	prefix.PhysicalEntity
Gene_or_gene_product	prefix.PhysicalEntity
Gene_product	prefix.PhysicalEntity
Gene	prefix.Gene
Ion	prefix.PhysicalEntity
Protein	prefix.Protein
Receptor	prefix.PhysicalEntity
RNA	prefix.Rna
Simple_molecule	prefix.SmallMolecule
Simple_chemical	prefix.SmallMolecule
Tag	prefix.PhysicalEntity

prefix = org.biopax.paxtools.model.level3

Table 3: Mapping of Annotation-type to BioPAX term.

Step 3: Create bp:Interactions Each event is mapped to the corresponding `bp:Interaction` class in BioPAX. `bp:Interaction` is a superclass used to describe reactions and the relationship between the `bp:PhysicalEntity` elements. Depending on the type of the event, an appropriate sub class of the `bp:Interaction` is chosen. The mapping is described in Table 4. The created `bp:Interaction` is assigned a unique id which is the same as the annotation-ID of the event in the standoff. Additionally, all interaction which have the `bp:ConversionDirection` attribute, are set to `bp:LEFT_TO_RIGHT`.

Standoff Event	BioPAX Class
Conversion	prefix.Conversion
Acetylation	prefix.BiochemicalReaction
Deacetylation	prefix.BiochemicalReaction
Methylation	prefix.BiochemicalReaction
Demethylation	prefix.BiochemicalReaction
Phosphorylation	prefix.BiochemicalReaction
Dephosphorylation	prefix.BiochemicalReaction
Ubiquitination	prefix.BiochemicalReaction
Deubiquitination	prefix.BiochemicalReaction
Gene_expression	prefix.TemplateReaction
Transcription	prefix.TemplateReaction
Translation	prefix.TemplateReaction
Catalysis	prefix.Catalysis
Degradation	prefix.Degradation
Catabolism	prefix.Degradation
Protein_catabolism	prefix.Degradation
Association	prefix.ComplexAssembly
Binding	prefix.ComplexAssembly
Dissociation	prefix.ComplexAssembly
Regulation	prefix.Control
Positive_regulation	prefix.Catalysis
Activation	prefix.Control
Negative_regulation	prefix.Control
Inactivation	prefix.Control
Localization	prefix.Transport
Transport	prefix.Transport

prefix = org.biopax.paxtools.model.level3

Table 4: Mapping of annotation-type to BioPAX interaction class.

Step 4: Add participants Events relevant for this paper fall into 3 categories 1) `bp:TemplateReaction` (for transcription ,translation and Gene_expression events), 2) `bp:Conversion` (for conversion events including phosphorylation, dephosphorylation etc., transport events, binding events and dissociation events) and 3) `bp:Control` (for regulation, positive regulation, activation, negative regulation and inactivation events).

Gene Expression, Transcription and Translation events are modeled as a `bp:TemplateReaction`. If the `Theme` of a transcription event is of type RNA, then it is mapped to the `bp:product` property of the `bp:TemplateReaction`. If the `Theme` is a DNA, then it is added as `bp:template` property. Similarly, if the `Theme` of a Gene expression event (Translation or Transcription) is of type Protein, then the corresponding `bp:PhysicalEntity` is set as the `bp:product` of the `bp:TemplateReaction`. If the `Theme` of a Translation event is an RNA, then it is set as the `bp:template` property.

Conversion events are easily mapped to BioPAX elements. Conversion events are all modeled as `bp:BiochemicalReaction`. The `bp:PhysicalEntity` corresponding to `Theme` is set to the `bp:left` of the `bp:BiochemicalReaction`. `Site` information is encoded into the suitable `bp:sequenceSite` property.

For instance, in the case of a Phosphorylation event, the reaction corresponds to `Theme` becoming phosphorylated. For this a new `bp:PhysicalEntity` is created which has the same properties as `Theme`, except that it has an additional `bp:ModificationFeature`, which corresponds to the phosphorylated residue. This new entity is then set to `bp:right` of the `bp:BiochemicalReaction`. If these reactions have the `Cause` entity, then, a new `bp:Control` interaction is created with the `Cause` entity as the `bp:controller` and the created `bp:BiochemicalReaction` as the `bp:controlled`.

Similarly, Binding, Dissociation and Degradation events map from their definitions onto the BioPAX setting.

Localization and transport events with the `ToLoc` and `FromLoc` roles are handled differently. The `ToLoc` and `FromLoc` entities are added as compartments in the BioPAX model. We then model a `bp:Transport` reaction with the `Theme` entity transported from the `FromLoc` compartment to the `ToLoc` compartment. Localization events with `AtLoc` role are not explicitly modeled as reaction. Only the compartment of the corresponding `Theme`'s `bp:PhysicalEntity` in the BioPAX model is appropriately set. Additionally, the annotation-ID of the event is appended as a comment to the corresponding element in BioPAX.

Control events are more complex since they can involve another event as a `Theme` or `Cause`. Positive/Negative Regulation, Activation and Inactivation events where `Theme` is mapped to a `bp:PhysicalEntity` are modeled as a `bp:BiochemicalReaction`. Here the entity is converted from an active/inactive form to an inactive/active form. Next, a corresponding `bp:Control` interaction is created (see Table 4). If the `Cause` is also an entity then it is added as the `bp:controller` to the `bp:Control` interaction. However, in case `Cause` is an event, then the right side entity (or product) of the Interaction encoded by the `Cause` event is derived and added as the `bp:controller`. The previously created `bp:BiochemicalReaction` is then added as the `bp:controlled` element for the `bp:Control` interaction. The `bp:controlType` property is set to `bp:ACTIVATION` and `bp:INHIBITION` for the Positive Regulation/Activation and Negative Regulation/Inactivation events respectively.

Regulation, Positive Regulation and Negative Regulation can also have events in the `Theme` role. In this case, the `Interaction` corresponding to the `Theme` is searched, and added as the `bp:controlled` element of a new `bp:Control` interaction. Should there be a `Cause` entity or event then it is handled as described previously.

Step 5: Optional Postprocessing Operations The software for BioPAX supports post processing similar to the SBML converter: 1) Unused entities can be removed, 2) interactions completed and 3) interactions without reactants and products removed. Additionally, we can assign a unique identifier to BioPAX entities by querying external databases like UniProt, this information is encoded into the `bp:Xref` class using either `bp:RelationshipXref` or `bp:UnificationXref`.

4.1 Discussion

The conversion from standoff to BioPAX is relatively straightforward. The finer grained options to represent different types of information makes it more naturally suited to translate annotations from standoff format. Nevertheless, issues highlighted in the SBML conversion exist in the BioPAX conversion too. For example, certain events such as Localization events with an `AtLoc` role do not end up as `bp:Interaction` etc.

4.2 Implementation

The algorithm is implemented in python. It uses the Java Paxtools 4.2.1 toolkit (Demir et al., 2013) to encode and manipulate models into the BioPAX format. JPype is used as the bridge to connect python to the Paxtools library. The other components of the implementation (such as the standoff-parser) are the same as used in the SBML implementation.

5 Results and Evaluation

For initial evaluation of our software we used the mTOR pathway event corpus also used in a related study on converting pathway models to standoff format (Ohta et al., 2011). The corpus consists of 60 PubMed abstracts and the same number of files of hand-annotated standoff files. The 60 abstracts contain 11960 words. The hand-annotated data contains 1284 events, 1483 Protein, 1 Entity, 201 Complexes (which gives a total of 2970 text bound annotation triggers). In total the annotations contain 1228 `Theme` roles, 19 `Product` roles, 205 `Causes`, 139 `Site`, 8 `atLoc`, 4 `fromLoc`, 16 `toLoc` and 51 `participant`. The conversion run on the hand-annotated data correctly translates entities and events to SBML and BioPAX according to the mapping described in the previous sections.

In order to check our software with state-of-the-art event extraction systems we applied an unaltered, freshly downloaded Turku Event Extraction System/TEES Version 2.1 (Björne et al., 2013) to the 60 PubMed abstracts. The resulting TEES/60 corpus contains 1472 text bound triggers (in a1) and 783 text bound triggers (in a2). TEES extracted 1473 Proteins which were all successfully translated to SBML and BioPAX. 20 entities were detected, 3 of which were translated into compartments (based on their usage in Localization), 10 were used as site and translated into site comments. In total 1126 events were detected by TEES of which the majority was translated. The exception were 30 localization events of which 1 was a localization with an `AtLoc` role (translated into a compartment). 29 Localization events were only annotated with a theme and therefore were ignored. 270 regulation events have an event based theme. Only 99 of those are also cause annotated and handled as sbml:reaction. The remaining 171 disappear since the extracted information from TEES is not enough to establish links in the models (both BioPAX and SBML).

Importantly, the failure to translate some of these events into SBML/BioPAX is caused by incomplete information provided by the NLP event extraction system. For instance, Localization events which only have a `Theme` role do not provide enough information to be added to the model. Obviously this is one of the areas where hand-annotated data provides better conversion results. Nevertheless, these kind of results are encouraging because the translation into biological knowledge allows for further processing and cleaning of automatically extracted data and potentially may lead to better extraction systems by providing additional learning signals.

Working with Natural language is never easy. Natural language is full of underspecification, ambiguities and context-dependencies. Standoff formats represent a compromise between exact specifications such as SBML and BioPAX that come with their own design approach and assumptions. Trying to map from one world into the other we noticed a few problems

Coarse type granularity of biological players: Coarse granularities such as "Gene or gene product", which encompass genes, RNA and protein, make it difficult to assign a type for the entity. This is important for reactions such as Gene expression, where the decision whether something is a `sbml:product` or `sbml:modifier` depends on exact distinctions.

Underspecification of event types: The event type Regulation refers to any process (`Cause`) that modulates any attribute of another process (`Theme`). In the pathway representation context, it is more natural that the process that gets modulated be an event (which is modeled as a `sbml:reaction` in SBML and `bp:Interaction` in BioPAX). It is not clear how to correctly represent the scenario when the process that gets modulated is an entity (modeled as `sbml:species` in SBML and `bp:PhysicalEntity` in BioPAX). However, the event specification allows Theme (that which is regulated) to be either an entity or an event.

Underspecification of roles: Event extraction systems try to extract as much as possible but often are not able to extract all necessary information. For example, the following says there is a Positive_regulation on Theme T23, but no information is available on the process that is regulating it (no `Cause`).

```
E13 Positive_regulation:T36
Theme:T23
```

In such cases the converter is unable to extract SBML and BioPAX information.

6 Conclusion

In this paper we proposed and discussed a scheme to convert NLP event representations to standard biomedical pathway data formats (SBML and BioPAX). This is important for several reasons. The system allows curators to integrate event extraction data into their normal work flow. For instance, the extracted information can give curators a base template, which can be further edited in their favorite drawing tool. The integration into graphical annotation tools could provide the basis to later capture the curator's changes. These changes could in turn be used to generate new human annotations and to improve current event extraction systems. Together with other tools that support the conversion of SBML models into NLP standoff representations (Ohta et al., 2011), our system bridges the gap between biological modeling and automatic event extraction and opens the way to a more tight interaction between the two fields.

Tight integration of NLP and biomedical research is a recent trend (Huang and Lu, 2015) with a number of groups moving in this direction (Wei et al., 2013; Cejuela et al., 2014; Miwa et al., 2013, for example). For pathway curation, it is important that the results of event extraction technologies become part of curation applications/workflows. To achieve this we will have to overcome problems inherent in the design of formats such as SBML/BioPAX and/or standoff formats. For instance, SBML wa s primarily developed as process-based transition notation that cannot faithfully capture all known biochemistry. Popular software like CellDesigner add a layer of custom XML annotations to resolve this. For our tools to be used in CellDesigner we have to add such information in the conversion process. Another layer of information can be provided by automatic annotation using UniProt. For the future it will be important to integrate other databases and external references.

Lastly, we plan to perform a more thorough evaluation of the conversion by reconstructing a complete known pathway (e.g. the mTOR pathway, for which high quality maps are already available). We are also performing a large scale evaluation of the software on the EVEX event database – a text mining resource of PubMed abstracts and full texts (Van Landeghem et al., 2011)

References

[Ananiadou et al.2010] Sophia Ananiadou, Sampo Pyysalo, Jun'ichi Tsujii, and Douglas B Kell. 2010. Event extraction for systems biology by text mining the literature. *Trends in biotechnology*, 28(7):381–90, July.

[Björne et al.2013] Jari Björne, Suwisa Kaewphan, and Tapio Salakoski. 2013. Uturku: Drug named entity detection and drug-drug interaction extraction using svm classification and domain knowledge. In *Proceedings of the 7th International Workshop on Semantic Evaluation (SemEval 2013)*.

[Cejuela et al.2014] Juan Miguel Cejuela, Peter McQuilton, Laura Ponting, Steven J Marygold, Raymund Stefancsik, Gillian H Millburn, Burkhard Rost, et al. 2014. tagtog: interactive and text-mining-assisted annotation of gene mentions in plos full-text articles. *Database*, 2014:bau033.

[Demir et al.2010] Emek Demir, Michael P Cary, Suzanne Paley, Ken Fukuda, Christian Lemer, Imre Vastrik, Guanming Wu, Peter D'Eustachio, Carl Schaefer, Joanne Luciano, et al. 2010. The biopax community standard for pathway data sharing. *Nature biotechnology*, 28(9):935–942.

[Demir et al.2013] Emek Demir, Özgün Babur, Igor Rodchenkov, Bülent Arman Aksoy, Ken I Fukuda, Benjamin Gross, Onur Selçuk Sümer, Gary D Bader, and Chris Sander. 2013. Using biological pathway data with paxtools. *PLoS computational biology*, 9(9):e1003194.

[Funahashi et al.2008] Akira Funahashi, Yukiko Matsuoka, Akiya Jouraku, Mineo Morohashi, Norihiro Kikuchi, and Hiroaki Kitano. 2008. Celldesigner 3.5: a versatile modeling tool for biochemical networks. *Proceedings of the IEEE*, 96(8):1254–1265.

[Huang and Lu2015] Chung-Chi Huang and Zhiyong Lu. 2015. Community challenges in biomedical text mining over 10 years: success, failure and the future. *Briefings in Bioinformatics*, page bbv024.

[Hucka et al.2003] Michael Hucka, Andrew Finney, Herbert M Sauro, Hamid Bolouri, John C Doyle, Hiroaki Kitano, Adam P Arkin, Benjamin J Bornstein, Dennis Bray, Athel Cornish-Bowden, et al. 2003. The systems biology markup language (sbml): a medium for representation and exchange of biochemical network models. *Bioinformatics*, 19(4):524–531.

[Junker et al.2006] Björn H Junker, Christian Klukas, and Falk Schreiber. 2006. Vanted: a system for advanced data analysis and visualization in the context of biological networks. *BMC bioinformatics*, 7(1):109.

[Kim et al.2011] Jin-Dong Kim, Sampo Pyysalo, Tomoko Ohta, Robert Bossy, Ngan Nguyen, and Jun'ichi Tsujii. 2011. Overview of bionlp shared task 2011. In *Proceedings of the BioNLP Shared Task 2011 Workshop*, pages 1–6. Association for Computational Linguistics.

[Le Novere et al.2009] Nicolas Le Novere, Michael Hucka, Huaiyu Mi, Stuart Moodie, Falk Schreiber, Anatoly Sorokin, Emek Demir, Katja Wegner, Mirit I Aladjem, Sarala M Wimalaratne, et al. 2009. The systems biology graphical notation. *Nature biotechnology*, 27(8):735–741.

[Miwa et al.2013] Makoto Miwa, Tomoko Ohta, Rafal Rak, Andrew Rowley, Douglas B Kell, Sampo Pyysalo, and Sophia Ananiadou. 2013. A method for integrating and ranking the evidence for biochemical pathways by mining reactions from text. *Bioinformatics*, 29(13):i44–i52.

[Nédellec et al.2013] Claire Nédellec, Robert Bossy, Jin-Dong Kim, Jung-jae Kim, Tomoko Ohta, Sampo Pyysalo, and Pierre Zweigenbaum. 2013. Overview of bionlp shared task 2013. *ACL 2013*, page 1.

[Novere et al.2005] Nicolas Le Novere, Andrew Finney, Michael Hucka, Upinder S Bhalla, Fabien Campagne, Julio Collado-Vides, Edmund J Crampin, Matt Halstead, Edda Klipp, Pedro Mendes, et al. 2005. Minimum information requested in the annotation of biochemical models (miriam). *Nature biotechnology*, 23(12):1509–1515.

[Ohta et al.2011] T. Ohta, S. Pyysalo, and J. Tsujii. 2011. From pathways to biomolecular events: Opportunities and challenges. In *Proceedings of the BioNLP 2011 Workshop*, pages 105–113, Portland, Oregon, USA, June. Association for Computational Linguistics.

[Ohta et al.2013] T. Ohta, S. Pyysalo, R. Rak, A. Rowley, H.-W. Chun, S. J. Jung, S. P. Choi, and S. Ananiadou. 2013. Overview of the pathway curation (pc) task of bionlp shared task 2013. In *Proceedings of the BioNLP Shared Task 2013 Workshop*, pages 67–75. Association for Computational Linguistics, August.

[Van Landeghem et al.2011] Sofie Van Landeghem, Filip Ginter, Yves Van de Peer, and Tapio Salakoski. 2011. Evex: a pubmed-scale resource for homology-based generalization of text mining predictions. In *Proceedings of BioNLP 2011 Workshop*, pages 28–37. Association for Computational Linguistics.

[Wei et al.2013] Chih-Hsuan Wei, Hung-Yu Kao, and Zhiyong Lu. 2013. Pubtator: a web-based text mining tool for assisting biocuration. *Nucleic acids research*, page gkt441.

Shallow Training is cheap but is it good enough? Experiments with Medical Fact Coding

Ramesh Nallapati, Radu Florian
IBM T. J. Watson Research Center
1101 Kitchawan Road,
Yorktown Heights, NY 10598, USA
{nallapati,radu}@us.ibm.com

Abstract

A typical NLP system for medical fact coding uses multiple layers of supervision involving fact-attributes, relations and coding. Training such a system involves expensive and laborious annotation process involving all layers of the pipeline.

In this work, we investigate the feasibility of a shallow medical coding model that trains only on fact annotations, while disregarding fact-attributes and relations, potentially saving considerable annotation time and costs. Our results show that the shallow system, despite using less supervision, is only 1.4% F1 points behind the multi-layered system on Disorders, and contrary to expectation, is able to improve over the latter by about 2.4% F1 points on Procedure facts. Further, our experiments also show that training the shallow system using only sentence-level fact labels with no span information has no negative effect on performance, indicating further cost savings through weak supervision.

1 Introduction

Medical fact coding is the joint task of recognizing the occurrences of medical facts from electronic patient medical records expressed in natural language, and linking each occurrence of a fact to a specific code in a medical taxonomy such as SNOMED[1].

A representative sentence from a medical record along with its annotated facts is shown in Figure 1. In the parlance of traditional natural language processing, this task is roughly equivalent to the tasks of named-entity recognition (Nadeau and Sekine, 2007) and entity-linking[2] rolled into one.

Several open evaluations such as *ShARe-CLEF* (Pradhan et al., 2013) and *Semeval* (Pradhan et al., 2014) have been run recently to address the twin problems of fact recognition (recognizing occurrences of medical facts in text) and fact-coding (linking each occurrence of a fact to a pre-assigned code). These evaluations report performance numbers on both the tasks separately.

Often times, facts that occur in a medical text may not correspond to any pre-assigned codes, and are referred to as *CUI-less* facts in the *Semeval* evaluation. In the aforementioned evaluations, the systems are expected to output and are evaluated against CUI-less facts as well. However, in typical end-user applications such as medical billing, one does not care about the occurrences of unrecognized, non-billable facts. This work is targeted at such end applications where discovering only the occurrences of fact-codes recognized by a medical taxonomy is desirable. Consequently, CUI-less facts are ignored in our evaluation framework.[3]

In this work, we will focus only on the fact types of Disorders and Procedures, and use SNOMED as our medical taxonomy. We also use *Linkbase*[4] as our knowledge-base for descriptions of the fact codes.

2 Multi-layered Models for Fact Coding

Some of the unique characteristics of medical fact coding compared to the traditional entity recognition are as follows:

1. Unlike traditional entities, medical facts can be non-contiguous.

2. Unlike traditional entities, medical facts can be overlapping.

[1] http://www.nlm.nih.gov/research/umls/Snomed/snomed_main.html

[2] http://www.nist.gov/tac/2013/KBP/EntityLinking/index.html

[3] In the official data of the Semeval task, it is reported that at least a quarter of the annotated facts are CUI-less (Pradhan et al., 2014). Hence ignoring these facts essentially renders a comparison of our evaluation numbers with the official Semeval numbers meaningless.

[4] http://www.nuance.com/for-healthcare/resources/clinical-language-understanding/ontology/index.htm

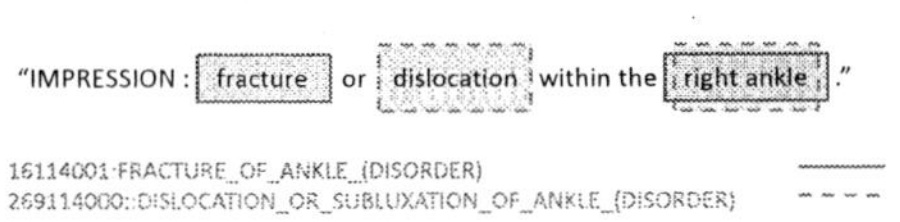

Figure 1: An example sentence containing two non-contiguous and mutually overlapping facts: Fact 1 is composed of the words 'fracture', 'right', and 'ankle' while Fact 2 comprises the words 'dislocation', 'right' and 'ankle'. Note that both facts are non-contiguous, since there is a break between 'fracture' and 'right ankle' as well as between 'dislocation' and 'right ankle'. Likewise, both facts are overlapping with each other since they share the tokens 'right' and 'ankle'.

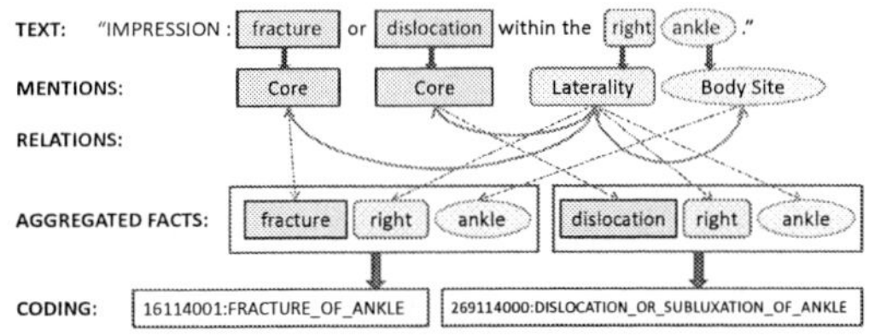

Figure 2: Output of various stages of the multi-layered pipeline on the example sentence in Figure 1. The mentions stage produces mentions of contiguous attributes, while the relations stage ties them together to produce larger, potentially non-contiguous entities. The final coding stage compares the fact-text to a database of fact-codes and their descriptions, and outputs the predicted medical codes.

The example sentence in Figure 1 satisfies these two unique criteria. Since entities may not occur contiguously, a BIO (Begin-Inside-Outside) style sequence tagger is no longer directly applicable (Bodnari et al., 2013). Therefore, some researchers have used BIOT (Begin-Inside-Outside-BeTween) style coding to model the non-contiguous nature of the entities (Cogley et al., 2013), while others have attempted the approach of breaking down the entities into attributes that satisfy the contiguousness requirement of the BIO style taggers, and then reconstructing the original non-contiguous entities by tying the mentions of attributes together using relations (Gung, 2013). The former approach of BIOT tagging addresses the problem of non-contiguous entities but does not address the problem of overlapping entities, while the latter can address both the problems. Hence, in this work, we will use the latter approach as our *multi-layered* baseline system.

An example output produced by various stages of the multi-layered system for the example sentence of Figure 1 is shown in Figure 2. In this example, a Disorder fact is broken down into *attributes* such as *Disorder-Core*, *Body-site* and *Laterality*, whose occurrences are always contiguous. The mentions of these attributes are identified by a BIO-style sequence tagger such as the CRF (Lafferty et al., 2001). Next, a relations classifier is run on all pairs of attribute mentions in a given sentence. Finally, all attribute-mentions connected by relations are aggregated to produce fact-mentions, which are then lexically compared to a database of fact-descriptions to output the code of each mentioned fact, if one exists in the taxonomy.

Although the above mentioned strategy of a multi-layered system is very effective, annotating the training data required for all stages of the pipeline can be quite laborious and expensive. The motivating question for us in this work is whether we can eliminate some of the stages such as attribute-mentions and relations and still deliver comparable fact-coding performance. We call our approach *shallow coding* as it aims to reduce the number of layers of supervised data needed to train the model.

3 Shallow Coding Model

The multi-layered model uses a bottom-up approach starting from attribute-mentions and incrementally building all the way to fact-codes. In contrast, the shallow coding model uses a top-down approach wherein the entire text of the sentence is used as a query to retrieve matching facts directly. Note that this model does not use any sequence tagger to identify relevant spans of the text before matching them with the fact-descriptions. Since using the entire sentence for matching results in retrieval of many spurious facts, they are further analyzed in subsequent stages to output the final set of predicted facts.

This approach detects occurrences of only the facts with pre-assigned codes in the taxonomy since the retrieved candidate facts are those that already exist in the taxonomy. In contrast, the multi-layered model can also detect facts that have no pre-assigned codes since the fact-recognition step is independent of the taxonomy. Since our final objective in this work is to generate recognizable fact codes, the shallow coding model is an appro-

Fact Code (Fact-Type)	Description	Source
49436004 (Disorder)	atrial fibrillation af afib	Linkbase DocID-131 DocID-236
195080001 (Disorder)	atrial fibrillation and flutter atrial flutter	Linkbase DocID-567

Table 1: A sample of the database of fact codes and their descriptions collected from a union of Linkbase and training data annotations.

priate candidate for the task. It is however not an appropriate model for the *ShARe-CLEF* and *Semeval* evaluations that also care about unrecognizable (CUI-less) facts. Hence we are unable to evaluate this model using the official evaluations. We however, compare the shallow model with our own implementation of the multi-layered approach as a baseline.[5]

The rest of the section discusses in detail, the various stages of the shallow coding system in the order of their execution.

3.1 Information Retrieval (IR) Stage

An inverted-index of codes and their corresponding concept-descriptions, as provided in the Linkbase knowledge-base is first created. The index is also augmented with fact annotations from training data, treating each fact-mention as an additional description for the corresponding fact-code. Such augmentation with training annotations is necessary since the language used in SNOMED descriptions differs significantly from that used in medical reports.[6] To prevent overfitting at training time, we use a leave-one-out strategy where for each sentence in the training set, the retrieval results exclude fact-annotations from the document that the sentence belongs to. A few example descriptions augmented with training data annotations are shown in Table 1.

During the retrieval process for a given sentence, the sentence is first filtered for all punctuation and stop-words, and an initial search is performed using a sliding-window of length 3 words and the retrieved descriptions over all the window searches are pooled together by their fact-codes. The reason for using a sliding-window search is that it minimizes spurious long-distance matches with the sentence. Any facts that span longer than the sliding window size may be ranked lower in the initial search, but are boosted in the re-ranking stage as described below.

The pooled descriptions are then pruned by their retrieval scores to a maximum of 10 descriptions per code. We then re-rank the retrieved facts by the maximum of the inclusion scores of their retrieved descriptions computed with respect to the entire sentence:

$$\text{incl-score}(f, s) = \max_{d \in f}\left(\frac{\sum_{w \in (d \cap s)} \text{TF}(w, f)\text{IDF}(w)}{\sum_{w \in d} \text{TF}(w, f)\text{IDF}(w)}\right), \quad (1)$$

where f is a fact-code, s is a sentence, d is a description pooled into f, and w is a word-token in the description obtained after removing stop-words and stemming the remaining words. The inverse-document-frequency (IDF) weights are computed from the index of descriptions and not from the training documents, and term-frequency $\text{TF}(w, f)$ is computed as the proportion of all descriptions in the fact f that contain the specific word w. The inclusion score is simply the IDF-weighted fraction of the description tokens contained in the sentence.

Further, to ensure that a single instance of the sliding window query does not dominate the search results, we also introduce a redundancy based penalty term into the inclusion score in Eqn. 1 where each word w in the numerator is discounted by $\log(1 + c(w))$, where $c(w)$ is the count of the number of times the word w is seen in the retrieved descriptions in the original ranking thus far.

The number of top ranking facts we return per sentence is a variable based on the sentence-length:

$$\text{n}(s) = \max(25, \min(3 \times \text{len}(s), 50)), \quad (2)$$

where $\text{n}(s)$ is the number of facts returned for the sentence s, and $\text{len}(s)$ is the number of processed tokens in s.

Note that we use unstemmed tokens in the initial search, but stemmed tokens for re-ranking, as this has been empirically found to improve performance by a small amount. In all our experiments, the initial search is performed us-

[5] The multi-layered approach should in fact be considered an upper-bound since it has access to more layers of labeled data.

[6] For example, one of the descriptions for Disorder code 49436004 is '*Atrial fibrillation*'. However, in medical reports, doctors typically use the short form '*afib*' to represent the same fact. Such variations can only be captured if we include training annotations as additional descriptions in the index.

Component	Recall
Lucene Search only	89.10
+ Inclusion-score-reranking	95.68
+ Redundancy penalty	96.12

Table 2: Contribution of various components towards the performance of the IR system. The numbers reported are on our Integris development set on Disorders facts.

ing default-ranking function as implemented in *Lucene*.[7]

Table 2 lists the contribution of Lucene search, re-ranking using inclusion score and using redundancy based penalty on our development set. The results indicates that while re-ranking is very critical towards achieving high recall, using a redundancy-based penalty to encourage diversity of results also is incrementally useful.

3.2 Alignment

All the descriptions retrieved from the previous step are then independently aligned word-to-word with the sentence text. For each description, we compute the alignment that has the minimum span but maximal possible matching, using a dynamic programming implementation that has a quadratic complexity in sentence length. We allow non-contiguous alignments in keeping with the fact a medical fact may consist of non-contiguous words. If multiple alignments satisfy this criterion, we return all such alignments. Note that the matches are computed using stemmed tokens, and order of matching is disregarded in computing the alignment. An example alignment where a single fact matches twice in a sentence via multiple descriptions is displayed in Figure 3.

For each description d aligned with the sentence s, an alignment score is computed as follows:

$$\text{Align-score}(a(d,s)) = \text{incl-score}(d,s) \times \text{tightness-ratio}(a(d,s)) \times \left(\sum_{w \in d \cap s} (\log(1.0 + \text{IDF}(w)))\right), \quad (3)$$

where $a(d,s)$ is the alignment of the description with the sentence, incl-score(d,s) is computed as shown in Eqn. 1, and tightness-ratio$(a(d,s))$ is computed as follows:

$$\text{tightness-ratio}(a(d,s)) = \frac{\sum_{w \in d \cap s}(1)}{\text{span-len}(a(d,s))}, \quad (4)$$

[7]http://lucene.apache.org/

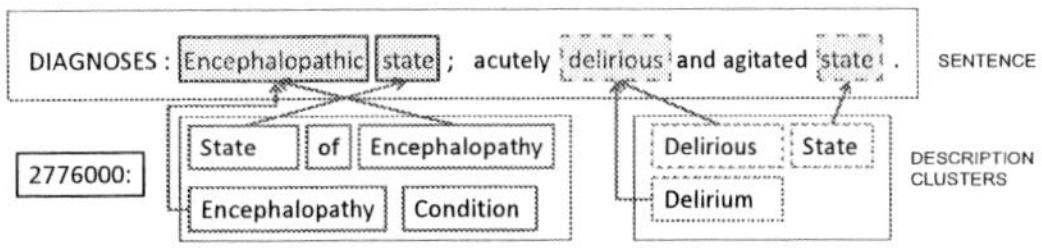

Figure 3: The Disorder code 2776000 occurs twice in the same sentence expressed as 'encephalopathic state' and 'delirious state'. Note that word order is ignored in computing the alignment. The words 'encephalopathy' and 'encephalopathic' match with each other due to matching of their stems. Also worth noting is the observation that the word 'state' in the description 'state of encephalopathy' could have been aligned with the last word in the sentence, but it does not happen since the alignment algorithm prefers maximal matches that have minimal span. The descriptions 'state of encepahlopathy' and 'encephalopathy condition' overlap over the word 'encephalopathy' and therefore form a cluster of descriptions. Likewise, the descriptions 'delirious state' and 'delirium' overlap over the word 'delirious' in the sentence, and form another cluster with the same fact code. These two clusters represent two distinct occurrences of the fact-code 2776000 in the sentence.

where span-len$(a(d,s))$ is the difference between the sentence-positions of right-most word in the alignment and the left-most word. Tightness ratio is higher for contiguous alignments than otherwise. As Eqn. 3 indicates, alignments that have a high inclusion score, tight alignment and a number of 'important' aligned words (as measured by their IDF scores) get high alignment-scores.

Since each fact can have multiple descriptions each of which may align in one or more ways with the sentence, we cluster the alignments of each fact based on their alignment positions in the sentence. In other words, each alignment-cluster c for a given fact contains all the descriptions that have at least one aligned position in common with another description in the cluster. Each such alignment-cluster constitutes an example, that goes to classifier-stages for further analysis. The alignment of a cluster with respect to a sentence is given by the alignment of the description in the cluster that has the best alignment score as given by Eqn. 3.

$$a(c,s) = \arg \max_{a(d,s)\ \forall\ d \in c} \text{Align-score}(a(d,s)) \quad (5)$$

3.3 Match Classifier

As mentioned above, each alignment cluster is treated as an example that is analyzed by the Match-classifier. At training time, the clusters are first mapped to positively annotated facts, such that each cluster is aligned with a positive fact in a greedy manner on a one-to-one basis. All the clusters mapped to positively annotated facts are considered positive examples, and the rest, negative.

Further, for training the Match-classifier, we only use those negative examples whose alignments do not overlap with those of any positive examples. This is done so that the Match-classifier accurately captures the semantics of similarity between the sentence and retrieved facts. The negative examples that overlap with the positive ones may have been annotated as negative for one of the following two reasons: (i) the retrieved fact is not related to the sentence, and (ii) the retrieved fact is related but is overruled because some other retrieved fact applies more accurately to the sentence. The Match-classifier is designed to deal with only case (i) above, hence we ignore the negative facts that overlap with any of the positive facts for training purposes. These facts will be handled separately by the Overlap-classifier in the next stage.

At test time, all the examples are run through the Match-classifier and classified as positive or negative for a given sentence. If the alignment of a given positively classified example does not overlap with that of any other example, it is directly output as positive for the given sentence. Else, it is sent to the subsequent stages for further analysis.

The following is the full list of features used in the Match-classifier.

Similarity features:

Unigrams: number of words and proportion of words in the description that are matched in the sentence, as well as the IDF-weighted versions of these two features.

Bigrams: number and proportion of bigrams in the description matched, as well as IDF-weighted versions of these features, where the IDF of a bigram is computed as the average of the IDFs of the pair of words in it.

Unordered bigrams: same as above, but ignoring the ordering of the bigrams.

Character-trigram features: each word in the description is mapped to a word in the sentence that has the highest number of character-level trigrams in common, and its similarity to the mapped word is measured in terms of the proportion of its character-trigrams matched. As features, we use the number and proportion of words in the description mapped, weighted by the character-trigram similarity scores.

Edit-distance based features: similar to character-trigram features, we map each word in the description to a word in the sentence using minimum edit-distance as the criterion. Next, we compute number and proportion of words matched using (1-edit-distance)/(word-length) as the similarity weight.

Synonym features: each word in the description is replaced with one of its synonyms from a dictionary[8], and computed unigram features with the replaced words, as above. The maximum value of the features over all synonyms is used as the final feature value.

For each of the above features, we compute its maximum value over all descriptions in the cluster and it as the final feature value.

Lexical features:

Matched and unmatched words: the matched words and their bigrams in the best alignment of the cluster, conjoined with the code, as well as the unmatched words within the span of the alignment conjoined with the code.

POS features: the parts-of-speech categories of matched words and their bigrams in the best alignment of the cluster, conjoined with the code, as well the POS categories of unmatched words within the span of the alignment, conjoined with the code.

Context words: Two words to the left and two words to right of the alignment, conjoined with the code of the description, used both as unigrams and bigrams.

Other features:

Alignment-based features: the tightness ratio (see Eqn. 4 above) of the best alignment for the cluster, average distance between the words in the align-

[8]The synonyms are generated in an unsupervised fashion based on descriptions that co-occur in a fact but differ by a single word, e.g.: 'lung cancer', and 'pulmonary cancer' are used to describe the same fact, hence 'lung' and 'pulmonary' are considered synonymous.

ment, and the number of unmatched words in the span of the alignment.

Prior features: the number and fraction of times the best aligned description in the cluster has been annotated with the given code in the training set.

Header features: the section-header name of the current sentence (E.g.: Diagnosis, History of illnesses, Discharge Summary, etc.) conjoined with the code of the matching description.

3.4 Overlap Classifier

All the examples classified as positive by the Match classifier that overlap with at least one other positively classified example are input to the Overlap classifier, that further analyzes these examples. The Overlap classifier uses all the features used in the Match-classifier as well as additional features based on the type of overlap between the two examples, and hierarchy relationship in SNOMED taxonomy between the two overlapping facts. We compute these features for each example with respect to all other examples that overlap with it. For a given example, even if the same feature fires with multiple overlapping examples, we do not add up the counts since we consider it as a binary feature.

Overlap features: For each example, a binary feature is computed to characterize whether its alignment (a) is subsumed by the alignment of the other example, (b) subsumes the alignment of the other example, (c) exactly equals the alignment of the other example or (d) overlaps without any of the three properties above. Other variants of this feature also include the feature conjoined with the overlapping words, and conjoined with the fact codes of the two examples.

Hierarchy features: For each example, we define a binary feature to characterize whether an example's fact-code is (a) a descendant, (b) an ancestor, (c) a sibling or (d) a co-parent of the other overlapping example's code in the taxonomy. Variants of this feature also include the feature conjoined with words in the overlap, and the fact codes of the two examples.

We only use positive examples that overlap with at least one other example, and negative examples that overlap with at least one positive example for training the classifier. This kind of sub-sampling of the training data allows the Overlap classifier to learn the semantics of how certain facts overrule other facts although both facts may be equally related to the sentence in question.[9]

Component	F1
Match Classifier only	78.65
+ Overlap Classifier	81.73
+ Rejection Rules	83.07

Table 3: Contribution of the two classifiers and the rejection rules towards the performance of the Shallow coding system. All numbers are reported on Disorder facts on the Integris development set.

At test time, each example is classified in an I.I.D. manner[10] , and the positively classified examples are then input to the final stage as described below.

3.5 Rejection Rules

In the final stage of the Shallow coding model, we apply two rules to potentially reject inconsistently classified examples from the previous stage. The two rules are listed below:

Rejection of subsumed examples: If the alignment of a positively classified example A strictly subsumes that of another positively classified example B, then example B is rejected and labeled as negative, since example A, with its longer alignment, is usually the more reliable and more specific fact. E.g.: Fact#195080001 with alignment 'atrial fibrillation and flutter' overrules Fact#49436004 that aligns with only 'atrial fibrillation', as its alignment is subsumed by the former's.

Rejection of ancestors: If the alignment of a positively classified example A overlaps with that of another positively classified example B and A is an ancestor of B, then the example A is rejected, since B, being the descendant is a more specific fact than A.

Note that the above rules are applied to pairs of positively classified and overlapping examples A and B, where A's confidence score as given by the Overlap classifier is higher than that of B.

[9]For example, if the medical text contains the phrase 'atrial fibrillation and flutter', it would match against both the facts shown in Table 1. However, Fact#195080001, being the more specific match is the correct fact and therefore overrules Fact#49436004.

[10]It is easy to see that the interactions of the overlapping examples may be modeled by a joint model such as the CRF. We have tried using CRFs in our experiments. Since the structure of the CRF can be arbitrary depending on the overlapping structure in a sentence, exact inference is hard. Hence, we used pseudo-likelihood for training the CRF and Gibbs sampling for testing, but it has not produced better results than the I.I.D. classifier using the features listed above. Hence we do not report the CRF's performance.

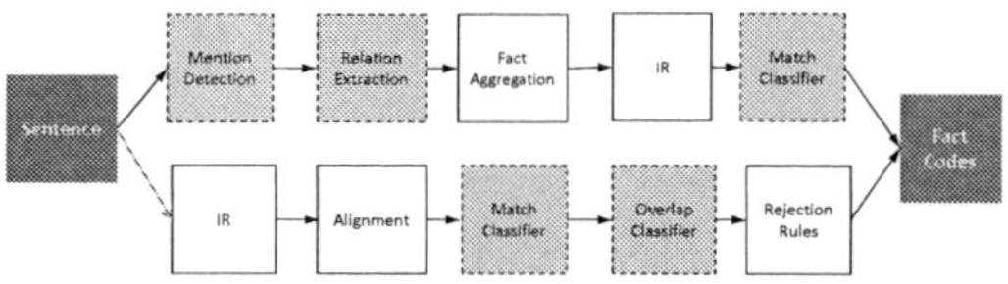

Figure 4: Comparison of various stages in the multi-layered pipeline vs the Shallow coding pipeline: the boxes with broken borders in either pipeline represent the stages that require labeled data. In the Shallow coding pipeline, Match-classifier and Overlap-classifier are the only stages that need training data, and they both use different slices of the same fact-span data for training. In contrast, the deep pipeline needs separate training data for mentions, relations and coding.

Dataset (FactType)	Subset	nDoc	nSent	nFact
Integris (Disorders)	train	409	14,218	10,906
	test	384	28,408	7,807
Mult-inst (Disorders)	train	12,370	484,822	204,124
	test	1,530	99,564	27,307
Proc-notes (Procedures)	train	1,624	71,151	17,996
	test	201	8,915	2,996

Table 4: Statistics of the datasets and corresponding fact-types used in our experiments. Integris is used purely as a development dataset on which we developed and tuned our models. We trained and evaluated Disorders on the multi-institution train and test datasets respectively. Similarly, we trained and evaluated our models on Procedures on the Proc-notes train and test splits. In the table, *nDoc* stands for number of documents, *nSent* for number of sentences, and *nFact* for number of facts.

Table 3 reports the incremental contribution of each classifier component to the overall performance of the shallow coding system. The numbers show that each component makes a significant contribution towards the overall performance.

Figure 4 compares the various stages involved in the multi-layered pipeline to the shallow coding system. The number of stages that need annotated data for training are indicated by boxes with broken edges in the figure, and is much less for the shallow system. In fact, both the stages that need training data in the shallow system, namely the Match classifier and Overlap classifier use different slices of the same training data, as described earlier.

Dataset	Model	Prec.	Rec.	F1
Integris	Mult-layer	82.76	84.51	**83.63**
	Shallow	85.27	80.98	83.07
Multi-Inst	Mult-layer	86.36	87.72	**87.03**
	Shallow	86.68	84.54	85.60
Proc-notes	Mult-layer	38.31	55.27	45.25
	Shallow	44.79	50.90	**47.65**

Table 5: Performance comparison: the shallow coding system is only about 0.6% F1 points below the multi-layered one on Disorders on the development set. On the unseen data of Multi-institution using the same fact-types, it is about 1.4% F1 behind the multi-layered model. On Proc-notes data involving Procedure facts, the shallow system is able to outperform the multi-layered architecture by 2.4% F1 points.

4 Experiments and Results

For tuning and developing our model, we used medical reports from an institution called *Integris*, which are partitioned into training and test sets. We tuned our model only on Disorder facts and evaluated them on both Disorders and Procedures. For evaluating the model on Disorders, we used another dataset from multiple institutions with its own train and test partitions which we call the *Multi-inst* dataset. For evaluating procedures, we used a dataset consisting of Procedure Notes documents with its own train and test partitions. The statistics of the datasets are summarized in Table 4.

The results of our experiments are summarized in Table 5. The shallow coding model is only about 1.4% F1 points behind the traditional multi-layered supervised model on Disorder facts, making it attractive for situations where cost savings are critical. On the more complex medical fact-types of Procedures, the shallow coding system outperforms the multi-layered system by 2.4 % F1 points. The fact that Procedure facts are harder is evident from the performance numbers of either system on Procedures compared with those on Disorders. A a few example Procedure facts, along with their attribute level annotations are displayed in Figure 5.

On complex fact-types involving long distance relations between the attributes, errors accumulate over the layers of the multi-layered system resulting in poorer performance.[11] In such a scenario, the shallow model may be more attractive.

[11]We are unable to show detailed comparison of the errors of the two models as our datasets are proprietary.

4.1 Weakly supervised training

Further, our experiments on both Disorders and Procedures showed that the performance of the shallow system practically remains unchanged even if it is provided with only sentence-level fact labels at training time, omitting their actual spans. The exact span of each fact in a training sentence is not needed since the model's alignment stage computes this information reasonably accurately, as long as it knows that the fact exists in the sentence. There was however, a caveat in our experiments: we retained the fact descriptions in the retrieval index that were created from the fact-spans in training sentences (see Section 3.1). Without these augmented descriptions, the performance of the system degrades considerably. Although this fact-span information was used only in the IR stage, it essentially means that the system did ultimately have access to fact-spans, and therefore is not a strict weakly-supervised model. Despite this important caveat, we believe that there is promise in a *weakly-supervised system* for medical fact coding, where facts are annotated only at sentence level without the exact span information, which may yield additional annotation cost savings. Note that such a weakly supervised model will not be applicable in the context of a sequence tagger that annotates mentions of facts or attributes first (such as the multi-layered model described in this paper or the ones described in (Bodnari et al., 2013) and (Gung, 2013)), since these models demand availability of annotated mention spans at training time.

Weakly supervised training has been successful in other information extraction tasks such as relation extraction (Surdeanu et al., 2012; Weston et al., 2013), but has not been used in the context of entity recognition, to the best of our knowledge. This may have been due to the fact that in traditional entity recognition, entities tend to be contiguous and non-overlapping, and therefore annotating entity spans may cause no significant overhead over annotating only sentences with entity-labels. Since these two properties do not hold true in medical fact recognition, weak supervision may be more attractive here. We hope this work paves the way for more future work in this direction.

5 Conclusions and Future Work

In this work, we propose a new shallow coding model that learns to annotate medical facts that are overlapping and non-contiguous without using any attribute level annotations and relations annotations. Our work shows that this approach, while not being too far behind on Disorders, actually outperforms a more sophisticated and more deeply supervised model on Procedures.

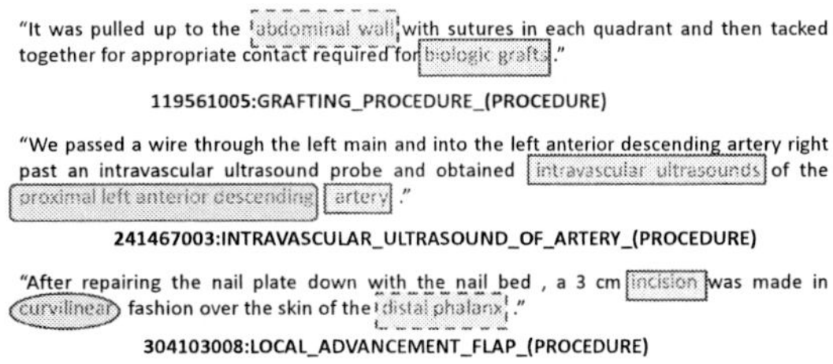

Figure 5: Examples of Procedure facts along with their attributes: the rectangle with sharp edges are *Procedure-Cores*, ones with broken edges are *Body-sites*, rectangles with rounded edges are *Lateralities*, and the ovals are *Approaches*. Note that the attributes for Procedures are more complicated and exhibit long-distance relations among themselves.

As part of future work, we plan to investigate the feasibility of a weakly-supervised system that trains on only sentence-level fact labels. We believe that optimal performance may be achieved by a hybrid system that uses a small number of annotated training facts for generating an augmented retrieval index, and a large number of sentences with fact-labels but without span information, for training the classifiers. This would further reduce the annotation costs substantially.

We implemented a basic system combination of the shallow coding and the multi-layered models where the predictions of the multi-layered system are re-ranked based on the prediction of the shallow model for facts that are aligned between the two systems. However, such combination did not result in any significant improvement. As part of future work, we plan to build a meta-classifier that learns to effectively combine the outputs of the two systems using more sophisticated features, hopefully further improving over either system.

Acknowledgments

We are extremely thankful to *Nuance, Inc.* for supporting this work, and for providing all the resources including annotated medical reports for training and evaluation, and for providing access to the knowledge-bases needed in this work.

References

A. Bodnari, L. Deleger, T. Lavergne, A. Neveol, and P. Zweigenbaum. 2013. A supervised named-entity extraction system for medical text. In *Online Working Notes of the CLEF 2013 Evaluation Labs and Workshop*.

J. Cogley, N. Stokes, and J. Carthy. 2013. Medical disorder recognition with structural support vector machines. In *Online Working Notes of the CLEF 2013 Evaluation Labs and Workshop*.

J. Gung. 2013. Using Relations for Identification and Normalization of Disorders: Team CLEAR in the ShARe CLEF 2013 eHealth Evaluation Lab. In *Online Working Notes of the CLEF 2013 Evaluation Labs and Workshop*.

J. D. Lafferty, A. McCallum, and F. C. N. Pereira. 2001. Conditional random fields: Probabilistic models for segmenting and labeling sequence data. In *Proceedings of the Eighteenth International Conference on Machine Learning*, pages 282–289.

D. Nadeau and S. Sekine. 2007. A survey of named entity recognition and classification. In *Linguisticae Investigationes*, number 30, pages 3–26.

S. Pradhan, N. Elhadad, B. R. South, D. Martinez, L. Christensen, A. Vogel, H. Suominen, W. W. Chapman, and G. Savova. 2013. Task 1: ShARe/CLEF eHealth Evaluation Lab. In *Online Working Notes of the CLEF 2013 Evaluation Labs and Workshop*.

S. Pradhan, N. Elhadad, W. Chapman, S. Manandhar, and G. Savova. 2014. SemEval-2014 Task 7: Analysis of Clinical Text. In *Proceedings of the 8th International Workshop on Semantic Evaluation (SemEval 2014)*, pages 54–62.

Mihai Surdeanu, Julie Tibshirani, Ramesh Nallapati, and Christopher D. Manning. 2012. Multi-instance multi-label learning for relation extraction. In *Proceedings of the 2012 Joint Conference on Empirical Methods in Natural Language Processing and Computational Natural Language Learning*, EMNLP-CoNLL '12, pages 455–465, Stroudsburg, PA, USA. Association for Computational Linguistics.

Jason Weston, Antoine Bordes, Oksana Yakhnenko, and Nicolas Usunier. 2013. Connecting language and knowledge bases with embedding models for relation extraction. *CoRR*, abs/1307.7973.

Stacked Generalization for Medical Concept Extraction from Clinical Notes

Youngjun Kim
School of Computing
University of Utah
Salt Lake City, UT 84112
youngjun@cs.utah.edu

Ellen Riloff
School of Computing
University of Utah
Salt Lake City, UT 84112
riloff@cs.utah.edu

Abstract

The goal of our research is to extract medical concepts from clinical notes containing patient information. Our research explores stacked generalization as a meta-learning technique to exploit a diverse set of concept extraction models. First, we create multiple models for concept extraction using a variety of information extraction techniques, including knowledge-based, rule-based, and machine learning models. Next, we train a meta-classifier using stacked generalization with a feature set generated from the outputs of the individual classifiers. The meta-classifier learns to predict concepts based on information about the predictions of the component classifiers. Our results show that the stacked generalization learner performs better than the individual models and achieves state-of-the-art performance on the 2010 i2b2 data set.

1 Introduction

Clinical notes (or electronic medical records) contain important medical information related to patient care management. Health care professionals enter a patient's medical history and information about their care at a health care provider. A patient's diseases, symptoms, treatments, and test results are often encoded in these notes in an unstructured manner.

In the last two decades, Natural Language Processing (NLP) techniques have been applied to clinical notes for medical concept extraction. Medical concept extraction typically consists of two main steps: detection of the phrases that refer to medical entities, and classification of the semantic category for each detected medical entity. Medical domain knowledge and sophisticated information extraction methods are required to achieve high levels of performance. Medical concept extraction is a fundamental problem that can also serve as the stepping stone for higher level tasks, such as recognizing different types of relationships between pairs of medical concepts.

The main goal of our research is to explore the use of stacked generalization learning for the medical concept extraction task. Stacked learning (Wolpert, 1992) is a meta-learning ensemble-based method that regulates the biases of multiple learners and integrates their diversities. An ensemble of individual classifiers is created and then another classifier (the meta-classifier) sits on top of the ensemble and trains on the predictions of the component classifiers. A key advantage of stacked generalization is that the meta-classifier learns how to weight and combine the predictions of the individual classifiers, allowing for a fully automated ensemble system. New component classifiers can be easily added without the need for manual intervention. Voting-based ensembles are another strategy for combing multiple classification models, and they often perform well. But they can require manual adjustment of the voting threshold when new components are added, and they do not automatically learn how to weight different components. Stacked generalization provides a more easily extensible and adaptable framework.

In the next sections, we discuss related work, describe our individual classifiers for medical concept extraction, and present the stacked generalization learning framework. Finally, we present experimental results on the 2010 i2b2 data set and compare our results with state-of-the-art systems.

2 Related Work

In early natural language processing (NLP) research for clinical notes, most systems used rule-based approaches. MedLEE (Friedman et al., 1994) uses a rule-based system that extracts med-

Proceedings of the 2015 Workshop on Biomedical Natural Language Processing (BioNLP 2015), pages 61–70,
Beijing, China, July 30, 2015. ©2015 Association for Computational Linguistics

ical concepts by performing a shallow syntactic analysis and using semantic lexicons. SymText was developed by Haug et al. (1995; 1997) and evolved into MPlus (Christensen et al., 2002). This system was used to extract medical findings, diseases, and appliances from chest radiograph reports. HITEx (Zeng et al., 2006) is a pipelined system with multiple preprocessing modules and has been used to extract family history information, principal diagnosis, comorbidity and smoking status from clinical notes. MetaMap (Aronson and Lang, 2010) was developed to recognize Metathesaurus concepts from biomedical texts by utilizing the UMLS (Unified Medical Language System).

Recently, statistical learning approaches have received more attention because of the manual effort typically required to create rule-based systems. Most current information extraction (IE) systems in clinical NLP use statistical machine learning approaches that often achieve better performance than rule-based approaches. Our work is also closely related to Named Entity Recognition (NER). For both newswire and biomedical texts, machine learning models have achieved good results for extracting specific types of entities (e.g., (Collier et al., 2000; Lafferty et al., 2001; Collins, 2002; Zhou and Su, 2002; McDonald and Pereira, 2005)).

Our research focuses on the medical concept detection task that was introduced in 2010 for the *i2b2 Challenge Shared Tasks* (Uzuner et al., 2011). These challenge tasks included: (a) the extraction of medical problems, tests, and treatments, (b) classification of assertions made on medical problems, and (c) relations between medical problems, tests, and treatments. The best performance on the 2010 i2b2 concept extraction task (a) was achieved by de Bruijn et al. (2011) with 83.6% recall, 86.9% precision, and 85.2% F_1 score. They integrated many features commonly used in NER tasks including syntactic, orthographic, lexical, and semantic information (from various medical knowledge databases). Jiang et al. (2011) trained a sequence-tagging model that consisted of three components in a pipeline: concept taggers with local features and outputs from different knowledge databases, post-processing programs to determine the correct type of semantically ambiguous concepts, and a voting ensemble module to combine the results of different taggers. Their system achieved an 83.9% F_1 score. Subsequent research by Tang et al. (2013) showed that clustering and distributional word representation features achieved an higher F_1 score of 85.8%.

Ensemble methods that combine multiple classifiers have been widely used for many NLP tasks and generally yield better performance than individual classifiers. For protein/gene recognition, Zhou et al. (2005) used majority voting from multiple classifiers to achieve better performance than any single classifier. Finkel et al. (2005) combined the outputs of forward and backward (reversing the order of the words in a sentence) sequence labelling, which improved recall. Similarly, Huang et al. (2007) integrated the outputs of three models for gene mention recognition. They applied intersection to the outputs of forward and backward labeling SVM (support vector machine) models and then union with the outputs of one CRF (conditional random fields) model. Doan et. al (2012) showed that a voting ensemble of rule-based and machine learning systems obtained better performance than individual classifiers for medication detection. For medical concept detection, Kang et al. (2012) used majority voting between seven different systems for performance improvement.

Our research explores an ensemble method called stacked generalization (Wolpert, 1992; Breiman, 1996), which has been shown to produce good results for several NLP tasks. Stacking is an ensemble-based method for combining multiple classifiers by training a meta-classifier using the outputs of the individual classifiers. Ting and Witten (1999) showed that stacked generalization using confidence scores from the predictions of multiple classifiers obtained better results than the individual systems. Džeroski and Zeno (2004) showed good performance for stacked learning on a collection of 21 datasets from the UCI Repository of machine learning databases (Blake and Merz, 1998). Nivre and McDonald (2008) applied stacked learning to dependency parsing by integrating two different models (graph-based models and transition-based models). Recently, some research has used stacked learning in the bioinformatics domain. Wang et al. (2006) used stacked learning with two base learners for predicting membrane protein types. Netzer et al. (2009) applied stacked generalization to identify breath gas marker and reported improved classification accuracy. For NLP from clinical texts, Kilicoglu et al. (2009) used stacked learning for document level classification to identify rigorous, clinically relevant studies.

Stacked learning is similar to weighted majority voting (Littlestone and Warmuth, 1994) and Cascading learning (Gama and Brazdil, 2000). However, weighted majority voting only determines a voting weight for each individual classifier, while stacked learning can assign different weights to different types of predictions. Training in cascading learning requires multiple rounds of learning, while stacked learning typically consists of just two stages. Also, cascading learning does not need multiple base learners. Tsukamoto et al. (2002) employed cascaded learning using a single algorithm that improved performance on an NER task.

Our stacked generalization framework is different from weighted majority voting or cascading learning. Our stacked learning architecture trains a meta-classifier using features derived from the predictions and confidence scores of a set of diverse component classifiers. To the best of our knowledge, this research is the first to use stacked generalization with a rich set of meta-features for medical concept extraction from clinical notes.

3 Stacked Generalization with Multiple Concept Extraction Models

The goal of our research is to investigate stacked generalization learning for medical concept extraction with a diverse set of information extraction models. We will first describe each individual model and then present the stacked learning framework.

3.1 Information Extraction Models

Our ensemble consists of four types of individual component systems, which are described below.

MetaMap: We use a widely-used knowledge-based system called MetaMap (Aronson and Lang, 2010). MetaMap is a rule-based program that assigns UMLS Metathesaurus semantic concepts to phrases in natural language text. Unlike our other IE systems, MetaMap is not trained with machine learning so it is not dependent on training data. Instead, MetaMap is a complementary resource that contains a tremendous amount of external medical knowledge.

We encountered one issue with using this resource for our task. MetaMap can assign a large set of semantic categories, many of which are not relevant to the i2b2 concept extraction task. However it is not obvious how to optimally align the MetaMap semantic categories with our task's semantic categories because their coverage can substantially differ. Therefore we built a statistical model based on the concepts that MetaMap detected in the training data. We collected all of MetaMap's findings in the training data, aligned them with the gold standard medical concepts, and calculated the probability of each MetaMap semantic category mapping to each of our task's three concept types ("problem", "treatment", and "test"). We then assigned a MetaMap semantic type to one of our concept types if the semantic type is ranked among the top 30% of semantic types based on Prob(concept_type | sem_type). For example, "sosy" ("Sign or Symptom" in MetaMap) was mapped to the "problem" concept type because it had a high probability of being aligned with labeled problems in the data set. Table 1 shows the semantic types that we ultimately used for concept extraction.[1]

Category	MetaMap semantic types
Problem	acab, anab, bact, celf, cgab, chvf, dsyn, inpo, mobd, neop, nnon, orgm, patf, sosy
Treatment	antb, carb, horm, medd, nsba, opco, orch, phsu, sbst, strd, topp, vita
Test	biof, bird, cell, chvs, diap, enzy, euka, lbpr, lbtr, mbrt, moft, phsf, tisu

Table 1: MetaMap semantic types used for concept extraction.

Rules: We used the training data to automatically create simple rules. The idea is to exploit the training data to create a simple rule-based system without any manual effort. For each phrase labeled as a medical concept in the training data, we created a rule that maps the phrase to the concept type that it was most frequently assigned to in the training data. Similar to the MetaMap model above, we then computed P(concept_type | phrase) using frequency counts.

To generate phrase matching rules, we applied

[1]Refer to `http://metamap.nlm.nih.gov/Docs/SemanticTypes_2013AA.txt` for the mapping between abbreviations and the full semantic type names.

two thresholds to each rule: a minimum probability threshold (θ_P) and a minimum frequency threshold (θ_F). First, we extracted annotated phrases from the training data. Next, for each phrase we computed its overall frequency and P(concept_type | phrase) for each of the 3 concept types. We then selected the phrases that passed the two thresholds and assigned them to the corresponding concept type. In cases where one phrase subsumed another phrase, such as "disease" and "coronary disease", and both phrases pass the thresholds, we only chose the longer phrase. We then created a rule for each phrase that labels all instances of that phrase as the concept type (e.g., "diabetes" $\rightarrow$ $Problem$). A concept was extracted when the candidate phrase occurs more than two times (θ_F) in the training data and the rule's probability is over 60% (θ_P).

Contextual Classifier (SVM): We created a supervised learning classifier with contextual features. We applied the Stanford CoreNLP tool (Manning et al., 2014) to our data sets for tokenization, lemmatization, part-of-speech (POS) tagging, and Named Entity Recognition (NER). We trained a Support Vector Machine (SVM) classifier with a linear kernel using the LIBLINEAR (Library for Large Linear Classification) software package (Fan et al., 2008) for multi-class classification.

We reformatted the training data with IOB tags (B: at the beginning, I: inside, or O: outside of a concept). We defined features for the targeted word's lexical string, lemma, POS tag, affix(es), orthographic features (e.g. Alphanumeric, HasDigit), named entity tag, and pairwise combinations of these features. Also, we used the predictions of MetaMap as additional features. Table 2 shows the complete feature set used to create the SVM model, as well as the CRF models described below. We set the cost parameter to be $c = 0.1$ (one of LIBLINEAR's parameters) after experimenting with different values by performing 10-fold cross validation on the training set.

Sequential Classifier (CRF): We trained several sequential taggers using linear chain Conditional Random Fields (CRF) supervised learning models. In contrast to the contextual classifier mentioned above, the CRF classifiers use a structured learning algorithm that explicitly models transition probabilities from one word to the next. Our CRF models also use the features in

Feature	Description
Word	w_0 (current word), w_{-1} (previous word), w_1 (following word), w_{-2} (second previous word), w_2 (second following word)
Bi-grams of words	$[w_{-2}, w_{-1}]$, $[w_{-1}, w_0]$, $[w_0, w_1]$, $[w_1, w_2]$
Lemmas	$l_{-3}, l_{-2}, l_{-1}, l_1, l_2, l_3$
Affixes	prefixes and suffixes, up to a length of 5
Orthographic	15 features based on regular expressions for w_0, w_{-1}, w_1
POS tags	$p_0, p_{-1}, p_1, p_{-2}, p_2$
Bi-grams of POS tags	$[p_{-2}, p_{-1}]$, $[p_{-1}, p_0]$, $[p_0, p_1]$, $[p_1, p_2]$
Lemma + POS	$[l_0, p_0]$
NER class	n_0
MetaMap semtype	m_0, m_{-1}, m_1, $[m_{-1}, m_0]$, $[m_0, m_1]$

Table 2: Feature set for SVM and CRF models.

Table 2. We used Wapiti (Lavergne et al., 2010), which is a simple and fast discriminative sequence labeling toolkit, to train the sequential models. As with the SVM, 10-fold cross validation was performed on the training set to tune the Wapiti's CRF algorithm parameters. We set the size of the interval for the stopping criterion to be $e = 0.001$. For regularization, $L1$ and $L2$ penalties were set to 0.005 and 0.4 respectively.

Post processing: The concepts annotated by the i2b2 annotation guidelines[2] include modifying articles, pronouns, and prepositional phrases. For treatments such as medications, the amount, dose, frequency, and mode are included in the annotation only when they occur as pre-modifiers. However, when they are part of *signatura*, which explains how to use the medication for the patient, they are excluded from concept boundaries. For example,

800 mg ibuprofen
Lasix 20 mg b.i.d. by mouth

[2]`https://www.i2b2.org/NLP/Relations/assets/ConceptAnnotationGuideline.pdf`

"800 mg ibuprofen" is annotated as a treatment concept, while only "Lasix" is annotated in the second example.

When applying MetaMap to the training set, we observed that there is a huge difference between the i2b2 annotations and MetaMap's concept boundary definition, especially with respect to articles and pronouns. MetaMap typically excludes modifying articles, pronouns, and prepositional phrases. For example, for "a cyst in her kidney", only "cyst" was extracted by MetaMap.

Therefore we added a post-processing step that uses three simple heuristics to adjust concept boundaries to reduce mismatch errors. Although these rules were originally compiled for use with MetaMap, we ultimately decided to apply them to all of the IE models. The three heuristic rules are:

1. We include the preceding word contiguous to a detected phrase when the word is a quantifier (e.g., "some"), pronoun (e.g., "her"), article (e.g., "the'), or quantitative value (e.g., "70%").
2. We include a following word contiguous to a detected phrase when the word is a closed parenthesis (")") and the detected phrase contains an open parenthesis ("(").
3. We exclude the last word of a detected phrase when the word is a punctuation mark (e.g., period, comma).

3.2 Ensemble Methods

We explored two types of ensemble architectures that use the medical concept extraction methods described above as components of the ensemble. We created a Voting Ensemble, as a simple but often effective ensemble method, and a Stacked Generalization Ensemble, which trains a meta-classifier with features derived from the outputs of its component models. Both architectures are described below.

Voting Ensemble Method: We implemented the majority voting strategy suggested by Kang et al. (2012) with a simple modification to avoid labeling concepts with overlapping text spans. When two different concepts have overlapping text spans, the concept that receives more votes is selected. For overlapping concepts with identical vote counts, we used the normalized confidence scores from the individual classifiers and select the concept with the higher confidence score. Each confidence score, $s \in S$ (the set of all confidence scores), was normalized by Z-score as:

$$Nor(s) = \frac{s - E(S)}{std(S)} \text{ where}$$

$$E(S) = \text{the mean of the scores}$$

$$std(S) = \text{the standard deviation of the scores}$$

Stacked Generalization Method: We created a meta-classifier by training a SVM classifier with a linear kernel based on the predictions from the individual classifiers. Figure 1 shows the architecture of our stacked learning ensemble. First, to create training instances for a document, all of the concept predictions from the individual IE models are collected. We then use a variety of features to consider the degree of agreement and consistency between the IE models. Each concept predicted by an IE model is compared with all other concepts predicted in the same sentence. For each pair of concepts, the following eight matching criteria are applied to create eight features:

- If the text spans match
- If the text spans partially match (any word overlap)
- If the text spans match and concept types match
- If the text spans partially match and the concept types match
- If the text spans have the same start position
- If the text spans have same end position
- If one text span subsumes the other
- If one text spans is subsumed by the other

Features are also defined that count how many different models produced a predicted concept, and features are defined for predictions produced by just a single model (indicating which model produced the predicted concept).

In addition, we created a feature for the confidence score of each predicted concept. When multiple components predicted a concept, the highest score was used. We also created a feature that counts how many times the same phrase was predicted to be a concept in other sentences in the same document. The number of word tokens in a prediction, and whether the prediction contains a conjunction or prepositional phrase, were also used as features.

We performed 10-fold cross validation on the training set to obtain predictions for each classifier. These predictions were used to train the meta-classifier.

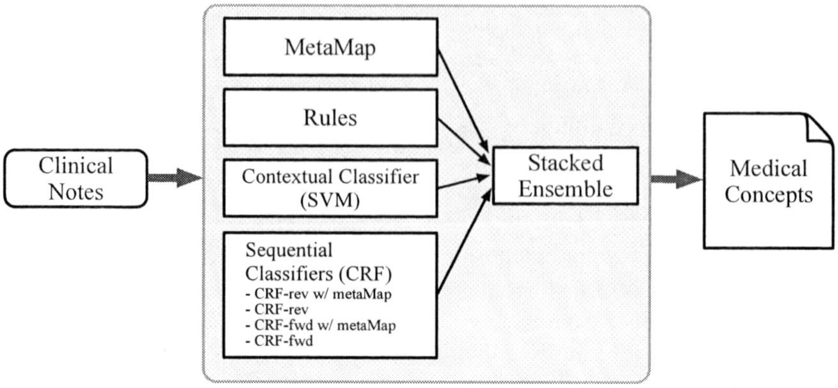

Figure 1: Stacked Learning Ensemble Architecture

4 Experimental Results

We present experimental results for each of our concept extraction components individually, as well as for each of the two ensemble methods: voting and stacked generalization learning.

4.1 Data

The 2010 i2b2 Challenge corpus was used for evaluation. The corpus consists of discharge summaries from Partners HealthCare (Boston, MA) and Beth Israel Deaconess Medical Center, as well as discharge summaries and progress notes from the University of Pittsburgh Medical Center (Pittsburgh, PA). It contains 349 clinical notes as training data and 477 clinical notes as test data. 18,550 problems, 13,560 treatments and 12,899 tests (for a total of 45,009 medical concepts) are annotated as the semantic concepts in the test data.

4.2 Performance of Individual Models

We used the i2b2 Challenge evaluation script to compute recall, precision, and F_1 scores. In this paper, we present the results of class exact match: both the text span and semantic category must exactly match the reference annotation.

MetaMap: We used MetaMap 2013v2 with the 2013AB NLM relaxed database.[3] As we mentioned in Section 3.1, we only used a subset of MetaMap's semantic types based on statistics collected by aligning MetaMap's findings with the medical concepts in the labeled training data.[4] We selected the top 30% of its semantic types (shown in Table 1) based the collected probabilities. The first row of Table 3 shows the results for MetaMap using these semantic categories. As explained before, MetaMap suffers from boundary mismatch errors due to the difference between the i2b2 annotations and MetaMap's concept boundary definition. In spite of our added post-processing rules to address this issue, we could not eliminate this problem especially for concepts containing many pre-modifiers or prepositional phrases. We also observed that MetaMap often did not recognize acronyms and abbreviations in the clinical notes.

Method	Rec	Pr	F
MetaMap	36.1	47.4	41.0
Rules	18.5	72.6	29.5
SVM	81.2	77.5	79.3
CRF-fwd	81.5	86.2	83.8
CRF-fwd w/ MetaMap	82.5	86.7	84.5
CRF-rev	82.4	86.5	84.4
CRF-rev w/ MetaMap	82.9	87.0	84.9
Voting ensemble	83.5	88.2	85.8
Stacked ensemble	**83.5**	**88.6**	**86.0**

Table 3: Recall (Rec), Precision (Pr), and F_1 score (F) of each method on the 2010 i2b2 Challenge test data.

Rules: The second row of Table 3 shows the results of matching with the rules that we extracted from the training data. This simple approach obtained fairly good precision of 72.6%, but low

[3] We used the following MetaMap options: -C -V NLM -y -i -g --composite_phrases 3 --sldi

[4] Using all of MetaMap's semantic types produces extremely low precision.

recall. This method relies entirely on common words found in the training data, so unseen words in the test data were not recognized. In addition, pre-modifiers were often missed. For example, only "embolization" was extracted from text mentioning "coil embolization".

SVM: The SVM context-based classifier achieved an F_1 score of 79.3% (third row in Table 3) with its rich contextual features. A subsequent analysis revealed that this classifier excels at recognizing concepts that consist of a single word, achieving recall of 89.3% for these cases, about 2.3% higher than the sequential classifiers (CRFs) perform on these cases.

CRF: We implemented four different variations of sequential classifiers. We trained CRF classifiers with both forward and backward tagging (by reversing the sequences of words) (Kudo and Matsumoto, 2001; Finkel et al., 2005). As a result, each medical concept had different IOB representations. For example, the IOB tags of "positive lymph nodes" by forward and backward tagging were *"positive/B-problem lymph/I-problem nodes/I-problem"* and *"positive/I-problem lymph/I-problem nodes/B-problem"*, respectively. For each of these forward (CRF-fwd) and backward (CRF-rev) taggers, we created versions both with and without MetaMap output as features. Overall, the CRF models performed better than the other IE methods. Among the four sequential models, backward tagging with MetaMap features obtained the best results, which are shown in row 7 of Table 3, with an F_1 score of 84.9%. A subsequent analysis revealed that this classifier excels at recognizing multi-word concepts, achieving a recall of 79.8% (about 5% higher than the SVM) and a precision of 82.8% (about 7.4% higher than the SVM) for medical concepts with multiple words.

4.3 Performance of Ensembles

Finally, we evaluated the performance of the two ensemble architectures described in Section 3.2.

Voting Ensemble: We created a Voting ensemble consisting of all seven individual IE models: the rules, MetaMap, the contextual classifier, and all four sequential tagging models. The 8th row in Table 3 shows the results with a voting threshold of three (i.e. three votes are needed to label a concept). This voting ensemble obtained better performance than any of the individual classifiers,

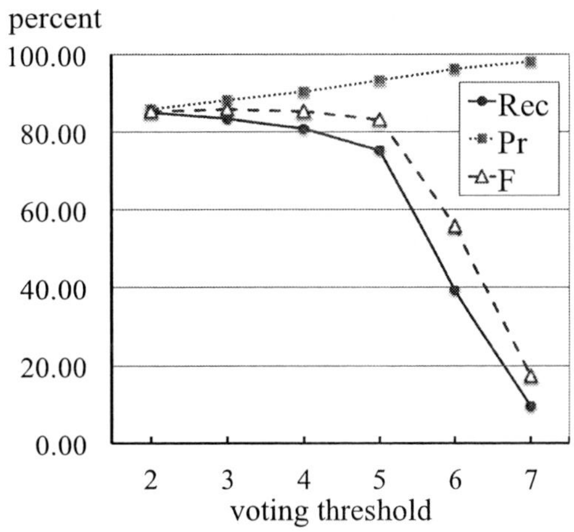

Figure 2: Recall (Rec), Precision (Pr), and F_1 score (F) of the voting ensemble for varying voting thresholds.

reaching an F_1 score of 85.8%.

The voting threshold is a key parameter for Voting Ensembles that can dramatically affect performance. The voting threshold can serve as a recall/precision knob to obtain different trade-offs between recall and precision. In Figure 2, we show results for voting thresholds ranging from two to seven. The curves show that precision increases as the threshold gets higher, but recall drops simultaneously. When the voting threshold exceeds five, recall drops precipitously.

Stacked Generalization: We evaluated the Stacked Generalization Ensemble using the same set of seven individual IE models used in the Voting Ensemble. The last row of Table 3 shows that the Stacked Ensemble achieved slightly higher precision than the Voting Ensemble, overall producing 83.5% recall, 88.6% precision, and an 86.0% F_1 score. Using a paired t-test across the F_1 scores for all test documents (i.e., each F_1 score was calculated for each document, and then averaged across all test documents), the Stacked Ensemble performed significantly better than all of the individual IE models ($p < 10^{-4}$), but not significantly better than the Voting Ensemble ($p = 0.0849$).

We performed ablation tests for both the Voting and Stacked Generalization Ensembles to evaluate the impact of each IE model on the ensembles. An ablated ensemble was tested by removing a single model from the ensemble. Table 4 shows the F_1 score for each ablated ensemble and the differ-

Method	Voting		Stacked	
	F_1 score	Impact	F_1 score	Impact
MetaMap	85.69	-0.10	85.81	-0.20
Rules	85.76	-0.02	85.93	-0.08
SVM	85.51	-0.28	85.70	-0.31
CRF-fwd	85.56	-0.23	85.84	-0.17
CRF-fwd w/ MetaMap	85.56	-0.22	85.83	-0.18
CRF-rev	85.41	-0.37	85.76	-0.25
CRF-rev w/ MetaMap	85.41	-0.37	85.77	-0.24

Table 4: The ablation tests of Voting and Stacked Generalization Ensembles

ence from the F_1 score of the original (complete) ensemble. As shown in Table 4, every IE model contributed to the performance of both the Voting and Stacked Ensembles. Removing the Rules component had a very small impact, presumably because the machine learning models also acquire information from the training data. All of the other IE models appear to have played a valuable role. For the voting ensemble, the F_1 score dropped the most when the CRF-rev or CRF-rev w/ MetaMap models were removed. For Stacked Generalization, removing the SVM model had the biggest impact.

Overall, our results confirm that ensemble architectures consistently outperform individual IE models. Although the Stacked Ensemble and Voting Ensemble produce similar levels of performance, Stacked Generalization has a significant practical advantage over Voting Ensembles. Adding new models to an ensemble is easy, but Voting Ensembles require a voting threshold that must be adjusted when the number of component models changes. Consequently, it can be difficult to assess the overall impact of adding new models (e.g., adding twice as many models may require a higher voting threshold, which may yield higher precision but substantially lower recall). A simple count-based voting threshold is coarse, so small changes can sometimes produce dramatic effects. In contrast, Stacked Generalization uses a meta-classifier to automatically learn how to best weight and use the components in its ensemble. Consequently, adding new models to a Stacked Ensemble only requires re-training of the meta-classifier.

To demonstrate this advantage over voting, we added a second copy of the MetaMap component as an additional system in our ensemble. Voting between the eight systems using our original threshold of three dropped the F_1 score by -0.3%. Adding a third copy of the MetaMap component (producing nine component systems) decreased the F_1 score by -6.8% (absolute). In the same scenarios, the Stacked Learning Ensemble proved to be much more robust, showing almost no change in performance (-0.2% and -0.3% with eight and nine systems respectively).

Table 5 shows the performance of other state-of-the-art systems for medical concept extraction alongside the results from our Stacked Learning Ensemble. The Stacked Ensemble produces higher precision than all of the other systems. Overall, the F_1 score of the Stacked Ensemble is comparable to the F_1 score of the best previous system by Tang et al. (2013). Our Stacked Ensemble achieves slightly higher precision, while the the Tang et al. system produces slightly higher recall.

System	Rec	Pr	F
de Bruijn et al. (2011)	83.6	86.9	85.2
Kang et al. (2012)	81.2	83.3	82.2
Tang et al. (2013)	84.3	87.4	85.8
Stacked Ensemble	83.5	88.6	86.0

Table 5: Recall (Rec), Precision (Pr), and F_1 score (F) of other state-of-the-art systems and our Stacked Ensemble.

5 Analysis

We did manual error analysis to better understand the nature of the mistakes made by our system. Many of the errors revolved around incorrect boundaries for extracted concepts. When allowing a ±1 boundary error for the outputs of the

Stacked Ensemble, the F_1 score went up to 87.9%. Most of these boundary errors on the test set were due to omitting a premodifier or incorrectly including a preceding verb. The first row of Table 6 shows examples of false negatives that fell into this category. The reference annotations appear in **boldface** and the system outputs are surrounded by brackets.

Boundary	Examples
±1	*positive* [***lymph nodes***] [*repeat* ***the echocardiogram***]
±2	[***overdosing***] ***on insulin*** [***head wound*** *remain dry*] ***1000 ml*** [***fluid restriction***]
Others	***active source of*** [***bleeding***] [*careful monitoring of* ***heart rate***]

Table 6: Examples of boundary errors by the Stacked Ensemble.

When allowing for ±2 boundary word errors, the F_1 score increased to 89.4%. The omission of a prepositional phrase or a pre-modifying phrase and the incorrect inclusion of a verb phrase were frequently observed in these errors. For broader boundaries, the errors are similar to ±2 cases but caused by longer pre-modifying phrases.

We also analyzed false negatives that did not contain any words in common with the outputs of the Stacked Learning Ensemble. For about 34% of the false negative concepts that were missed, none of the words in the concept appeared in the training data.

6 Conclusion

We demonstrated that a Stacked Generalization Ensemble achieves high precision and overall performance comparable to the state-of-the-art for the task of medical concept extraction from clinical notes. Stacked learning offers the advantage of being able to easily incorporate any set of individual concept extraction components because it automatically learns how to combine their predictions to achieve the best performance. We believe that Stacked Generalization offer benefits for many problems in medical informatics because it allows for easy, flexible, and robust integration of multiple component systems, including rule-based systems, external dictionaries and knowledge bases, and machine learning classifiers.

Acknowledgments

This research was supported in part by the National Science Foundation under grant IIS-1018314.

References

Alan R. Aronson and François-Michel Lang. 2010. An overview of MetaMap: historical perspective and recent advances. *J Am Med Inform Assoc*, 17:229–36.

Catherine L. Blake and Christopher J. Merz. 1998. UCI Repository of Machine Learning Databases.

Leo Breiman. 1996. Stacked regressions. *Machine Learning*, 24:49.

Lee M. Christensen, Peter J. Haug, and Marcelo Fiszman. 2002. MPLUS: a probabilistic medical language understanding system. In *Proc. ACL-02 Work. Nat. Lang. Process. Biomed. Domain.* pages 29–36.

Nigel Collier, Chikashi Nobata, and Jun-ichi Tsujii. 2000. Extracting the names of genes and gene products with a hidden markov model. In *Proceedings of the 18th conference on Computational linguistics*. pages 201–207.

Michael Collins. 2002. Discriminative training methods for hidden markov models: theory and experiments with perceptron algorithms. In *Proceedings of the ACL-02 conference on EMNLP*. 1–8.

Berry de Bruijn, Colin Cherry, Svetlana Kiritchenko, Joel Martin, and Xiaodan Zhu. 2011. Machine learned Solutions for Three Stages of Clinical Information Extraction: the State of the Art at i2b2 2010. *J Am Med Inform Assoc*, 18(5):557-562.

Son Doan, Nigel Collier, Hua Xu, Pham Hoang Duy, and Tu Minh Phuong. 2012. Recognition of medication information from discharge summaries using ensembles of classifiers. *BMC Med Inform Decis Mak*, 12:36.

Saso Džeroski and Bernard Ženko. 2004. Is combining classifiers with stacking better than selecting the best one? *Machine Learning*, 54(3):255–273.

Rong-En Fan, Kai-Wei Chang, Cho-Jui Hsieh, Xiang-Rui Wang, and Chih-Jen Lin. 2008. LIBLINEAR: A Library for Large Linear Classification. *Journal of Machine Learning Research*, 9:1871–1874.

Jenny Finkel, Shipra Dingare, Christopher D Manning, Malvina Nissim, Beatrice Alex, and Claire Grover. 2005. Exploring the Boundaries: Gene and Protein Identification in Biomedical Text *BMC Bioinformatics*, 6(S1):S5.

Carol Friedman, Philip O. Alderson, John H. M. Austin, James J. Cimino, and Stephen B. Johnson. 1994. A general natural language text processor for clinical radiology. In *J Am Med Inform Assoc.*, 1(2):161–174

João Gama and Pavel Brazdil. 2000. Cascade generalization. *Machine Learning*, 41(3):315–343.

Peter J. Haug, Spence Koehler, Lee Min Lau, Ping Wang, Roberto Rocha, and Stanley M. Huff. 1995. Experience with a mixed semantic/syntactic parser. In *Proc Annu Symp Comput Appl Med Care*, pages 284–288.

Peter J. Haug, Lee Christensen, Mike Gundersen, Brenda Clemons, Spence Koehler, and Kay Bauer. 1997. A natural language parsing system for encoding admitting diagnoses. In *Proc AMIA Annu Fall Symp*, pages 814–818.

Han-Shen Huang, Yu-Shi Lin, Kuan-Ting Lin, Cheng-Ju Kuo, Yu-Ming Chang, Bo-Hou Yang, I-Fang Chung, and Chun-Nan Hsu. 2007. High-recall gene mention recognition by unification of multiple backward parsing models. In *Proceedings of BioCreative II*, pages 109–111.

Min Jiang, Yukun Chen, Mei Liu, S Trent Rosenbloom, Subramani Mani, Joshua C Denny, and Hua Xu. 2011. A study of machine-learning-based approaches to extract clinical entities and their assertions from discharge summaries. *J Am Med Inform Assoc*, 18(5):601–606.

Ning Kang, Zubair Afzal, Bharat Singh, Erik M. Van Mulligen, and Jan A. Kors. 2012. Using an ensemble system to improve concept extraction from clinical records. *J. of Biomedical Informatics*, 45(3):423–428.

Halil Kilicoglu, Dina Demner-Fushman, Thomas C. Rindflesch, Nancy L. Wilczynski, and R. Brian Haynes. 2009. Towards automatic recognition of scientifically rigorous clinical research evidence. *J Am Med Inform Assoc*, 16: 25–31.

Taku Kudo and Yuji Matsumoto. 2001. Chunking with support vector machines. *Proceedings of NAACL-2001*, pages 1–8.

John D. Lafferty, Andrew McCallum, and Fernando C.N. Pereira. 2001. Conditional Random Fields: Probabilistic Models for Segmenting and Labeling Sequence Data. In *Proceedings of the International Conference on Machine Learning*. 282–289.

Thomas Lavergne, Olivier Cappé, and François Yvon. 2010. Practical very large scale CRFs. In *Proceedings of ACL-2010*. pages 504–513.

Nick Littlestone and Manfred K. Warmuth. 1994. The Weighted Majority Algorithm. *Information and Computation*, 108:212–261.

Christopher D. Manning, Mihai Surdeanu, John Bauer, Jenny Finkel, Steven J. Bethard, and David McClosky. 2014. The Stanford CoreNLP Natural Language Processing Toolkit. In *Proceedings of ACL-2014: System Demonstrations*. pages 55–60.

Ryan McDonald and Fernando Pereira. 2005. Identifying Gene and Protein Mentions in Text Using Conditional Random Fields. *BMC Bioinformatics*, 6(S1):S6.

M. Netzer, G. Millonig, M. Osl, B. Pfeifer, S. Praun, J. Villinger, W. Vogel, and C. Baumgartner. 2009. A new ensemble-based algorithm for identifying breath gas marker candidates in liver disease using ion molecule reaction mass spectrometry. *Bioinformatics*, 25(7):941–947.

Joakim Nivre and Ryan McDonald. 2008. Integrating graph-based and transition-based dependency parsers. In *Proc. of ACL 2008*. pages 950–958.

Buzhou Tang, Hongxin Cao, Yonghui Wu, Min Jiang, and Hua Xu. 2013. Recognizing clinical entities in hospital discharge summaries using Structural Support Vector Machines with word representation features. *BMC Med Inform Decis Mak*, 13(S1):S1.

Kay Min Ting and Ian H. Witten. 1999. Issues in stacked generalization, *Journal of Artificial Intelligence Research*, 10:271-289.

Koji Tsukamoto, Yutaka Mitsuishi, and Manabu Sassano. 2002. Learning with multiple stacking for named entity recognition. In *Proc. of COLING-02*. pages 1–4.

Özlem Uzuner, Brett R South, Shuying Shen, and Scott L DuVall. 2011. 2010 i2b2/VA challenge on concepts, assertions, and relations in clinical text. *J Am Med Inform Assoc*, 18:552–556.

Shuang-Quan Wang, Jie Yang, and Kuo-Chen Chou. 2006. Using stacked generalization to predict membrane protein types based on pseudo-amino acid composition. *Journal of Theoretical Biology*, 242(4):941–946.

David H. Wolpert. 1992. Stacked generalization. *Neural Networks*, 5:241–259.

Qing T Zeng, Sergey Goryachev, Scott Weiss, Margarita Sordo, Shawn N Murphy, and Ross Lazarus. 2006. Extracting principal diagnosis, co-morbidity and smoking status for asthma research: evaluation of a natural language processing system. *BMC Med Inform Decis Mak*, 6:30.

GuoDong Zhou, Dan Shen, Jie Zhang, Jian Su, and SoonHeng Tan. 2005. Recognition of Protein/Gene Names from Text Using an Ensemble of Classifiers. *BMC Bioinformatics*, 6(S1):S7.

GuoDong Zhou and Jian Su. 2002. Named entity recognition using an hmm-based chunk tagger. In *Proceedings of ACL 2002*, pages 473–480.

Extracting Disease-Symptom Relationships by Learning Syntactic Patterns from Dependency Graphs

Mohsen Hassan, Olfa Makkaoui, Adrien Coulet and Yannick Toussaint
LORIA (CNRS, Inria, Université de Lorraine),
Campus scientifique, Vandoeuvre-lès-Nancy, F-54506, France
{mohsen.sayed,olfa.makkaoui,adrien.coulet,yannick.toussaint}@loria.fr

Abstract

Disease-symptom relationships are of primary importance for biomedical informatics, but databases that catalog them are incomplete in comparison with the state of the art available in the scientific literature. We propose in this paper a novel method for automatically extracting disease-symptom relationships from text, called SPARE (standing for Syntactic PAttern for Relationship Extraction). This method is composed of 3 successive steps: first, we learn patterns from the dependency graphs; second, we select best patterns based on their respective *quality* and *specificity* (their ability to identify only disease-symptom relationships); finally, the patterns are used on new texts for extracting disease-symptom relationships. We experimented SPARE on a corpus of 121,796 abstracts of PubMed related to 457 rare diseases. The quality of the extraction has been evaluated depending on the pattern *quality* and *specificity*. The best F-measure obtained is 55.65% (for $specificity \geq 0.5$ and $quality \geq 0.5$). To provide an insight on the novelty of disease-symptom relationship extracted, we compare our results to the content of phenotype databases (OrphaData and OMIM). Our results show the feasibility of automatically extracting disease-symptom relationships, including true relationships that were not already referenced in phenotype databases and may involve complex symptom descriptions.

1 Introduction

Disease-Symptom (D-S) relationships are of major importance for biomedical informatics since they provide a fine-grained description of disease that could be used to guide medical diagnosis in clinical care. However, biomedical databases that catalog D-S relationships such as OrphaData and OMIM are incomplete in comparison with the state of the art available in the scientific literature (Köhler et al., 2014). In addition, extracting this information manually from the literature by experts requires a lot of time and effort, which motivates the need for developing automatic methods.

Our study focuses on extracting symptoms in relation with rare diseases (RDs). These are diseases that affect a small percentage of the population, ranging from 1/1,000 to 1/200,000. As their number is relatively important (between 6,000 and 8,000 (Mazzucato et al., 2014)), RDs have received a particular attention in the medical domain.

In this context, we propose an automatic method, called SPARE (Syntactic PAttern for Relationship Extraction), for D-S relationship extraction based on shortest path patterns generated from the dependency graphs (DGs) of texts. We applied SPARE to the extraction of D-S relationships associated with rare diseases. Because symptoms associated with rare diseases may be uncommon and complex (*i.e.,* they can not be expressed with one word or a simple expression), we particularly focus on enabling the recognition of symptoms that are not listed in phenotype databases or ontologies.

As a result, objectives of this work are threefold: (1) learning patterns specific for diseases-symptom relationships extraction; (2) identifying symptom description that is pointed by specific pattern; and (3) extracting D-S relationships.

This article is organized as follow: we introduce D-S relationship relative issues in section 2. Section 3 presents main methods for relationship extraction. In section 4, we detail the SPARE method. Section 5 describes experiments and re-

Proceedings of the 2015 Workshop on Biomedical Natural Language Processing (BioNLP 2015), pages 71–80,
Beijing, China, July 30, 2015.

sults. Finally, we discuss and conclude about the results described in the article.

2 Disease-Symptom Relationships

OrphaData and OMIM are two examples of databases that catalog D-S relationships. OrphaData[1] is the database accessible from Orphanet, the portal for rare diseases and orphan drugs. It includes description of symptoms (clinical signs) of rare disease. OMIM[2] (Online Mendelian Inheritance in Man) is a database for genetic diseases. It contains disease descriptions that include a list of symptoms named "clinical synopsis".

Due to the fact that their content is manually curated by experts, OrphaData and OMIM are high quality resources. However, these resources do not contain a complete list of relationships between diseases and symptoms that exist in the biomedical literature. As shown in Table 1, among the 8,644 diseases listed by OrphaData only 2,689 diseases (31.11%) are associated with clinical signs and symptoms. Indeed, one can use cross references between OrphaData and OMIM[3] to associate OrphaData diseases to symptoms described in OMIM. Nevertheless, even when considering these additional symptoms, only 4,856 (56.18%) OrphaData diseases have symptoms. The rest, 3,788 OrphaData diseases, is not related to any symptom. This motivates us to extract these relations from the literature.

	#Diseases	#Diseases associated with symptoms	#Symptoms	#D-S Relations
OrphaData	8,644	2,689	1,273	52,503
OMIM	23,929	23,910	46,369	432,760

Table 1: Information about OrphaData and OMIM databases

Recognizing diseases and symptoms in texts is a preliminary step for D-S relationships extraction. Previous work on disease recognition achieved good results (Leaman and Lu (2014) obtained 78.25% F-Measure, 76.3% recall and 80.3% precision). Less works aimed at recognizing symptoms. Their performances are low in comparison with those of disease recognition. For example, Martin *et al.* (2014) used HPO[4] (Köhler et al., 2014) for symptom extraction and obtained 36.8% F-Measure, 23.7% recall and 82.2% precision.

Extracting D-S relationships automatically is a challenging task mainly due to the following two reasons: first, there is no complete dictionary of symptoms to guide their recognition; second, symptoms are complex entities that are hard to recognize in text. Indeed, HPO, which contains 11,021 phenotypes terms, covers only symptoms related to genetic diseases. Thus, a simple "exact match" approach to recognize HPO symptoms in text would give a low recall: "serositis" in example 2.1 is not known as a symptom in HPO.

In addition, Named Entity Recognition (NER) tools recognize symptoms with low recall. This is the case of MetaMap (Aronson, 2001), a tool that annotates texts with concepts from UMLS (Bodenreider, 2004). In example 2.2 MetaMap annotates "Familial Mediterranean Fever" as disease but does not annotate "fever" or "attacks of fever" as a symptom.

Ex 2.1. *"<disease>Familial Mediterranean Fever</disease> is characterized by serositis"*

Ex 2.2. *"<disease>Familial Mediterranean Fever</disease> (FMF) is an autosomal recessive disorder characterized by attacks of fever"*

Recognizing the full description of symptoms is another challenge for symptom recognition, in particular with rare diseases where symptom description can be complex phrases. Some cases of partial annotations occur when HPO or MetaMap annotates only a part of the entity. For instance, example 2.3 shows that "pure spasticity of the lower limbs" is a symptom but MetaMap annotates only "spasticity".

Ex 2.3. *"One patient with <disease>Krabbe disease</disease> presented with pure <symptom>spasticity</symptom> of the lower limbs"*

The ambiguity between diseases and symptoms is another factor of complexity as diseases play, in some situations, the role of symptoms. For instance, example 2.4 shows that "muscle wasting" is recognized by MetaMap as a disease. However, it can be considered as a symptom for "Duchenne muscular dystrophy".

Ex 2.4. *"<disease>Duchenne muscular dystrophy</disease> is characterized by <disease>muscle wasting</disease>"*

[1]OrphaData website: http://www.orphadata.org/

[2]OMIM website: http://www.omim.org/

[3]4,162 OrphaData diseases have cross references to OMIM diseases.

[4]HPO (The Human Phenotype Ontology) provides a structured and controlled vocabulary for the phenotypic features of diseases.

3 Related Works

Various works have proposed methods to extract relationships from text. They are based on different approaches such as statistics, pattern-based or rule-based, and machine learning.

A co-occurrence method is a simple method to identify relationships between two entities that co-occur in the same sentence (Bunescu et al., 2006). It is based on the hypothesis that if two entities are mentioned frequently together, they are likely to be in a relation. Approaches based on co-occurrences of entities do not employ NER techniques. The type and the direction of relationships are not captured by these methods. Various statistical measures are used to decide whether the two entities co-cited together are in relation or not (Lee et al., 2007; Ramani et al., 2005). Examples of these measures are Pointwise Mutual Information, Chi-Square or Log-Likelihood Ratio (Manning and Schütze, 1999), which use the co-occurrence statistics of the two entities to hypothesize about the existence of a relationship between them. Ramani *et al.* (2005) use random co-citation model based on the hypergeometric distribution. Co-occurrence methods have been successfully applied to the automated construction of networks of biomolecules such as gene-protein and gene regulatory networks (Šarić et al., 2006; Friedman et al., 2001).

Pattern- and rule-based methods generate symbolic patterns or rules to extract relationships, with advantage that they are easy to interpret (Agichtein and Gravano, 2000). These patterns or rules can be generated manually (Divoli and Attwood, 2005) or automatically by learning from annotated corpus (Hakenberg et al., 2005). They are based on different levels of linguistic information like lexical, syntactic or dependency information and different levels of structures like sequences, trees and graphs. These methods tend to have a high precision but a low recall (Cellier et al., 2010; Béchet et al., 2012; Liu et al., 2013; Martin et al., 2014; Hassan et al., 2014).

Liu *et al.* (Liu et al., 2013) proposed a graph-based approach to learn rules for event extraction (that can be compared to relationship extraction). The rules are represented by the information on the shortest path between entities in an undirected DG. Béchet *et al.* (2012) and Cellier *et al.* (2010) proposed a method based on sequential pattern mining to extract disease-gene and gene-gene relationships. As the number of their patterns is very large, they introduced constraints for patterns filtration to reduce them. Close to our objectives, Martin *et al.* (2014) used sequential patterns for recognizing unidentified symptoms. Also, Hassan *et al.* (2014) proposed a pattern-based method for D-S relationship extraction, where diseases and symptoms are previously recognized and annotated by a NER tool. The patterns are learned from shortest paths between diseases and symptoms in directed DGs.

Machine Learning (ML) methods consider a relationship extraction task as a classification problem. Two ML techniques are mainly employed: feature-based and kernel-based methods. Feature-based methods such as support vector machines or conditional random fields have been employed by (Krallinger et al., 2008; Bundschus et al., 2008) for relationship extraction. Kernel methods use a kernel function to measure the similarity between a large amount of features *e.g.,* sub-sequences, trees, graphs (Zelenko et al., 2003; Zhang et al., 2008; Airola et al., 2008).

Bunescu and Mooney (2005) proposed a shortest path kernel method that uses the shortest path between two entities in an undirected DG for relationship extraction. This work is based on the hypothesis that the relationship between two entities in the same sentence is typically captured by the shortest path between them in the DG. Chowdhury *et al.* (2012) proposed a hybrid kernel that uses different types of information (*e.g.,* syntactic, contextual, semantic) and their different representations (*i.e.,* flat features, tree structures and graphs). This hybrid kernel helps improving the results of relationship extraction.

4 Method

We describe in this section the SPARE method for D-S relationship extraction. This method is composed of three steps: first, learning patterns out of DGs that include both a disease and a symptom; second, selecting patterns in regard to their quality (*i.e.,* their capacity to identify true relationships) and their specificity (*i.e.,* their capacity to identify only D-S relationships); third, using selected patterns to extract D-S relationships from text.

The originality of the SPARE method relies on measuring how syntactic patterns between diseases and symptoms are specific to D-S relationships. Using highly specific patterns allow us to consider the case where symptoms are not recognized by NER tools, which consequently offers the opportunity to discover new symptom descriptions that can be potentially rare and complex.

SPARE is inspired from various previous works such as using the shortest path between entities of a DG as described by Bunescu et Mooney (2005), then applied by Chowdhury *et al.* (2012) and Liu *et al.* (2013). Similarly to Liu *et al.* (2013), we extract patterns represented by the whole subgraph (*i.e.,* all nodes and edges in the shortest path), but unlike them, we keep edge directions. Hassan *et al.* (Hassan et al., 2014) proposed a pattern-based method for D-S relationship extraction. They assume that diseases and symptoms are initially recognized by a NER tool. Here we relax patterns, similarly to Blohm *et al.* (2011), and use specificity to consider cases of unrecognized symptoms.

The following subsections detail the three steps of SPARE.

4.1 Learning Syntactic Patterns from DG

For pattern learning, only DGs of sentences that contain at least one disease and one symptom are considered as we are interested in extracting D-S relationships. DGs are explored to find the shortest paths between diseases and symptoms. Because one sentence can mention several diseases and symptoms, several shortest paths may be found.

Ex 4.1. *"A 15-month-old girl with <disease>propionic acidemia</disease> presented <symptom>muscular hypotonia </symptom>"*

Ex 4.2. *"A 25-year-old woman with <disease>cystic fibrosis</disease> developed <symptom>hemoptysis</symptom>"*

Figures[5] 1(a) and 1(c) show the DGs generated from sentences of examples 4.1 and 4.2 after the replacement of the annotated entities (*i.e.,* diseases and symptoms) by generic words (*i.e.,* DISEASE and SYMPTOM) and other words by their lemmas. Figures 1(b) and 1(d) show the shortest paths extracted from associated DGs. The whole shortest path is kept, including all nodes, edges and directions.

Next, patterns are generated on the basis of shortest paths, using a generalization process. In this process, two shortest paths (or more) can be merged and represented in one generalized pattern. Different shortest paths are aggregated to a

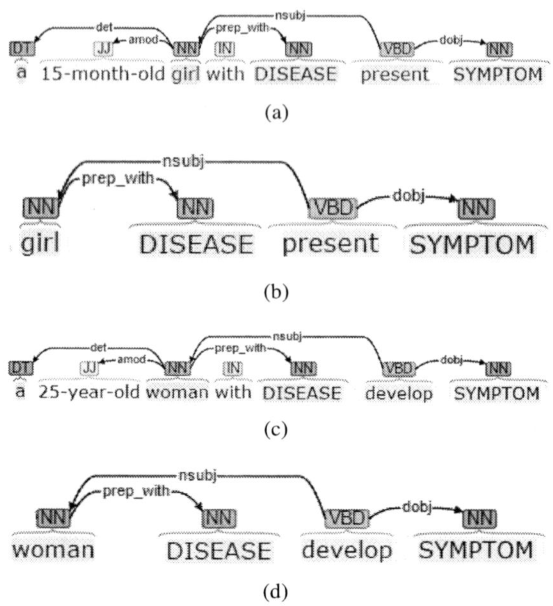

Figure 1: (a,c) DGs and (b,d) Shortest paths between disease and symptom respectively extracted from sentences of examples 4.1 and 4.2

pattern if those share the same edges and directions. Figure 2 illustrates this generalization process considering the shortest paths obtained from examples 4.1 and 4.2. If the values of the nodes in the pattern are different, then they are replaced by "*" (*i.e.,* matching any token). A list of values observed for each node is kept but for pattern documentation purpose only. The frequency of patterns is measured by their *support*, *i.e.,* how many sentences in our learning corpus match this pattern.

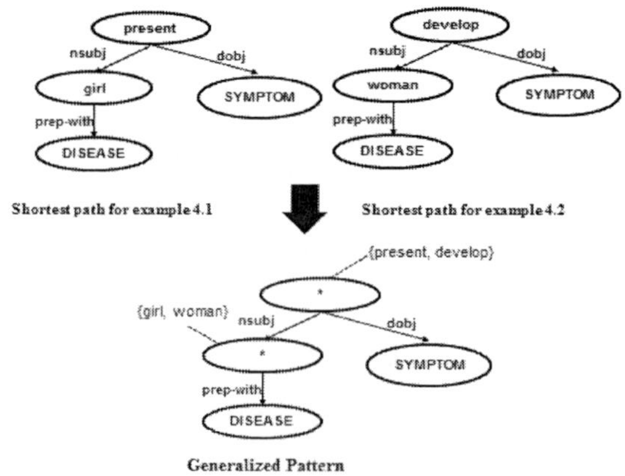

Figure 2: Example of pattern generation from two shortest paths

[5]DGs are processed by the Stanford Parser and drawn with the Brat tool at http://nlp.stanford.edu:8080/corenlp/

This generalization affects the precision and the recall of the patterns. Replacing the node value in the shortest path by using "*" (*i.e.,* any token) makes the pattern more generic, and has the consequence of increasing the recall of the patterns. On the other side, we assume that edges (*i.e.,* dependency types of DGs) and directions of the pattern guarantee its precision.

4.2 Pattern Selection

4.2.1 Quality-Based Selection

We classify patterns into two classes: positive and negative patterns. This classification relies both on the frequency and on the quality of patterns. The *quality* of patterns requires an evaluation procedure, on the basis of an annotated corpus, to be computed. The *quality* of a pattern is defined as:

$$quality = \frac{|T|}{|A|} \quad (1)$$

where T is the set of all true relationships and A is the set of all (true and false) relationships that are identified by the pattern. A relationship is qualified as true if it is annotated in the corpus, *i.e.,* if the sentence is actually mentioning the relationship. A pattern is considered positive if its *support* is greater than or equal to a minimum support denoted *min_support* and its *quality* is greater than or equal to a minimum quality denoted $min\_quality$.

4.2.2 Specificity-Based Selection

In order to measure how much a pattern is specific to D-S relationships and not to other relationships, a *specificity* measure of the pattern is defined. To measure this specificity, we performed a new evaluation task for which we consider: *(i)* a novel set of annotated sentences, not including one disease and one symptom but including one disease and another entity (*e.g.,* a symptom, a gene, a treatment or a living being); *(ii)* patterns from which we removed the constraint on the symptom node (*i.e.,* SYMPTOM is replaced by "*"). The pattern specificity is computed by the following formula:

$$specificity = \frac{|DS|}{|A|} \quad (2)$$

where DS is the set of true D-S relationships extracted by the pattern and A is the set of all (true and false) relationships that are extracted by the pattern (including D-S, disease-gene, disease-treatment and disease-living being relationships). For example, if the pattern extracts 23 true D-S relationships and 7 disease-any entity relationships, then pattern specificity is 23/30. The specificity measure is used to select the patterns that are the most specific to D-S relationships by selecting those that have a specificity greater than or equal to a minimum specificity denoted $min\_specificity$.

Both quality and specificity are associated with the precision of patterns (the ratio of true relationships on all extracted relationships, see formula 3) but are used in different contexts. The quality of a pattern is calculated based on the extracted relationships (D-S relationships only) of the training corpus. In order to keep the patterns that are able to extract true relationships with high precision, we restrict the pattern on disease and symptom constraints. In contrast, the specificity of a pattern is calculated using the relationships (D-S or disease-any entity relationships) in the whole corpus when the pattern is relaxed on the symptom constrain. Specificity is used to keep the patterns that are specific to D-S relationships only.

4.3 Relationship Extraction

4.3.1 Pattern Relaxation for Unknown Symptoms

Patterns with $specificity \geq min\_specificity$ are relaxed on the symptom constraint, meaning that one entity must be annotated as a disease, but there is no requirement for the second entity to be annotated as a symptom. This enables us to identify symptoms that are not recognized by NER tools. Similarly to the learning phase, DGs are generated from the text to explore for D-S relationships. Then, a pattern matching between DGs and the pattern set is applied to extract D-S relationships.

4.3.2 Extraction of Complex Symptoms

During pattern matching, the word that matches with the node of the second entity (not constrained) is considered to be a symptom. Indeed, it is considered to be a symptom if this word is a leaf of the DG, but is considered as the "head" of a more complex symptom description if it is not a leaf. To extract the complete description of the symptom, we explore the subtree that has, as a head, the node that matched as a symptom. For example, if a pattern matching is applied to the sentence provided in example 4.3, we obtain a match with the pattern presented in Figure 2. In this case, the word that is considered to extract the symptom description is "attack". Exploring the subtree represented in Figure 3(b) enables us to reconstruct the full symptom that is involved in the relationship. This reconstruction uses every word of the subtree, dependency types plus the initial order of words to reconstruct the symptom description, "acute attack of fever" in our example. This example illustrates the usefulness of DGs in identifying and representing complex entities like symptoms.

Ex 4.3. *"A patient with <disease>Familial Mediterranean Fever</disease> suffered acute attacks of fever"*

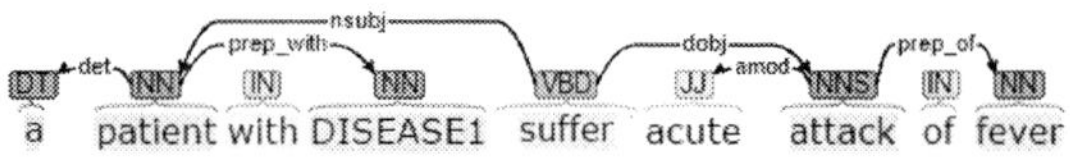

(a) DG of the sentence in example 4.3

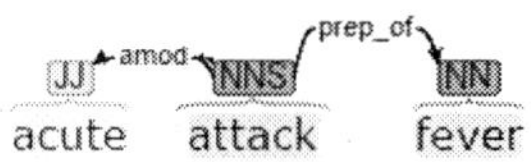

(b) The subtree of a complex symptom description

Figure 3: An example of complex symptom extraction

5 Experiments

5.1 Data Preparation

5.1.1 Rare Disease Corpus

Our rare disease corpus is composed of 121,796 PubMed abstracts obtained by querying PubMed with 457 rare diseases of OrphaData.[6] These diseases are selected because they fulfill following criteria: (1) they are associated with symptoms (namely "clinical signs") in OrphaData; (2) they can be mapped to an OMIM disease throught UMLS CUI; (3) their corresponding OMIM reference is annotated with symptoms (namely "clinical synopsis") in OMIM. This enables having a corpus of a reasonable size and guarantees that the selected diseases are associated with symptoms in both OrphaData and OMIM. This set of diseases and associated symptoms are used in subsection 5.5 to compare our relationships with the content of OrphaData and OMIM.

5.1.2 Preprocessing

The 121,796 abstracts are first split into 907,088 sentences using LingPipe[7]. These sentences are then annotated by MetaMap in order to label diseases, symptoms, genes, treatments and living beings with UMLS CUI. Finally, sentences that do not contain diseases are filtered out. Therefore, we obtained 301,599 sentences with at least one disease.

5.2 Pattern Learning

To learn patterns for D-S relationship extraction, 2,341 sentences with at least one disease and one symptom are kept. These sentences are split into a *learning corpus* made of 90% of sentences (randomly selected) and a *testing corpus* made of 10% of sentences. Both corpora are manually annotated by only one person to identify true and false relationships: the annotation task mainly requires linguistics and NLP skills. A true relationship is counted when a pair D-S is found and a relationship between them is actually mentioned in the text; whereas a false relationship is listed when the pair is found but no relationship is mentioned. The sentence in example 5.1 shows instances of both true and false relationships. "Schwartz Jampel syndrome"-"blepharospasm" is a true relationship, while "rare neuromuscular disorder"-"blepharospasm" is false. Table 2 shows the size of the learning and testing corpora in term of number of sentences, and of true and false relationships in each corpus. We use the Stanford parser to generate a DG for each sentence (de Marneffe et al., 2006). Shortest paths are computed from the 2,107 prepared DGs to generate 1,049 patterns. Figure 4 presents 7 examples of patterns generated.

Ex 5.1. *"<disease>Schwartz Jampel syndrome</disease> is a <disease>rare neuromuscular disorder</disease> characterized by <symptom>blepharospasm</symptom>"*

Corpus	#Sentences	#True Relations	#False Relations
learning	2,107	2,680	2,294
testing	234	330	326

Table 2: Size and content of the learning and testing corpora used for pattern learning and selection.

5.3 Pattern Selection

5.3.1 Quality-based Selection

Increasing the *min_support* value from 1, to 2, then to 3, reduces the number of patterns from 1,049, to 257, then to 118. To avoid rare patterns that can result from parser errors or complex sentences, we fixed *min_support* = 2.

[6]The list of 457 rare diseases is available at https://sourceforge.net/projects/spare2015/files/457-diseases

[7]LingPipe website: http://alias-i.com/lingpipe/

pattern	support	quality	specificity
DISEASE ←nsubj— * —vmod→ * —agent→ SYMPTOM	60	1	0.95
DISEASE ←prep_with— * ←nsubj— * —prep_with→ SYMPTOM	18	1	0.98
DISEASE ←prep_with— * ←nsubj— * —dobj→ SYMPTOM	17	1	0.92
DISEASE ←prep_with— * —rcmod→ * —dobj→ SYMPTOM	10	1	0.97
DISEASE ←prep_with— * —rcmod→ * —prep_with→ SYMPTOM	9	1	1
DISEASE ←prep_of— * ←nsubj— * —dobj→ SYMPTOM	6	1	0.97
DISEASE ←nsubj— * —prep_of→ * —vmod→ characterize —agent→ SYMPTOM	5	1	1

Figure 4: 7 examples of patterns from our pattern set and their *support*, *quality* and *specificity*.

We fixed $min\_quality = 0.5$ to reduce our selected patterns to 235. This choice is guided by the *F-measure* that we computed for each *quality* threshold, as presented in Figure 5. This optimal *F-measure* is 56.97% (*precision* 87.97%, *recall* 42.12%) on the testing corpus. If used on the testing corpus, these 235 patterns extract 139 true relationships and 19 false relationships on a total of 330 relationships. Formulas for precision, recall and F-measure are recalled hereafter:

$$precision = \frac{all\ true\ extracted\ relations}{all\ extracted\ relations} \quad (3)$$

$$recall = \frac{all\ true\ extracted\ relations}{all\ relevant\ relations} \quad (4)$$

$$F-measure = 2 * \frac{precision * recall}{precision + recall} \quad (5)$$

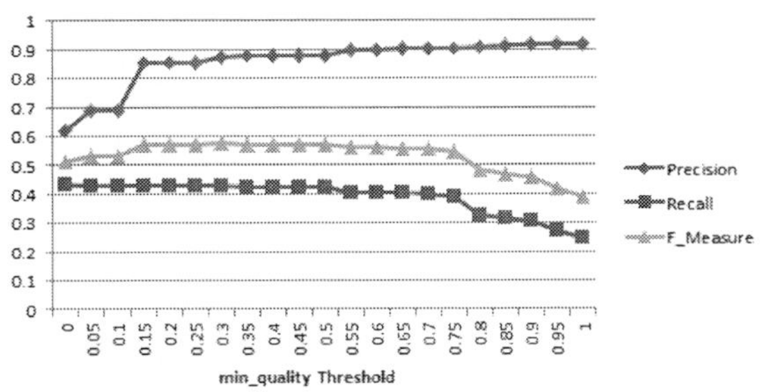

Figure 5: The effect of quality threshold on precision, recall and F-Measure values

5.3.2 Specificity-based Selection

For computing the pattern specificity, all sentences that contain at least one disease and another UMLS entity are selected. This produces 9,233 sentences. Then, all the 235 previously selected patterns are applied to the DGs of these sentences, resulting in the extraction of 5,197 D-S relationships and 391 disease-non symptom relationships (182 disease-gene, 182 disease-treatment, 27 disease-living being relationships). Finally, the specificity of each pattern is computed (see formula 2). Figure 6 shows that using $min\_specificity = 0.5$ achieves the best *F-measure*, 55.65% ($precision = 89.87\%$ and $recall = 40.3\%$), on the testing corpus. Finally we keep 220 patterns with $quality \geq 0.5$ and $specificity \geq 0.5$.[8]

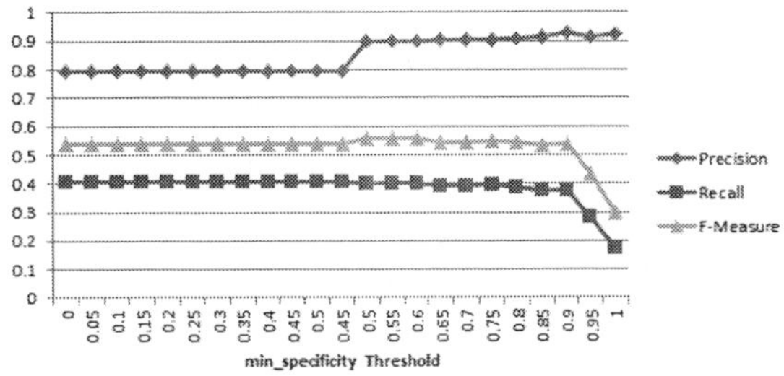

Figure 6: The effect of specificity threshold on precision, recall and F-Measure values

5.4 Application of Relationships Extraction

We applied selected patterns to the whole corpus[9] (301,599 sentences with at least one disease). The extracted relationships are divided into two groups. The first group contains 4,886 D-S relationships where symptoms were previously recognized by MetaMap. The second group contains 6,572 D-S relationships where symptoms were not recognized by MetaMap. After manual checking[10], these extractions achieved respectively 90.69% and 83.13% *precision*. The number of distinct symptoms in the second group is 3,849.

5.5 Comparison with Phenotype Databases

5.5.1 Comparison Approach

The novelty of extracted relationships is evaluated based on the comparison with D-S relationships available in OrphaData, and in OMIM. Results of the comparison are categorized into 3 groups: matched, partial matched and new relationships. To realize this comparison, it is required to map

[8]The list of 220 patterns is available at http://sourceforge.net/projects/spare2015/files/220Patterns

[9]The whole corpus is used (including the training and testing corpus) because the purpose of this task is to extract as much as possible D-S relationships and then, to compare them to the content of phenotype databases.

[10]The manual checking is done by only one person.

diseases of extracted relationships to OMIM diseases. Indeed, MetaMap provides, for each extracted disease, a UMLS CUI that may be mapped to OMIM.

For symptom mapping, we implemented a similarity measure to evaluate the similarity between the extracted symptom and those referenced in OMIM clinical synopsis and HPO. Our similarity value is based on the Jaccard index and is computed following formula:

$$Jaccard\ Index = \frac{text\_words \cap symp\_words}{text\_words \cup symp\_words} \quad (6)$$

where $text\_words$ are the words of the extracted symptom string and $symp\_words$ are the words that are describing a symptom defined either in OMIM or HPO.

Before computing the Jaccard index, each word in the extracted symptom and HPO (or OMIM) symptom is replaced by its lemma, stop words[11] are removed and a list of synonyms from WordNet (Fellbaum, 1998) is associated with each word. The synonym list of a word is used in case of the word does not match with any other word. The similarity value is then computed by the Jaccard index. For each symptom, the first three closest symptoms found in OMIM and HPO (six in total) are manually checked to select the best match if exists. A label "exact", "partial" or "new" is assigned to express if the match is exact or partial, or if the symptom is not listed in OMIM and HPO, thus considered as new.

5.5.2 Comparison Results

The relationships in the first group are compared automatically to Orphadata and OMIM relationships (because both their disease and symptom are associated with a UMLS CUI). The number of true D-S relationships is 4,431, including 803 relationships available in OrphaData and 646 available in OMIM. The union of these 2 sets counts up to 1,074 distinct D-S relationships already listed in OrphaData, OMIM or in both. Consequently, about 3,357 D-S relationships are potentially new and must be added to phenotype databases.

Regarding the relationships in the second group, the extracted symptoms are mapped to symptoms in HPO and OMIM. In this step, 3,236 symptoms (from 3,849 distinct symptoms in the relationships) are mapped to HPO and OMIM symptoms. The extracted relationship pairs are then compared to relationships in HPO and OMIM, which results in 1,422 matched relationships. As a result, we identified 613 $(3,849 - 3,236)$ new symptoms descriptions that may be of interest in rare disease studies and 4,041 $(5,463 - 1,422)$ potentially new D-S relationships[12][13].

6 Discussion

In SPARE, the choice of $min\_quality$ and $min\_specificity$ have important consequences on the results of the relationship extraction. Figures 5 and 6 show how the quality of the extraction changes when these two values are changed. In both cases, we observe relatively few evolution of the F-Measure. In Figure 5, $min\_quality$ between 0.35 and 0.5 achieve the best *F-measure* of 56.97%. They give the same result because the number of extracted patterns with $min\_quality$ between 0.35 and 0.5 is the same (235 patterns). Consequently, we chose arbitrarily $min\_quality = 0.5$. As shown in Figure 6, we chose $min\_specificity = 0.5$ because it achieves the best F-Measure. The result of F-Measure is constant when $min\_specificity$ between 0 and 0.45 because the number of patterns in this interval is the same.

We obtain a relatively good precision but a low recall. We Consider that a larger corpus for learning patterns could enable us to increase the recall. Our learning corpus is annotated manually with true and false relationships and increasing its size would require annotating additional relationships.

The corpus used in the learning task is relatively small, subsequently it is not enough to train ML methods. We propose to increase the size of the annotated corpus in order to apply ML methods on this corpus and compare with the results of our SPARE method.

Studying the novelty of our extracted relationships requires the comparison with the relationships of phenotype databases. For now, this comparison is semi-automatic and partial matching relies on a rather naive similarity measure. We

[11] We considered stop words listed in http://xpo6.com/list-of-english-stop-words/

[12] A list of extracted D-S relationship examples is available at https://sourceforge.net/projects/spare2015/files/D-SRelationsExamples.csv

[13] A list of extracted symptom examples is available at https://sourceforge.net/projects/spare2015/files/symptom-examples

would like to develop a more systematic approach by enabling a fine-grained comparison of phenotype descriptions. This could be achieved by normalizing then, comparing DGs of symptom descriptions.

SPARE method is a supervised classification process, in which threshold is selected manually. This selection can be computed automatically by considering the best F-Measure value.

7 Conclusion

In this paper, we proposed a pattern-based method that we call SPARE for extracting D-S relationships. The patterns are learned from shortest paths observed between the entities of interest (diseases and symptoms) within DGs. Using only the shortest path is simple and it captures the most important features required to describe the relationship between two entities. For extracting relationships involving rare or complex symptoms, we selected a subset of patterns that are specific to D-S relationships. In turn, a DG is helpful to extract and define complex symptoms, that are not recognized by other tools such as MetaMap. The novelty of relationship extracted has been compared with relationships listed in OrphaData and OMIM. This shows the ability of the SPARE to discover existing and potentially new relationships and the ability to identify new and complex symptom as well.

Acknowledgments

This work is supported by the ANR (French National Research Agency), project Hybride ANR-11-BS02-002.

References

Eugene Agichtein and Luis Gravano. 2000. Snowball: Extracting relations from large plain-text collections. In *Proceedings of the Fifth ACM Conference on Digital Libraries*, DL '00, pages 85–94, New York, NY, USA. ACM.

A. Airola, S. Pyysalo, J. Björne, T. Pahikkala, F. Ginter, and T. Salakoski. 2008. All-paths graph kernel for protein-protein interaction extraction with evaluation of cross-corpus learning. *BMC bioinformatics*, 9 Suppl 11.

A. R. Aronson. 2001. Effective mapping of biomedical text to the umls metathesaurus: the metamap program. *Proc AMIA Symp*, pages 17–21.

Nicolas Béchet, Peggy Cellier, Thierry Charnois, Bruno Crémilleux, and Marie-Christine Jaulent. 2012. Sequential pattern mining to discover relations between genes and rare diseases. In *CBMS*, pages 1–6.

Sebastian Blohm, Krisztian Buza, Philipp Cimiano, and Lars Schmidt-Thieme, 2011. *Relation Extraction for the Semantic Web with Taxonomic Sequential Patterns*, pages 185–209. Applied Semantic Web Technologies. Taylor and Francis Group.

Olivier Bodenreider. 2004. The unified medical language system (UMLS): integrating biomedical terminology. *Nucleic Acids Research*, 32(Database-Issue):267–270.

Markus Bundschus, Mathaeus Dejori, Martin Stetter, Volker Tresp, and Hans-Peter Kriegel. 2008. Extraction of semantic biomedical relations from text using conditional random fields. *BMC Bioinformatics*, 9(1):207.

Razvan C. Bunescu and Raymond J. Mooney. 2005. A shortest path dependency kernel for relation extraction. In *Proceedings of the Conference on Human Language Technology and Empirical Methods in Natural Language Processing*, HLT '05, pages 724–731, Stroudsburg, PA, USA. Association for Computational Linguistics.

Razvan Bunescu, Raymond Mooney, Arun Ramani, and Edward Marcotte. 2006. Integrating co-occurrence statistics with information extraction for robust retrieval of protein interactions from medline. In *Proceedings of the HLT-NAACL Workshop on Linking Natural Language Processing and Biology (BioNLP'06)*, pages 49–56, New York, NY, June.

Peggy Cellier, Thierry Charnois, and Marc Plantevit. 2010. Sequential patterns to discover and characterise biological relations. In A. F. Gelbukh, editor, *Computational Linguistics and Intelligent Text Processing (CICLing)*, LNCS 6008, pages 537–548. Springer.

Md. Faisal Mahbub Chowdhury and Alberto Lavelli. 2012. Combining tree structures, flat features and patterns for biomedical relation extraction. In *EACL*, pages 420–429.

Marie-Catherine de Marneffe, Bill MacCartney, and Christopher D. Manning. 2006. Generating typed dependency parses from phrase structure parses. In *IN PROC. INT'L CONF. ON LANGUAGE RESOURCES AND EVALUATION (LREC*, pages 449–454.

A. Divoli and T. K. Attwood. 2005. BioIE: extracting informative sentences from the biomedical literature. *Bioinformatics*, 21(9):2138–9.

Christiane Fellbaum, editor. 1998. *WordNet: an electronic lexical database*. MIT Press.

Carol Friedman, Pauline Kra, Hong Yu, Michael Krauthammer, and Andrey Rzhetsky. 2001. Genies: a natural-language processing system for the extraction of molecular pathways from journal articles. *Comput. Appl. Biosci.*, 17(suppl_1):S74–82, June.

Jörg Hakenberg, Conrad Plake, Ulf Leser, Harald Kirsch, and Dietrich Rebholz-schuhmann. 2005. Lll'05 challenge: genic interaction extraction – identification . . . with alignments and finite state automata. In *IN PROC LEARNING LANGUAGE IN LOGIC WORKSHOP (LLL05) AT THE 22ND INT CONF ON MACHINE LEARNING*, pages 38–45.

Mohsen Hassan, Adrien Coulet, and Yannick Toussaint. 2014. Learning subgraph patterns from text for extracting disease - symptom relationships. In *Proceedings of the 1st International Workshop on Interactions between Data Mining and Natural Language Processing co-located with The European Conference on Machine Learning and Principles and Practice of Knowledge Discovery in Databases, DMNLP@PKDD/ECML 2014, Nancy, France, September 15, 2014.*, pages 81–96.

Sebastian Köhler, Uwe Schoeneberg, Johanna C. Czeschik, Sandra C. Doelken, Jayne Y. Hehir-Kwa, Jonas Ibn-Salem, Christopher J. Mungall, Damian Smedley, Melissa A. Haendel, and Peter N. Robinson. 2014. Clinical interpretation of CNVs with cross-species phenotype data. *Journal of Medical Genetics*, pages jmedgenet–2014–102633+, October.

Martin Krallinger, Florian Leitner, Carlos Rodriguez-Penagos, and Alfonso Valencia. 2008. Overview of the protein-protein interaction annotation extraction task of BioCreative II. *Genome biology*, 9 Suppl 2(Suppl 2):S4+.

Robert Leaman and Zhiyong Lu. 2014. Disease named entity recognition and normalization with dnorm. In *Proceedings of the 5th ACM Conference on Bioinformatics, Computational Biology, and Health Informatics*, BCB '14, pages 587–587, New York, NY, USA. ACM.

Insuk Lee, Zhihua Li, and Edward M. Marcotte. 2007. An improved, bias-reduced probabilistic functional gene network of baker's yeast, saccharomyces cerevisiae. *PLoS ONE*, 2(10).

Haibin Liu, Karin Verspoor, Donald C. Comeau, Andrew MacKinlay, and W John Wilbur. 2013. Generalizing an approximate subgraph matching-based system to extract events in molecular biology and cancer genetics. In *Proceedings of the BioNLP Shared Task 2013 Workshop*, pages 76–85, Sofia, Bulgaria, August. Association for Computational Linguistics.

Christopher D. Manning and Hinrich Schütze. 1999. *Foundations of Statistical Natural Language Processing*. MIT Press, Cambridge, MA, USA.

Laure Martin, Delphine Battistelli, and Thierry Charnois. 2014. Symptom extraction issue. In *Proceedings of BioNLP 2014*, pages 107–111, Baltimore, Maryland, June. Association for Computational Linguistics.

Monica Mazzucato, Laura Visonà Dalla Pozza, Silvia Manea, Cinzia Minichiello, and Paola Facchin. 2014. A population-based registry as a source of health indicators for rare diseases: the ten-year experience of the Veneto Region's rare diseases registry. *Orphanet Journal of Rare Diseases*, $item.volume:37+, March.

A.K. Ramani, R.C. Bunescu, Raymond J. Mooney, and E.M. Marcotte. 2005. Consolidating the set of known human protein-protein interactions in preparation for large-scale mapping of the human interactome. *Genome Biology*, 6(5):r40.

Jasmin Šarić, Lars Juhl Jensen, Rossitza Ouzounova, Isabel Rojas, and Peer Bork. 2006. Extraction of regulatory gene/protein networks from medline. *Bioinformatics*, 22(6):645–650, March.

Dmitry Zelenko, Chinatsu Aone, and Anthony Richardella. 2003. Kernel methods for relation extraction. *J. Mach. Learn. Res.*, 3:1083–1106, March.

Min Zhang, GuoDong Zhou, and Aiti Aw. 2008. Exploring syntactic structured features over parse trees for relation extraction using kernel methods. *Inf. Process. Manage.*, 44(2):687–701, March.

Extracting Time Expressions from Clinical Text

Timothy A. Miller[1], **Steven Bethard**[2], **Dmitriy Dligach**[1], **Chen Lin**[1], and **Guergana K. Savova**[1]

[1] Boston Children's Hospital Informatics Program, Harvard Medical School
{firstname.lastname}@childrens.harvard.edu

[2] Department of Computer and Information Sciences, University of Alabama at Birmingham
steven.bethard@colorado.edu

Abstract

Temporal information extraction is important to understanding text in clinical documents. Temporal expression extraction provides explicit grounding of events in a narrative. In this work we provide a direct comparison of various ways of extracting temporal expressions, using similar features as much as possible to explore the advantages of the methods themselves. We evaluate these systems on both the THYME (Temporal History of Your Medical Events) and i2b2 Challenge corpora. Our main findings are that simple sequence taggers outperform conditional random fields on the new data, and higher-level syntactic features do not seem to improve performance.

1 Introduction

Temporal information is ubiquitous in clinical narratives, and accurately extracting temporal information has recently been the focus of a great deal of work in clinical natural language processing (NLP) (Raghavan et al., 2012; Miller et al., 2013; Sun et al., 2013). Relevant temporal information includes events, time expressions, and temporal relations between pairs of events and/or times. The accurate extraction of temporal information would be enabling technology for sophisticated downstream processing that requires temporal awareness of patient status. One promising application is question answering, where a physician can directly ask questions about a patient's medical record. Many question types of interest are explicitly temporal (*When was the patient's last colonoscopy?*), but almost all are implicitly temporal in the sense that every question needs to be understood relative to some time frame (*What drugs is the patient on?* cannot simply return all drugs in the record but has to understand the question itself is anchored in the present).

Time Class	Example
Date	February 2 2010, Friday morning
Time	5:30 PM, 20 minutes ago
Duration	For the next 24 hours, nearly 2 weeks
Quantifier	twice, three times
Prepostexp	postoperatively, post-surgery
Set	twice daily, weekly

Table 1: Time expression classes and two examples of each class.

This work focuses on the automatic identification of time expressions in clinical text. Time expressions are words and phrases that correspond to points or spans on a timeline, such as dates or times. Other temporal expression types include *Durations*, *Quantifiers*, *Sets*, and *Prepostexps*. Table 1 shows the time expression classes used in this work, with examples given of each class. The significant deviation from general domain methods is the *Prepostexp* type, which is specific to the clinical domain. Exemplified by terms like *postoperatively*, this type represents time spans relative to some event, often an operation.

Temporal information extraction has been a topic of a great deal of work both in the clinical and general NLP domains. In the general NLP domain, the TimeBank (Pustejovsky et al., 2003) spurred much of the early research by providing a manually annotated corpus of events, times and temporal relations. Shared tasks such as TERN[1], which focused on time expressions, and TempEval (Verhagen et al., 2007; Verhagen et al., 2010; UzZaman et al., 2013), which included events and temporal relations as well, helped build a com-

[1] http://www.itl.nist.gov/iad/mig/tests/ace/2004/

Proceedings of the 2015 Workshop on Biomedical Natural Language Processing (BioNLP 2015), pages 81–91,
Beijing, China, July 30, 2015. ©2015 Association for Computational Linguistics

munity around temporal information extraction. The community explored a wide variety of approaches, the best of which used either manually engineered databases of regular expression rules (Strötgen et al., 2013) or a supervised learning word classification paradigm (Bethard, 2013), and achieved precisions and recalls above 80% in the shared tasks.

In the clinical domain, temporal information extraction has seen a great deal of recent interest, with the i2b2 (Informatics for Integrating Biology and the Bedside) shared task on temporal information extraction (Sun et al., 2013) and the recent release of the THYME (Temporal History of Your Medical Events) corpus of clinical annotations (Styler IV et al., 2014). The i2b2 shared task contained a track explicitly focusing on extraction of temporal expressions. In that task, a variety of approaches were used for time expression extraction. The best performing system (Xu et al., 2013) used machine learning, with a conditional random field classifier (CRF) for finding spans and a support vector machine classifier for classifying attributes. Other top approaches used adapted regular expressions (Sohn et al., 2013) on top of the off the shelf Heideltime system (a general-domain NLP system for parsing time expressions) (Strötgen and Gertz, 2010). Another approach used a hybrid system where the output from a CRF-based system was combined with the output of a rule-based system (Kovačević et al., 2013).

In this work, we develop and evaluate several machine learning methods for extracting time expressions from clinical text. These methods include simple sequential classifiers, a sequential model (conditional random field), a constituency parser-based method, and an ensemble sequence method that attempts to leverage the differing performance of all the other models. The contributions of this work are the comparison and analysis of a large number of different machine learning models for this task, the first use of deep syntactic features for this task, and an evaluation on two different corpora, including the first evaluation of these methods on the THYME corpus.

	THYME (TempEval)	i2b2
Date	1271	1639
Time	54	69
Duration	195	406
Quantifier	61	n/a
Set	83	n/a
Prepostexp	149	n/a
Frequency	n/a	249

Table 2: Descriptive statistics of THYME and i2b2 corpora. Frequency in i2b2 is roughly the union of set and quantifier in THYME.

2 Materials and Methods

2.1 Corpora

We use two corpora for training and evaluating the methods described above. The first is the THYME corpus (Styler IV et al., 2014), which consists of clinical and pathology notes of patients with colon cancer from Mayo Clinic. The THYME corpus is split into training, development, and test sets based on patient number, with 50% in training and 25% each in development and test sets. For our experiments we use the same patient set as the upcoming TempEval 2015[2], patients 28-127. The training data contains 1874 time expressions, the development contains 1119, and the test set contains 1047. We used the development set for optimizing learning parameters, then combined it with the training set to build the system used for reporting results in Section 3.

The second corpus we use is the i2b2 2012 Challenge dataset (Sun et al., 2013). The i2b2 dataset contains discharge summaries from Partners Healthcare and Beth Israel Deaconess Medical Center. This data is split into a training and test set, with no predefined development set. We arbitrarily set aside filenames above 600 from the training set as a development set and for tuning parameters. Under this configuration, the i2b2 dataset contained 1856 training examples, 507 development examples, and 1820 test examples. Again, training and development examples were combined to build the system that is evaluated in Section 3.

Table 2 shows the distribution of the different time classes in the THYME and i2b2

[2]http://alt.qcri.org/semeval2015/task6/

corpora. While distribution is broadly similar, i2b2 had a higher percentage of *duration* expressions while THYME had many *prepost-exp* expressions, which in i2b2 were annotated as the *date* category.

2.2 Systems

We implemented a variety of systems in an attempt to empirically evaluate the best way to model the time span classification task. For all systems, the temporal expression extractor is implemented within Apache cTAKES[3] (clinical Text Analysis and Knowledge Extraction System) (Savova et al., 2011), making use of its components for feature generation as well as its interface to the source general-domain NLP system ClearTK (Bethard et al., 2014) which in turn interfaces with different machine learning libraries, including LibSVM (Chang and Lin, 2011) and CRFSuite (Okazaki, 2007).

2.2.1 Sequence Models

We developed three sequence-based models for this task, each with different perceived strengths. The first system is perhaps the simplest, a standard BIO (Begin-Inside-Outside) tagger using an off the shelf support vector machine (SVM) classifier (Cortes and Vapnik, 1995). BIO taggers work by labeling every token in a sentence as the beginning (B), inside (I), or outside (O) of some subsequence in the data (in this case a temporal expression). The tagger progresses left to right through a sentence, making a classification decision at each word, with features based on any information that would be available to a system at run time. After processing a sentence, tag sequences are converted to time expression spans and evaluated in the span format. The main benefit of this system is its efficiency, as it operates in a "greedy" fashion, getting a locally optimal labeling.

The second sequence system is a backwards BIO tagger. This system works just like the BIO tagger described above, except it starts at the end of the sentence and works its way forward. As mentioned above, this family of models is not globally optimal. In preliminary work, we found that the BIO tagger frequently left off the first word of a time expression, especially if it was a common word like 'the' or 'this.' Additionally, time expressions are often noun phrases, which typically carry a lot of meaning in the right-most word, so starting from the right has that advantage as well. For evaluation purposes, this model and the forward BIO tagger can be given exactly the same features, so there is a very clear evaluation of just the single difference in model strategy, going forwards or backwards.

The third system is a conditional random field (CRF) sequence labeler. Conditional random fields (Lafferty et al., 2001) are discriminatively-trained undirected graphical models that find the globally optimal labeling for a given configuration of random variables. We use a standard CRF architecture, the linear-chain CRF, where the random variables for sequence labels have only dependencies between the previous and next label, and random variables for arbitrary features of the observed evidence. Like the sequential taggers above, the CRF tagger assigns BIO tags to every word in a sequence, and time expressions are deterministically extracted from those assignments. The CRF tagger processes one sentence at a time, assigning labels to all tokens within that sentence simultaneously.

2.2.2 Constituency Model

The other system we developed is based on a constituency parse representation. Constituency trees represent the phrasal structure of a sentence, building up structure from the word level to a single tree which encloses the whole sentence. Our time expression classification model starts at the root of a tree and traverses it depth-first for a given sentence, and at each node in the tree classifies the enclosed span of words as a time expression or not a time expression, using a support vector machine classifier. During the depth-first traversal, further downward traversal is terminated with a positive classification (a time expression is found) or with a constituency spanning a single word. Figure 1 shows an example constituency tree with a time expression.

During training, the depth-first search will compare the span of every constituent in the search path with the spans of the gold standard, and any matching constituents are positive instances. Any non-matching spans are negative training instances. Features for each

[3]http://ctakes.apache.org

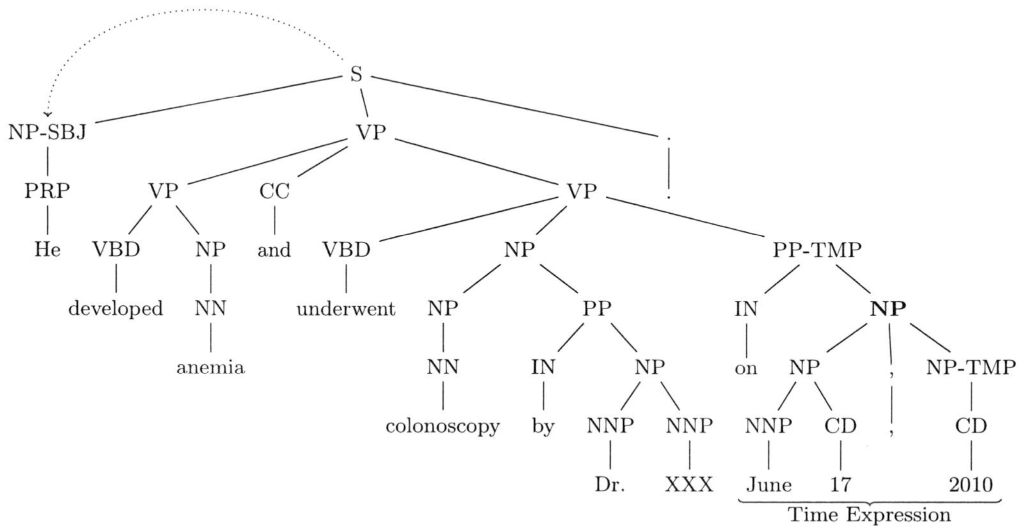

Figure 1: **Example constituency tree containing a time expression.** This sentence contains a single time-expression (*June 17, 2010*), spanned by the bolded NP in the figure. That NP is a single positive training instance, while all other constituents will be negative training instances.

positive or negative instance (described in detail below) can be extracted arbitrarily based on the position of the instance in the tree, but this representation obviously lends itself more to hierarchical, syntax-based features and makes sequence-based features more difficult (though not impossible) to represent.

The appeal of this approach is that time expressions will nearly always be constituents, so the classifier is constrained to select only constituent sequences. This also seems to combine advantages of the systems above, as it gets to consider whole spans at once (like the global optimization of the CRF), while using a simple binary classifier (like the SVM-based BIO taggers). One potential drawback is that it requires high accuracy parsing, at least for constituents composed of temporal expressions.

2.2.3 Ensemble Model

The final model we developed is an ensemble sequence model that is trained on features encapsulating the outputs of the four above systems and making predictions based on those features. The rationale for this model is that our other models differ enough to have varying strengths and weaknesses, and an ensemble system may be able to learn when to select which system. The features at each word in the sequence are the outputs of the component systems at a window of width n around the word. So, for systems i the features at position j in the document are the following set:

$$feats_j = \cup_i \cup_{j'=j-n}^{j+n} \{out_{j'}^i\} \quad (1)$$

where $out_i^j \in \{B, I, O\}$, indicating the output label of system i at position j.

We use a CRF-based tagger for this model – with a much smaller (and thus more learnable) feature space, our intuition is that a globally optimal model should have even more of an advantage over word-by-word discriminative taggers. In preliminary work we found that a window of $n = 1$ gave the best performance on the development set, so the final model was trained using that value.

2.3 Features

To make the comparison fair, we made an effort to have feature parity between systems as much as possible. For the three sequence-based models this was largely accomplished. For the constituency parser-based model the approach is so different that the features do not align perfectly with the other systems, but roughly the same information is present.

Feature Type	Features
Tokens	Word=June,Word=17,Word=COMMA,Word=2010
POS tags	POS=NNP,POS=CD,POS=COMMA,POS=CD
Character classes	Char=LuLlLlLl,CharCollapsed=LuLl,Char=NdNd,CharCollapsed=Nd, Char=Pc,CharCollapsed=Pc,Char=NdNdNdNd,CharCollapsed=Nd
Gazetteer	MonthOfYear,Number,Year
Parse	node=NP,parent=PP,prod=NP→NP-COMMA-NP,root=false,leaf=false

Table 3: Table representing features extracted from the time expression in Figure 1, organized by feature type. The comma character is represented as COMMA so it is not confused with the commas used to separate features. Character classes are explained in the main text.

2.3.1 Sequence Features

Sequence features include lexical features, gazetteer features, syntactic parse features, and section features. Lexical and gazetteer features are both token-based, and are defined for both the current token under consideration (i.e., the one the classifier is currently trying to label) and the three tokens on either side of the token under consideration. These features include token part of speech (POS) tags and two character based features. The part of speech tags are obtained from the cTAKES POS tagger (a clinical data-trained wrapper for the Apache OpenNLP[4] POS tagger). The character-based features map every character in the token to a unicode character category[5], for example, uppercase letter (*Lu*), lowercase letter (*Ll*), decimal digit (*Nd*), etc. This character-mapped token representation is then turned into two features, one in the unmodified format and one where repeats are collapsed. For example, the token "2004" would map to two features: one where its represented as four digit characters (*NdNdNdNd*) and one where the repeats are collapsed (*Nd*).

Gazetteer features rely on a lookup table that contains information about lexical items that are very likely to generalize. We define a small set of temporal word classes and created a gazetteer that maps lexical items to those classes. The set of classes with representative examples is: {Number (numbers up to 200), Year (four-digit numbers that could reasonably appear in current notes), Unit (second, minute), PartOfDay (morning), DayOfWeek (Monday), WeekendOfWeek (weekend), MonthOfYear (January, jan), SeasonOfYear (Summer), DecadeOfCentury (nineties), Time (noon), Age (teenager), TimeReference (previously), Frequency (monthly), Adjuster (next), Modifier (nearly), PrePost (postoperative), TimeSeparator (:)}. The full list of items is too long to list here but will be part of the open source release of this system.

Parse features for the sequence model are not as natural a fit as with a constituency node-based model, but some features can be derived based on spans. With the BIO tagger models (forward and backward) we define a *candidate span* to consider, defined in terms of the forward tagger but easily extendable to the backwards tagger. A candidate span for the current token we are classifying has as its rightmost token the current token, and its leftmost token as the start of the sequence that the current token would be a part of if it is classified as part of a temporal expression. This is simple to find in practice: if the previous token is O (not part of a temporal expression) then the current candidate span is only the current token; otherwise the candidate span starts at the most recent token labeled B (the start of a temporal expression). For the CRF sequence tagger, classification decisions have not yet been made, so the candidate span always covers only the current token.

Given this definition of a candidate span, we define several features. We have one feature for the category of the lowest constituent that dominates the current span, a feature for the parent category of the dominating node, a feature that indicates whether the dominating node is a leaf (preterminal) node, and a feature to indicate whether the dominating node matches the current span exactly.

[4]`http://opennlp.apache.org`

[5]`http://www.unicode.org/reports/tr44/#General_Category_Values`

We then have production-rule associated features. First, we simply represent the production rule of the dominating node as a string (e.g., "NP -> DT NN"). We also use "bag of children" features which represent each of the elements of the right hand side of the production rule, ignoring ordering.

The next feature type is based on surrounding classifications. Here the BIO taggers have access to the previous classification decision (B, I, or O). The CRF in the linear chain configuration can use labels on either side of the current word. While this represents a difference in features available to the systems, it is one that is inherent to the methodology (something that is only possible with CRFs and not with BIO taggers) so we consider this to not violate our goal of feature parity.

The last feature type is specific to the THYME corpus, as it is based on identifiers in the section headers of Mayo Clinic notes. These identifiers are easily extracted with regular expressions and are codes that indicate the purpose of a section (e.g., medications, allergies, etc.). For this feature we simply use the string representing the code for the section that encloses the token under consideration. These are intended to capture the fact that some sections may contain condensed narrative, and are likely to contain time expressions, while others have expressions that resemble time expressions but are not (*5/9* to mean five out of nine).

2.3.2 Constituency Features

The constituency parse-based system attempts to use similar features where possible – we will refer to the features above when possible and point out implementation differences.

First, the features for character class and part of speech for tokens are replicated, by applying them to all the tokens within the span of the current tree node being classified. Gazetteer features are replicated similarly – each word covered by the current tree node is mapped to its time class, if it exists. This is done without reference to ordering.

From the tree itself, we use several features similar to those above, but explicitly based on the tree rather than having to be mapped to the tree. For the current node and its parent, we have features for node category (e.g., NP). For the current node alone we use boolean features for whether it is the root node of the sentence and whether it is a leaf node. We have string features for the bag of children, as well as a feature representing the production rule. Table 3 shows the features that would be extracted for this classifier for the time expression in Figure 1.

2.4 Evaluation

Our evaluation looks at three variables – different machine learning methods, the usefulness of automatic parses at deriving syntactic features, and the domain of the data. For scoring the evaluation, we primarily use a simple scorer built into ClearTK that requires *exact* span matching. We also track partially overlapping spans and count them as correct for *overlapping* span matching. For comparability, we also use the 2012 i2b2 Challenge scoring tool for i2b2 data, which allows both exact and overlapping matching. We use exact span matching as our primary scoring method to conservatively estimate performance, in part because the output of these systems will typically be passed to a time normalization system, which may not be able to handle the variations in input. The metrics we use are precision $\left(\frac{\#correcttimespans}{\#predictedtimespans}\right)$, recall $\left(\frac{\#correcttimespans}{\#goldtimespans}\right)$, and F1 $\left(\frac{2*p*r}{p+r}\right)$.

We look first at multiple methods on two different corpora. In this experiment we are looking to see whether there is any method which is clearly superior to the others, especially across corpora. This experiment is important because methods like the CRF and the CRF-based ensemble have some nice theoretical properties (finding the globally optimal sequence), but as a result have slower run time, and to understand this tradeoff we need to measure performance differences. For the first four systems (the non-ensemble systems), we simply train each method on the combined training and development sets for each corpus, and test on the test set for that corpus.

For the ensemble system, we note that since it is trained on the outputs of other systems, we must do an internal cross-validation of the component systems before performing the tests, to ensure that the labels provided to the ensemble method are representative of what it

	THYME						i2b2 2012 Challenge					
	Exact			Overlapping			Exact			Overlapping		
	P	R	F	P	R	F	P	R	F	P	R	F
BIO	0.784	0.676	**0.726**	0.948	0.836	0.888	0.775	0.718	0.745	0.921	0.853	0.886
Backwards	0.770	**0.687**	**0.726**	0.948	**0.846**	0.894	0.786	**0.740**	**0.762**	0.917	**0.862**	**0.889**
CRF	**0.788**	0.584	0.671	0.961	0.712	0.818	**0.814**	0.617	0.702	**0.960**	0.728	0.828
Constituency	0.715	0.563	0.630	**0.989**	0.799	0.884	0.657	0.545	0.596	0.920	0.762	0.834
Ensemble	0.784	0.669	0.722	0.962	0.841	**0.897**	0.809	0.706	0.754	0.948	0.828	0.884
Xu et al.(2013)										*0.881*	*0.950*	*0.914*

Table 4: Precision (P), Recall (R), and F1-Score (F) for different systems and corpora. The highest score in each column is in bold. BIO=Begin-Inside-Outside tagger, Backwards=Reverse BIO tagger,CRF=Conditional Random Field tagger,Constituency=Constituency parser-based classifier, Ensemble=CRF-based ensemble classifier. Italicized results from Xu et al. indicated reported, not replicated, results.

will see on test data. We first perform a 5-fold cross validation on the training set, for each fold training the component on four folds and running the trained component on the fifth. The output on that fifth fold forms the training data that the ensemble method will see. By repeating that for each fold, the ensemble method obtains proper system-generated labels from the component system for the entire training set to use as its training data.

The second experiment looks at the importance of accurate syntactic parsing for generating features. For the syntax-focused experiments, we use only the THYME corpus, since it has a layer of gold standard treebank annotations. The tagger we evaluate is the best performing system on the first experiment, the Backwards BIO tagger. In this case we examine three different conditions: First, using gold standard treebank for feature extraction; second, using automatic parses from a THYME-trained parser; and finally, without any syntactic features at all.

The final experiment examines the domain-specificity of the systems and corpora. In this experiment we train the best performing system (Backwards BIO tagger) on THYME data and then test on i2b2 data, and vice versa.

	THYME		
	Exact		
	P	R	F
Gold	0.771	0.699	0.733
Automatic	*0.770*	*0.687*	*0.726*
No Syntax	0.773	0.690	0.729

Table 5: Precision (P), recall (R), and F1-Score (F) for different syntactic configurations of the *Backwards* BIO tagger system. Gold - Manually annotated trees from Treebank used for features. Automatic – parser trained on clinical text from THYME Treebank, italicized to denote that it is copied from Table 4 above. No Syntax – Backwards BIO tagger system with no syntactic features.

3 Results

Results are shown in Tables 4-6. Table 4 shows the results of the primary experiment – performance of the various systems on both THYME and i2b2 corpora. In most conditions, the Backwards BIO Tagger obtains the highest or tied for the highest F-score, while the regular BIO tagger and ensemble method obtain very competitive F-scores. The Backwards BIO tagger tends to have the best recall of all systems, while preserving precision at a relatively high level. The CRF, despite being theoretically globally optimal, is not competitive in terms of F-score with the SVM-based taggers. The ensemble CRF nominally obtains the best performance in the *Overlapping* metric on the THYME corpus, but the improvement is marginal.

The backwards BIO tagger achieved an F-score of 0.889 on the i2b2 Challenge data allowing for partial matches (the *Overlapping* column). The best performing system in the i2b2 Challenge (Xu et al., 2013) is shown in the last row, with an F1 score of 0.914, with an advantage on recall. Our best system performance would tie for 4th in the span matching part of that challenge, without tuning for that dataset. While we incorporated features based on the best-performing similar system (Xu et al., 2013), including punctuation information, prepositions, and chunk information, these did not improve performance. Their paper described a larger system and did not contain enough detail on time expression extraction to replicate exactly (e.g., the *Other Keywords* section in the online supplement is not exhaustive and probably is important to their result).

Table 5 shows the results of experiments examining the role of syntactic features on our best performing system, the Backwards BIO tagger. This experiment suggests syntactic features are not valuable for the test set. Neither using gold standard trees for extracting features, nor removing syntactic features altogether, changed performance meaningfully.

Finally, Table 6 shows results of cross-domain experiments, using the best-performing Backwards BIO tagger. Performance falls quite a bit relative to the in-domain trained experiments, even in the relaxed *Overlapping* condition. Training on THYME and testing on i2b2 results in the worst performance, with an exact span matching F1 Score of 0.422.

4 Discussion

While the results are competitive with the best systems at the i2b2 Challenge, they raise many interesting questions.

First, it is very interesting that the best performing systems are the simplest and fastest. Despite the theoretical advantages of the conditional random field's global sequence optimization, the BIO approaches using local classifiers typically obtain the best performance. This is also in contrast with results from the i2b2 Challenge, where the best performing system used a CRF approach. We extensively explored the parameter space for CRFs on development data and our sense was that throughout this entire space performance lagged SVM-based tagging systems.

Next, it is unfortunate but interesting that the ensemble method does not improve performance over the component systems. Error analysis for this system showed both examples where the first word was missed and examples where the last word was missed. The forwards and backwards BIO taggers should be obtaining complementary errors of these types. Thus it is not clear why the ensemble method is unable to take advantage of the information from multiple systems to improve performance.

The syntax-based system shows the biggest gain when switching from the scorer that considers exact spans to one that considers overlapping spans. In preliminary work using gold standard parses, the exact span scores were significantly higher. These two facts suggest that the primary reason for the low accuracy of this model on exact spans is parsing errors. This was, in fact, one motivation for incorporating parser features – if the parser cannot reliably find exact spans, perhaps it is still possible to use its output at word levels to find patterns that a sequence-based model could use.

The lack of improvements with syntactic features in these experiments is therefore somewhat confusing, as using a totally syntax-based system is able to obtain decent performance. One hypothesis for their lack of impact is that annotation consistency plays a role. We noticed that annotations of time expressions around prepositional phrases are inconsistent. For example, in the prepositional phrase *since Tuesday*, the time expression is only *Tuesday*, but in some cases the whole PP is annotated in the gold standard. This may help explain the large jump in performance when partially overlapping spans are included, as there are many errors that are off by only an added or dropped preposition at the start of the time expression. (Note that this explanation may also apply to the constituency parser model.)

The cross-corpus performance (training on THYME and testing on i2b2 and vice versa) is surprisingly low. While the annotation guidelines are similar, one major difference is the addition of the *prepostexp* class to THYME, for expressions like *postoperative*. Meanwhile, i2b2 challenge data annotates expressions like *postoperative day 5*, which do not occur in THYME data, as the *date* class. This affects both recall and precision on the THYME to i2b2 evaluation as expressions like *postoperative day 5* cause both recall errors (not getting the whole expression) and precision errors (predicting the first word of the expression). Additional errors in this direction are caused by unseen abbreviations in THYME that are common in i2b2 (*POD* for postoperative day,

		Exact			Overlapping		
Train Corpus	Test Corpus	P	R	F	P	R	F
THYME	THYME	*0.770*	*0.687*	*0.726*	*0.948*	*0.846*	*0.894*
i2b2	i2b2	*0.786*	*0.740*	*0.762*	*0.917*	*0.862*	*0.889*
THYME	i2b2	0.436	0.410	0.422	0.722	0.679	0.700
i2b2	THYME	0.589	0.432	0.498	0.860	0.629	0.727

Table 6: Precision (P), recall (R), and F1-Score (F) for cross-domain experiments. We use the best-performing system for each experiment (Backwards BIO), with automatic parse features from a THYME-trained parser. Italicized rows are copied from Table 4 above for ease of comparison.

x 2 for twice a day). In the other direction (train on i2b2, test on THYME), recall errors are worst because it does not correctly identify any of the prepostop expressions. Surprisingly, in both directions there are relatively simple date formats missed due to slight differences in convention – THYME data often uses month names (e.g. Jan 5, 2014) while i2b2 typically does not, while i2b2 uses MM-YY format (e.g., March 7 represented as 05-07) while THYME does not. This suggests that a better system could be obtained by training on both corpora, although this will require some reconciliation of the differences in time classes, primarily what THYME calls *prepostexp* time expressions.

Errors on the best-performing system are primarily those where the start or end of the time expression is off by one. As above, these may be partially due to inconsistent prepositional phrase annotation, and the effect of fixing this is roughly seen in the overlapping scoring criterion. The remaining errors probably represent the most opportunity for system improvement, so we focus on that briefly.

One common issue occurs with coordination – phrases like *2003 or 2004*. While these are annotated as a single span, the system will get the two individual years, resulting in one recall error but two precision errors. This type of error might be fixed by a second pass that joins together time expressions connected by coordinators. A modified syntactic approach that operates bottom-up instead of top-down might also correctly recognize such expressions. Another source of error is in expressions that are unusually expressed in a few instances, such as *times three* to mean something happened three times. While this is *in* the training data, it is not the primary way of indicating this meaning, and there are not enough instances to learn this modification. Similarly, sometimes punctuation is inserted or modified into an expression that slightly changes its representation to the classification algorithm (e.g., *one-year* with a dash rather than *one year*). Fixing this issue in a general way is a tricky problem, as it is related to the larger issue of there being many ways to instantiate any given concept (*half past 7*, *half of 8*, *7:30*, etc.). In the clinical domain one hopes that usage is a bit more constrained and that one might be able to get away with a simpler approach such as just ignoring punctuation.

In conclusion, we have presented and evaluated multiple machine learning methods for temporal expression extraction. Our results suggest that simpler and faster BIO sequence tagger methods are as good as more complex models or ensemble methods. We also show that deep syntax does not seem beneficial to this task. Finally, we show that there is significance performance degradation when applying to new corpora, despite similar annotation guidelines and domains.

5 Acknowledgments

The project described was supported by R01LM010090 (THYME) from the National Library Of Medicine. The content is solely the responsibility of the authors and does not necessarily represent the official views of the National Institutes of Health.

References

Steven Bethard, Philip Ogren, and Lee Becker. 2014. Cleartk 2.0: Design patterns for machine learning in uima. In Nicoletta Calzolari, Khalid Choukri, Thierry Declerck, Hrafn Loftsson, Bente Maegaard, Joseph Mariani, Asuncion Moreno, Jan Odijk, and Stelios Piperidis, editors, *Proceedings of the Ninth International*

Conference on Language Resources and Evaluation (LREC'14), pages 3289–3293, Reykjavik, Iceland, May. European Language Resources Association (ELRA).

Steven Bethard. 2013. Cleartk-timeml: A minimalist approach to tempeval 2013. In *Second Joint Conference on Lexical and Computational Semantics (*SEM), Volume 2: Proceedings of the Seventh International Workshop on Semantic Evaluation (SemEval 2013)*, pages 10–14, Atlanta, Georgia, USA, June. Association for Computational Linguistics.

Chih-Chung Chang and Chih-Jen Lin. 2011. LIBSVM: A library for support vector machines. *ACM Transactions on Intelligent Systems and Technology*, 2:27:1–27:27. Software available at `http://www.csie.ntu.edu.tw/~cjlin/libsvm`.

Corinna Cortes and Vladimir Vapnik. 1995. Support-vector networks. *Machine Learning*, 20(3):273–297.

Aleksandar Kovačević, Azad Dehghan, Michele Filannino, John A Keane, and Goran Nenadic. 2013. Combining rules and machine learning for extraction of temporal expressions and events from clinical narratives. *Journal of the American Medical Informatics Association*, 20(5):859–866.

John Lafferty, Andrew McCallum, and Fernando Pereira. 2001. Conditional random fields: Probabilistic models for segmenting and labeling sequence data. In *ICML*, pages 282–289.

Timothy Miller, Steven Bethard, Dmitriy Dligach, Sameer Pradhan, Chen Lin, and Guergana Savova. 2013. Discovering temporal narrative containers in clinical text. In *Proceedings of the 2013 Workshop on Biomedical Natural Language Processing*, pages 18–26, Sofia, Bulgaria, August. Association for Computational Linguistics.

Naoaki Okazaki. 2007. Crfsuite: a fast implementation of conditional random fields (CRFs). `http://www.chokkan.org/software/crfsuite/`.

James Pustejovsky, Patrick Hanks, Roser Sauri, Andrew See, Robert Gaizauskas, Andrea Setzer, Dragomir Radev, Beth Sundheim, David Day, Lisa Ferro, et al. 2003. The timebank corpus. In *Corpus linguistics*, volume 2003, page 40.

Preethi Raghavan, Eric Fosler-Lussier, and Albert M Lai. 2012. Temporal classification of medical events. In *Proceedings of the 2012 Workshop on Biomedical Natural Language Processing*, pages 29–37. Association for Computational Linguistics.

Guergana K. Savova, Wendy W. Chapman, Jiaping Zheng, and Rebecca S. Crowley. 2011. Anaphoric relations in the clinical narrative: corpus creation. *J Am Med Inform Assoc*, 18:459–465.

Sunghwan Sohn, Kavishwar B Wagholikar, Dingcheng Li, Siddhartha R Jonnalagadda, Cui Tao, Ravikumar Komandur Elayavilli, and Hongfang Liu. 2013. Comprehensive temporal information detection from clinical text: medical events, time, and tlink identification. *Journal of the American Medical Informatics Association*, pages amiajnl–2013.

Jannik Strötgen and Michael Gertz. 2010. Heideltime: High quality rule-based extraction and normalization of temporal expressions. In *Proceedings of the 5th International Workshop on Semantic Evaluation*, pages 321–324. Association for Computational Linguistics.

Jannik Strötgen, Julian Zell, and Michael Gertz. 2013. Heideltime: Tuning english and developing spanish resources for tempeval-3. In *Second Joint Conference on Lexical and Computational Semantics (*SEM), Volume 2: Proceedings of the Seventh International Workshop on Semantic Evaluation (SemEval 2013)*, pages 15–19, Atlanta, Georgia, USA, June. Association for Computational Linguistics.

William F. Styler IV, Steven Bethard, Sean Finan, Martha Palmer, Sameer Pradhan, Piet C de Groen, Brad Erickson, Timothy Miller, Chen Lin, Guergana Savova, and James Pustejovsky. 2014. Temporal annotation in the clinical domain. *Transactions of the Association for Computational Linguistics*, 2:143–154.

Weiyi Sun, Anna Rumshisky, and Ozlem Uzuner. 2013. Evaluating temporal relations in clinical text: 2012 i2b2 challenge. *Journal of the American Medical Informatics Association*, 20(5):806–813.

Naushad UzZaman, Hector Llorens, Leon Derczynski, James Allen, Marc Verhagen, and James Pustejovsky. 2013. Semeval-2013 task 1: Tempeval-3: Evaluating time expressions, events, and temporal relations. In *Second Joint Conference on Lexical and Computational Semantics (*SEM), Volume 2: Proceedings of the Seventh International Workshop on Semantic Evaluation (SemEval 2013)*, pages 1–9, Atlanta, Georgia, USA, June. Association for Computational Linguistics.

Marc Verhagen, Robert Gaizauskas, Frank Schilder, Mark Hepple, Graham Katz, and James Pustejovsky. 2007. Semeval-2007 task 15: Tempeval temporal relation identification. In *Proceedings of the Fourth International Workshop on Semantic Evaluations (SemEval-2007)*, pages 75–80, Prague, Czech Republic,

June. Association for Computational Linguistics.

Marc Verhagen, Roser Sauri, Tommaso Caselli, and James Pustejovsky. 2010. Semeval-2010 task 13: Tempeval-2. In *Proceedings of the 5th International Workshop on Semantic Evaluation*, pages 57–62, Uppsala, Sweden, July. Association for Computational Linguistics.

Yan Xu, Yining Wang, Tianren Liu, Junichi Tsujii, and Eric I-Chao Chang. 2013. An end-to-end system to identify temporal relation in discharge summaries: 2012 i2b2 challenge. *Journal of the American Medical Informatics Association : JAMIA*.

Exploiting Task-Oriented Resources to Learn Word Embeddings for Clinical Abbreviation Expansion

Yue Liu[1], **Tao Ge**[2], **Kusum S. Mathews**[3], **Heng Ji**[1], **Deborah L. McGuinness**[1]
[1]Department of Computer Science, Rensselaer Polytechnic Institute
{liuy30,jih,dlm}@rpi.edu
[2]School of Electronics Engineering and Computer Science, Peking University
getao@pku.edu.cn
[3]Departments of Medicine and Emergency Medicine, Icahn School of Medicine at Mount Sinai
kusum.mathews@mssm.edu

Abstract

In the medical domain, identifying and expanding abbreviations in clinical texts is a vital task for both better human and machine understanding. It is a challenging task because many abbreviations are ambiguous especially for intensive care medicine texts, in which phrase abbreviations are frequently used. Besides the fact that there is no universal dictionary of clinical abbreviations and no universal rules for abbreviation writing, such texts are difficult to acquire, expensive to annotate and even sometimes, confusing to domain experts. This paper proposes a novel and effective approach – exploiting task-oriented resources to learn word embeddings for expanding abbreviations in clinical notes. We achieved 82.27% accuracy, close to expert human performance.

1 Introduction

Abbreviations and acronyms appear frequently in the medical domain. Based on a popular online knowledge base, among the 3,096,346 stored abbreviations, 197,787 records are medical abbreviations, ranked first among all ten domains.[1] An abbreviation can have over 100 possible explanations[2] even within the medical domain. Medical record documentation, the authors of which are mainly physicians, other health professionals, and domain experts, is usually written under the pressure of time and high workload, requiring notation to be frequently compressed with shorthand jargon and acronyms. This is even more evident within intensive care medicine, where it is crucial that information is expressed in the most efficient manner possible to provide time-sensitive care to critically ill patients, but can result in code-like messages with poor readability. For example, given a sentence written by a physician with specialty training in critical care medicine, "STAT TTE c/w RVS. AKI - no CTA. .. etc", it is difficult for non-experts to understand all abbreviations without specific context and/or knowledge. But when a doctor reads this, he/she would know that although "STAT" is widely used as the abbreviation of "statistic", "statistics" and "statistical" in most domains, in hospital emergency rooms, it is often used to represent "immediately". Within the arena of medical research, abbreviation expansion using a natural language processing system to automatically analyze clinical notes may enable knowledge discovery (e.g., relations between diseases) and has potential to improve communication and quality of care.

In this paper, we study the task of abbreviation expansion in clinical notes. As shown in Figure 1, our goal is to normalize all the abbreviations in the intensive care unit (ICU) documentation to reduce misinterpretation and to make the texts accessible to a wider range of readers. For accurately capturing the semantics of an abbreviation in its context, we adopt word embedding, which can be seen as a distributional semantic representation and has been proven to be effective (Mikolov et al., 2013) to compute the semantic similarity between words based on the context without any labeled data. The intuition of distributional semantics (Harris, 1954) is that if two words share similar contexts, they should have highly similar semantics. For example, in Figure 1, "RF" and "respiratory failure" have very similar contexts so that their semantics should be similar. If we know "respiratory fail-

[1]www.allacronyms.com
[2]www.allacronyms.com/_medical/HD

Proceedings of the 2015 Workshop on Biomedical Natural Language Processing (BioNLP 2015), pages 92–97,
Beijing, China, July 30, 2015. ©2015 Association for Computational Linguistics

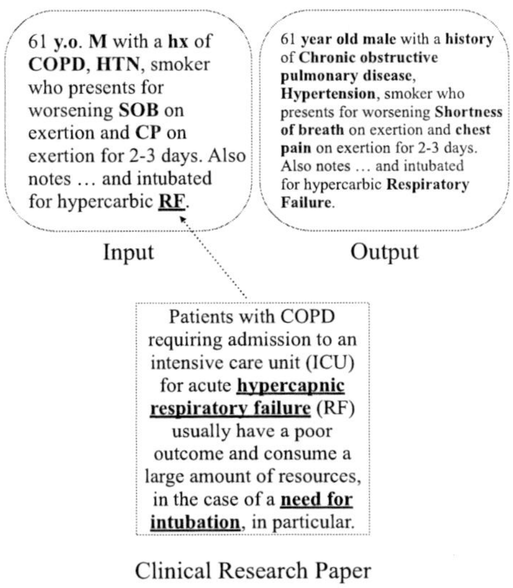

Figure 1: Sample Input and Output of the task and intuition of distributional similarity

ure" is a possible candidate expansion of "RF" and its semantics is similar to the "RF" in the intensive care medicine texts, we can determine that it should be the correct expansion of "RF". Due to the limited resource of intensive care medicine texts where full expansions rarely appear, we exploit abundant and easily-accessible task-oriented resources to enrich our dataset for training embeddings. To the best of our knowledge, we are the first to apply word embeddings to this task. Experimental results show that the embeddings trained on the task-oriented corpus are much more useful than those trained on other corpora. By combining the embeddings with domain-specific knowledge, we achieve 82.27% accuracy, which outperforms baselines and is close to human's performance.

2 Related Work

The task of abbreviation disambiguation in biomedical documents has been studied by various researchers using supervised machine learning algorithms (Liu et al., 2004; Gaudan et al., 2005; Yu et al., 2006; Ucgun et al., 2006; Stevenson et al., 2009). However, the performance of these supervised methods mainly depends on a large amount of labeled data which is extremely difficult to obtain for our task since intensive care medicine texts are very rare resources in clinical domain due to the high cost of de-identification and annotation. Tengstrand et al. (2014) proposed a distributional semantics-based approach for abbreviation expansion in Swedish but they focused only on expanding single words and cannot handle multi-word phrases. In contrast, we use word embeddings combined with task-oriented resources and knowledge, which can handle multiword expressions.

3 Approach

3.1 Overview

The overview of our approach is shown in Figure 2. Within ICU notes (e.g., text example in top-left box in Figure 2), we first identify all abbreviations using regular expressions and then try to find all possible expansions of these abbreviations from domain-specific knowledge base[3] as candidates. We train word embeddings using the clinical notes data with task-oriented resources such as Wikipedia articles of candidates and medical scientific papers and compute the semantic similarity between an abbreviation and its candidate expansions based on their embeddings (vector representations of words).

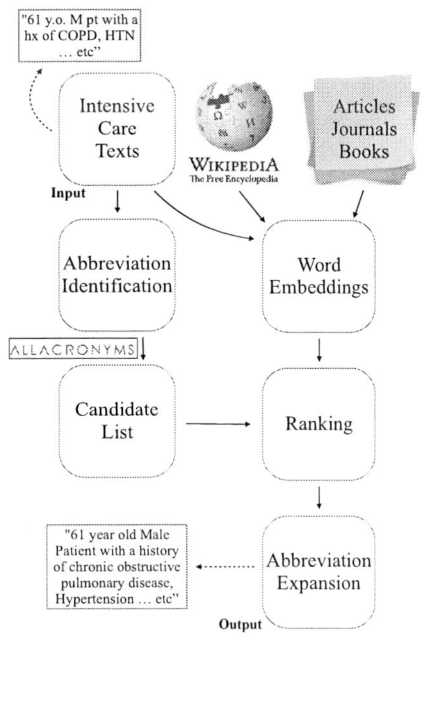

Figure 2: Approach overview.

[3]http://www.allacronyms.com

3.2 Training embeddings with task oriented resources

Given an abbreviation as input, we expect the correct expansion to be the most semantically similar to the abbreviation, which requires the abbreviation and the expansion share similar contexts. For this reason, we exploit rich task-oriented resources such as the Wikipedia articles of all the possible candidates, research papers and books written by the intensive care medicine fellows. Together with our clinical notes data which functions as a corpus, we train word embeddings since the expansions of abbreviations in the clinical notes are likely to appear in these resources and also share the similar contexts to the abbreviation's contexts.

3.3 Handling MultiWord Phrases

In most cases, an abbreviation's expansion is a multi-word phrase. Therefore, we need to obtain the phrase's embedding so that we can compute its semantic similarity to the abbreviation.

It is proven that a phrase's embedding can be effectively obtained by summing the embeddings of words contained in the phrase (Mikolov et al., 2013; Socher et al., 2013). For computing a phrase's embedding, we formally define a candidate c_i as a list of the words contained in the candidate, for example: one of *MICU*'s candidate expansions is *medical intensive care unit*=[*medical,intensive,care,unit*]. Then, c_i's embedding can be computed as follows:

$$\boldsymbol{x}(\boldsymbol{c_i}) = \sum_{t \in c_i} \boldsymbol{x}(\boldsymbol{t}) \tag{1}$$

where t is a token in the candidate c_i and $\boldsymbol{x}(\cdot)$ denotes the embedding of a word/phrase, which is a vector of real-value entries.

3.4 Expansion Candidate Ranking

Even though embeddings are very helpful to compute the semantic similarity between an abbreviation and a candidate expansion, in some cases, context-independent information is also useful to identify the correct expansion. For example, CHF in the clinical notes usually refers to "congestive heart failure". By using embedding-based semantic similarity, we can find two possible candidates – "congestive heart failure" (similarity=0.595) and "chronic heart failure"(similarity=0.621). These two candidates have close semantic similarity score but their popularity scores in the medical domain are quite different – the former has a rating score[4] of 50 while the latter only has a rating score of 7. Therefore, we can see that the rating score, which can be seen as a kind of domain-specific knowledge, can also contribute to the candidate ranking.

We combine semantic similarity with rating information. Formally, given an abbreviation b's candidate list $l(b) = \{c_1, c_2, ..., c_n\}$, we rank $l(b)$ based on the following formula:

$$score(c) = \lambda \frac{rating(c)}{\sum_{c_i \in l(b)} rating(c_i)} + (1-\lambda) \frac{\boldsymbol{x}(\boldsymbol{b}) \cdot \boldsymbol{x}(\boldsymbol{c})}{|\boldsymbol{x}(\boldsymbol{b})||\boldsymbol{x}(\boldsymbol{c})|} \tag{2}$$

where $rating(c)$ denotes the rating of this candidate as an expansion of the abbreviation b, which reflects this candidate's popularity, $\boldsymbol{x}(\cdot)$ denotes the embedding of a word. The parameter λ serves to adjust the weights of similarity and popularity[5]

4 Experiment Results

4.1 Data and Evaluation Metrics

The clinical notes we used for the experiment are provided by domain experts, consisting of 1,160 physician logs of Medical ICU admission requests at a tertiary care center affiliated to Mount Sanai. Prospectively collected over one year, these semi-structured logs contain free-text descriptions of patients' clinical presentations, medical history, and required critical care-level interventions. We identify 818 abbreviations and find 42,506 candidates using domain-specific knowledge (i.e., www.allacronym.com/_medical). The enriched corpus contains 42,506 Wikipedia articles, each of which corresponds to one candidate, 6 research papers and 2 critical care medicine textbooks, besides our raw ICU data.

We use word2vec (Mikolov et al., 2013) to train the word embeddings. The dimension of embeddings is empirically set to 100.

Since the goal of our task is to find the correct expansion for an abbreviation, we use *accuracy* as a metric to evaluate the performance of our approach. For ground-truth, we have 100 physician logs which are manually expanded and normalized by one of the authors Dr. Mathews, a well-trained

[4]All the rating information in this paper is from http://www.allacronyms.com. On this website, users are free to rate expansions of an abbreviation if they like the expansions. In general, a popular expansion has a high rating score.

[5]In the experiments, λ is empirically tuned to 0.2 on a separate development set.

domain expert, and thus we use these 100 physician logs as the test set to evaluate our approach's performance.

4.2 Baseline Models

For our task, it's difficult to re-implement the supervised methods as in previous works mentioned since we do not have sufficient training data. And a direct comparison is also impossible because all previous work used different data sets which are not publicly available. Alternatively, we use the following baselines to compare with our approach.

- Rating: This baseline model chooses the highest rating candidate expansion in the domain specific knowledge base.
- Raw Input embeddings: We trained word embeddings only from the 1,160 raw ICU texts and we choose the most semantically related candidate as the answer.
- General embeddings: Different from the Raw Input embeddings baseline, we use the embedding trained from a large biomedical data collection that includes knowledge bases like PubMed and PMC and a Wikipedia dump of biomedical related articles (Pyysalo et al., 2013) for semantic similarity computation.

4.3 Results

Table 1 shows the performance of abbreviation expansion. Our approach significantly outperforms the baseline methods and achieves 82.27% accuracy.

Approaches	Accuracy
Rating	21.32%
Raw input embeddings	26.45%
General embeddings	28.06%
Our Approach	**82.27%**

Table 1: Overall performance

Figure 3 shows how our approach improves the performance of a rating-based approach. By using embeddings, we can learn that the meaning of "OD" used in our test cases should be "overdose" rather than "out-of-date" and this semantic information largely benefits the abbreviation expansion model.

- 'OD'- rating-based: ['out-of-date', 'other diseases', 'on duty', 'once daily', 'optometry degree', 'organ donation', 'overdose', 'optic disc' ... etc.]
- 'OD'- our approach: ['overdose', 'osteochondritis dissecans', 'optic disc' ... ... etc.]

Figure 3: Ranking lists of expansions of "OD" by the rating-based method, our approach

Compared with our approach, embeddings trained only from the ICU texts do not significantly contribute to the performance over the rating baseline. The reason is that the size of data for training the embeddings is so small that many candidate expansions of abbreviations do not appear in the corpus, which results in poor performance. It is notable that general embeddings trained from large biomedical data are not effective for this task because many abbreviations within critical care medicine appear in the biomedical corpus with different senses.

- Output of general Embeddings on abbreviation 'OD': ['O.D.', 'optical density', 'OD450', 'O.D', 'OD570', 'absorbance', 'OD490', '600nm' ... etc.]

Figure 4: The output of general embeddings trained on large biomedical texts

For example, "OD" in intensive care medicine texts refers to "overdose" while in the PubMed corpus it usually refers to "optical density", as shown in Figure 4. Therefore, the embeddings trained from the PubMed corpus do not benefit the expansion of abbreviations in the ICU texts.

Moreover, we estimated human performance for this task, shown in Table 2. Note that the performance is estimated by one of the authors Dr. Mathews who is a board-certified pulmonologist and critical care medicine specialist based on her experience and the human's performance estimated in Table 2 is under the condition that the participants can not use any other external resources. We can see that our approach can achieve a performance close to domain experts and thus it is promising to tackle this challenge.

Groups	Accuracy
General readers	<40%
Nurses	40%
Mid-level provider (nurse practitioner or physician associate)	70%
General practicing physician	80%
Domain experts with additional training in Emergency Medicine or Critical Care Medicine	**>90%**

Table 2: Estimated human performance for abbreviation expansion

4.4 Error Analysis

The distribution of errors is shown in Table 3. There are mainly three reasons that cause the incorrect expansion. In some cases, some certain abbreviations do not exist in the knowledge base. In this case we would not be able to populate the corresponding candidate list. Secondly, in many cases although we have the correct expansion in the candidate list, it's not ranked as the top one due to the lower semantic similarity because there are not enough samples in the training data. Among all the incorrect expansions in our test set, such kind of errors accounted for about 54%. One possible solution may be adding more effective data to the embedding training, which means discovering more task-oriented resources. In a few cases, we failed to identify some abbreviations because of their complicated representations. For example, we have the following sentence in the patient's notes: " No n/v/f/c." and the correct expansion should be "No nausea/vomiting/fever/chills." Such abbreviations are by far the most difficult to expand in our task because they do not exist in any knowledge base and usually only occur once in the training data.

Type of error	Percentage
Out of Vocabulary	27%
Lack of training samples	54%
Unidentified representation	19%

Table 3: Error distribution

5 Conclusions and Future Work

This paper proposes a simple but novel approach for automatic expansion of abbreviations. It achieves very good performance without any manually labeled data. Experiments demonstrate that using task-oriented resources to train word embeddings is much more effective than using general or arbitrary corpus.

In the future, we plan to collectively expand semantically related abbreviations co-occurring in a sentence. In addition, we expect to integrate our work into a natural language processing system for processing the clinical notes for discovering knowledge, which will largely benefit the medical research.

Acknowledgements

This work is supported by RPI's Tetherless World Constellation, IARPA FUSE Numbers D11PC20154 and J71493 and DARPA DEFT No. FA8750-13-2-0041. Dr. Mathews' effort is supported by Award #1K12HL109005-01 from the National Heart, Lung, and Blood Institute (NHLBI). The content is solely the responsibility of the authors and does not necessarily represent the official views of NHLBI, the National Institutes of Health, IARPA, or DARPA.

References

Sylvain Gaudan, Harald Kirsch, and Dietrich Rebholz-Schuhmann. 2005. Resolving abbreviations to their senses in medline. *Bioinformatics*, 21(18):3658–3664.

Zellig S Harris. 1954. Distributional structure. *Word*.

Hongfang Liu, Virginia Teller, and Carol Friedman. 2004. A multi-aspect comparison study of supervised word sense disambiguation. *Journal of the American Medical Informatics Association*, 11(4):320–331.

Tomas Mikolov, Ilya Sutskever, Kai Chen, Greg S Corrado, and Jeff Dean. 2013. Distributed representations of words and phrases and their compositionality. In *Advances in Neural Information Processing Systems*, pages 3111–3119.

Sampo Pyysalo, Filip Ginter, Hans Moen, Tapio Salakoski, and Sophia Ananiadou. 2013. Distributional semantics resources for biomedical text processing. *Proceedings of Languages in Biology and Medicine*.

Richard Socher, Danqi Chen, Christopher D Manning, and Andrew Ng. 2013. Reasoning with neural tensor networks for knowledge base completion. In *Advances in Neural Information Processing Systems*, pages 926–934.

Mark Stevenson, Yikun Guo, Abdulaziz Al Amri, and Robert Gaizauskas. 2009. Disambiguation of biomedical abbreviations. In *Proceedings of the Workshop on Current Trends in Biomedical Natural Language Processing*, pages 71–79. Association for Computational Linguistics.

Lisa Tengstrand, Beáta Megyesi, Aron Henriksson, Martin Duneld, and Maria Kvist. 2014. Eacl-expansion of abbreviations in clinical text. In *Proceedings of the 3rd Workshop on Predicting and Improving Text Readability for Target Reader Populations (PITR)@ EACL*, pages 94–103.

Irfan Ucgun, Muzaffer Metintas, Hale Moral, Fusun Alatas, Huseyin Yildirim, and Sinan Erginel. 2006. Predictors of hospital outcome and intubation in copd patients admitted to the respiratory icu for acute hypercapnic respiratory failure. *Respiratory medicine*, 100(1):66–74.

Hong Yu, Won Kim, Vasileios Hatzivassiloglou, and John Wilbur. 2006. A large scale, corpus-based approach for automatically disambiguating biomedical abbreviations. *ACM Transactions on Information Systems (TOIS)*, 24(3):380–404.

Semantic Type Classification of Common Words in Biomedical Noun Phrases

Amy Siu
Max Planck Institute for Informatics
66123 Saarbrücken, Germany
siu@mpi-inf.mpg.de

Gerhard Weikum
Max Planck Institute for Informatics
66123 Saarbrücken, Germany
weikum@mpi-inf.mpg.de

Abstract

Complex noun phrases are pervasive in biomedical texts, but are largely underexplored in entity discovery and information extraction. Such expressions often contain a mix of highly specific names (diseases, drugs, etc.) and common words such as "condition", "degree", "process", etc. These words can have different semantic types depending on their context in noun phrases. In this paper, we address the task of classifying these common words onto fine-grained semantic types: for instance, "condition" can be typed as "symptom and finding" or "configuration and setting". For information extraction tasks, it is crucial to consider common nouns only when they really carry biomedical meaning; hence the classifier must also detect the negative case when nouns are merely used in a generic, uninformative sense. Our solution harnesses a small number of labeled seeds and employs label propagation, a semisupervised learning method on graphs. Experiments on 50 frequent nouns show that our method computes semantic labels with a micro-averaged accuracy of 91.34%.

1 Introduction

1.1 Motivation

In biomedical texts, entities are written as natural language expressions – often complex noun phrases. Previous works on information extraction in this domain have focused on short phrases that work well, for instance, with dictionary-based approaches. The most notable method is the MetaMap tool by Aronson and Lang (2010). Often, however, expressions are long and complex, mixing domain-specific names (of diseases, symptoms, drugs, etc.) with common nouns such as "condition", "degree" or "process". Examples for such complex phrases are:

1) monitoring of the carcinogenic process
2) development of processes for the prognosis of malaria.

In the first example, "process" is a vital part of the phrase and carries biomedical meaning, namely, denoting a body function. In the second example, "process" is used in the generic sense of the common noun and is relatively uninformative for the purpose of detecting biomedical entities in text. For information extraction tasks like entity discovery, relation mining and knowledge base population, it is crucial to distinguish these two situations. Moreover, in the first case, we would like to further annotate the common noun with a semantic type that captures the role of the word within the surrounding noun phrase.

This kind of semantic typing could be based on WordNet senses (Fellbaum, 1998), using techniques for word sense disambiguation (Navigli, 2009), or on UMLS entries. However, WordNet has limited coverage of the biomedical domain, and UMLS has rather coarse-grained and sometimes fuzzy types. Therefore, we devised a small collection of *fine-grained semantic types* ourselves. The novelty of our proposed semantic types lies in the explicit provision for non-biomedical types, as well as the uninformative type where applicable; Table 1 shows both of these elements in play for the target words *culture* and *degree*.

Our goal then is to automatically label common words in complex noun phrases with the most appropriate semantic type or inferring that the word is merely used in a generic sense without specific biomedical meaning. We focus on a judiciously chosen list of common nouns, referred to as *target words*, that frequently appear within long noun phrases in biomedical texts. The resulting annota-

Proceedings of the 2015 Workshop on Biomedical Natural Language Processing (BioNLP 2015), pages 98–103,
Beijing, China, July 30, 2015.

Target word	Semantic types
culture	medical sample social construct
degree	metric for temperature metric for bending stage in progression (e.g. second degree burn) academic degree degree of freedom in statistics edges out of a node in a graph generic, uninformative

Table 1: Semantic types for the target words *culture* and *degree*.

tions – for example, labeling "process" in "monitoring of the carcinogenic process" as body function – can in turn enhance the coverage and quality of information extraction tasks.

1.2 Approach and contribution

We develop a semisupervised method for labeling a target word, within a given noun phrase, with its most suitable semantic type or tagging it as biomedically unspecific and uninformative. Our method is based on label propagation over a graph that connects noun phrases and has a small number of manually labeled seed nodes. Each distinct noun phrase becomes a node, and an edge connects two nodes that share a target word with a weight reflecting the similarity between the contexts of the respective phrase occurrences. We then apply the MAD label propagation algorithm (Talukdar and Crammer, 2009) to infer the best type labels for the target words in the graph's nodes.

Experiments show that our method achieves 91.34% micro-averaged and 83.57% macro-averaged accuracy over 50 frequently appearing target words. Moreover, our method is capable of classifying both target words with and without an uninformative semantic type.

2 Related work

In general, the semantic interpretation of complex phrases is a long-studied problem in computational linguistics, and widely viewed as a very demanding task (see, e.g., Sag et al. (2002); Nakov and Hearst (2013)). For biomedical texts, however, complex phrases are an infrequently studied problem. Golik et al. (2013) propose to handcraft rules based on linguistic cues to identify longer noun phrases beyond dictionary entries. Similar to this paper, they are also motivated by the needs of a knowledge acquisition application. Their work makes a point in analyzing "semantically poor" terms, some of which essentially entail the uninformative semantic type we propose.

The problem setting closest to word usage detection is undoubtedly word sense disambiguation (WSD) of free text. For the general domain, the vast body of work has been surveyed by Navigli (2009), and mature software tools such as It Makes Sense (Zhong and Ng, 2010) covers most words. For the biomedical domain, the majority of previous works center around two WSD datasets (Weeber et al., 2001; Jimeno-Yepes et al., 2011) that together contain 253 ambiguous words, multi-word terms, and abbreviations. In addition, Stevenson et al. (2008), Fan et al. (2009), and Cheng et al. (2012) propose methods to generate labeled data. As for methodologies, vector space models (McInnes, 2008; Savova et al., 2008) are a common choice. Another common approach is to exploit the rich knowledge embedded in UMLS. Agirre et al. (2010) and Humphrey et al. (2006) leverage entity-entity relations and semantic type information in UMLS, respectively.

Entity disambiguation is another highly relevant research area. For the general domain, most efforts focus on named entities, and software systems such as AIDA (Hoffart et al., 2011) and Wikifier (Ratinov et al., 2011) are both robust and scalable. In contrast, for the biomedical domain, existing works target restricted scopes such as species (Wang et al., 2010) and acronyms (Harmston et al., 2012). Although MetaMap (Aronson and Lang, 2010) covers all the diverse entities in UMLS, its entity disambiguation functionality remains limited.

3 Methodology

3.1 Outline of methodology

Our method operates on one target word at a time. We collect noun phrases in our text corpus that contain the selected target word. On the one hand comes the manual preparation of the target semantic types and their seed phrases. On the other hand comes the automatic computation of similarities of noun phrase pairs. This similarity is based on *context* – a window of k words before and after the target word in a noun phrase (for clarity purposes, we denote by *context words* those words in the window surrounding the target). This context, in turn, is captured by three features, namely word occurrences, part-of-speech tags, and entity types (again for clarity purposes, we distinguish *context entity types* that are precomputed, from target semantic types that we want to classify). Using the seed phrases and context similarities, we cast the the noun phrases into a graph and apply the MAD label propagation algorithm.

In the following subsections, we describe how we construct each component.

3.2 Target semantic types

In our corpus, we observe that 90% of all noun occurrences come from 5000+ distinct nouns. Since it is infeasible to study so many of them, we pick 50 highly common but semantically ambiguous ones to be our target words. For each target word, we manually specify its applicable target semantic types based on two criteria. First, a target semantic type should have a discernible presence in the corpus. Second, the contexts of target semantic types should be amenable to a learning algorithm, i.e. they should be sufficiently distinct from each other. Recall that we would also like to identify the case when the target word is used in a generic, uninformative way. We facilitate this by adding a uninformative semantic type. We observe, however, that not all target words require this uninformative type. For instance, *culture* has two overwhelmingly dominant types (medical sample and social construct) such that the rest are negligible and do not need an explicit representation. This specification of target semantic types is based on manual observation, over both the corpus noun phrases and UMLS entries relevant to the target word.

Once the semantic types are set, we nominate a few representative phrases as seed phrases. This process is again manual, where we aim for phrases which are sufficiently prevalent, and which convey the target semantic type with high certainty. Table 2 shows all semantic types and all the seed phrases for the target word *activity*, and the complete list is available at `http://mpi-inf.mpg.de/~siu/bionlp2015/`. In our compilation, one target word has on average 3.78 target semantic types, which in turn has on average 2.68 seed phrases.

3.3 Context entity type estimation

We would like to assign an entity type to each context word. However, since a comprehensive entity disambiguation tool is not available, we estimate the entity types by a popularity-based approach that exploits the repetitiveness of thesauri entries and semantic assets in UMLS. First, take note of UMLS entity names that contain a single word. Next, for each distinct entity name, take note of the entities (distinct CUIs), as well as the number of occurrences (MRCONSO entries) represented. A few heuristics determine which entity is the most popular, and the corresponding CUI's UMLS semantic type[1] becomes the word's entity type. Taking *cat* as an example, it appears 16 times as a mammal, 3 times as the abbreviation for CAT scan, and 1 time as an enzyme. Therefore *cat*'s entity type is *Mammal*, the UMLS semantic type for CUI 0007450. In essence, this approach approximates the entity type with the largest prior distribution probability. Since biomedical word senses are often highly skewed (Jimeno-Yepes et al., 2011), we believe this approach is a reasonable interim substitute to a full-fledged entity disambiguation tool.

In addition to the 133 UMLS semantic types, we introduce an extra type to represent measurement units such as mg/kg and μmol.

We investigate two variants of entity type similarity. Under the hard variant, only the same entity type occurrences contribute towards context similarity (e.g. *Cell* and *Cell Component* would therefore be considered completely dissimilar). Under the soft variant, similar entity types also contribute (*Cell* and *Cell Component* now have a similarity of 0.9375). The similarity between two entity types A and B is:

$$0.5 \times group(A, B) + 0.5 \times lch(A, B)$$

where $group()$ returns 1 if A and B belong to the same UMLS semantic group, and 0 otherwise. $lch(A, B)$ is the similarity score between A, B in the UMLS semantic type hierarchy according to Leacock and Chodorow's method (1998), normalized to range between 0 and 1. The use of $group()$ is necessary because some semantic type pairs are highly similar but far apart in the hierarchy (e.g. *Body System* and *Tissue*).

3.4 Context similarity

We model the similarity between two phrases by calculating a similarity score between their contexts. Specifically, the similarity score is a linear combination of the contributions from the contexts' words, part-of-speech (POS) tags, and entity

[1]Not to be confused with the custom target semantic types in Section 3.2. They are used independently in this work.

Semantic type	Seed phrases	Sample classified noun phrases
physical activity	fetal activity	instruction in self-directed exercises and activity diaries
	physical activity	day-to-day household activities that create the backbone of healthy environments
body & protein process	catalytic activity	histochemically demonstrable esterase activity in the hypothalamus of the developing rat
	disease activity	lower insulin-stimulated GS activity in PCOS patients compared with controls
	inflammatory activity	plasma anti-pneumococcal polysaccharide antibody activity (serotypes 3, 6a and 23)
	kinase activity	polymerase activity relative to the wild-type protein
generic, uninformative	of activity of	dual activity of exploring karanjin isolation for medicinal purposes
	of activity in	the orchestration of a set of activities that should be executed in order to deliver an output

Table 2: Semantic types, seed phrases, and sample classified noun phrases for the target word *activity*.

types (either the hard or the soft variant):

$$sim(\text{context}_1, \text{context}_2) =$$
$$\alpha_1 \times J_w(\text{words}_1, \text{words}_2)$$
$$+ \alpha_2 \times J_w(\text{POS tags}_1, \text{POS tags}_2)$$
$$+ \alpha_3 \times J_w(\text{entity types}_1, \text{entity types}_2)$$

where $\alpha_1 + \alpha_2 + \alpha_3 = 1$, and $J_w()$ is the weighted Jaccard similarity function. Intuitively, $J_w()$ captures not only the overlap between two sets of items, but also the significance or weight of the items. In our setting, an item is a word, a POS tag, or a context entity type, and the weight depends on the item's distance to the target word – the smaller the distance, the higher the weight. Based on preliminary experiments, $1/d$ is found to be the best weighting scheme, where d is the distance between target and context words. For word, POS, and the hard variant of context entity type, only exact matches count towards $J_w()$ item overlap (singular/plural and American/British spellings of the same word qualify as exact matches). For the soft variant of context entity type, the $1/d$ weight is further scaled by the entity type-entity type similarity score.

3.5 MAD label propagation

Now we have all the ingredients to build a graph out of a collection of noun phrases. Take a phrase as a node. Compute the similarity score between two phrases' contexts, and make it the weight of the edge between the two corresponding nodes. A small number of nodes containing seed phrases become the seed nodes, and the seed phrase's semantic type is the label. Apply the MAD label propagation algorithm (Talukdar and Crammer, 2009) to label all the nodes, effectively classifying each node with the best target semantic type. Recall that each target word requires its own graph and hence separate application of MAD.

Label propagation, also known as belief propagation, is a semisupervised, iterative learning method on graphs. Some nodes, i.e. the seed nodes, in the graph are initially labeled. Informally, over the iterations, the seed nodes exert influence on their neighbors, whom in turn influence their neighbors, such that eventually all nodes become labeled. MAD is a state-of-the-art variant of the standard label propagation algorithm (Baluja et al., 2009), and it guarantees convergence. Based on preliminary experiments, $\mu_1 = 10 \times \mu_2 = 100 \times \mu_3$ were found to be the best hyperparameters for MAD. Since a graph with n nodes contains $O(n^2)$ edges, we prune low-weight edges to avoid excessively time consuming computations.

4 Results and discussion

4.1 The dataset

Our corpus consists of documents from a diverse set of biomedical free texts: PubMed abstracts and full-length articles, encyclopedic webpages from health portals, and online discussion forums. As a pre-processing step, each document is segmented into sentences by the LingPipe tool, and further tagged with POS and parsed into dependency graphs by the Stanford CoreNLP tool. We then extract the longest compound noun phrases from the sentences. Finally, for each target word, we make one collection by randomly selecting noun phrases containing that word. The average noun phrase length across collections are relatively uniform from 13 to 17 words.

4.2 Results

We tuned the method's parameters using a development dataset of 1,000 randomly selected nodes for each target word. Keeping the proportion of seed nodes at 5%, we obtained the best parameter setting (the α's in context similarity and window size k) for each individual word.

In the test dataset, each target word has a graph of 10,000 random nodes with also 5% seeds. On

Target word	#Types	Micro	Macro	Best context	Target word	#Types	Micro	Macro	Best context	Target word	#Types	Micro	Macro	Best context
activity	3	0.91	0.91	WPH	function	3	0.94	0.94	WPS	reaction	5	0.97	0.94	WP
administration	2	0.93	0.84	WPS	group	3	0.92	0.74	WPS	reduction	3	0.72	0.75	WPS
area	6	0.92	0.89	WP	information	4	0.95	0.95	WPH	region	4	0.90	0.50	WPS
body	4	0.96	0.94	WPH	line	5	0.89	0.85	WPS	report	2	0.99	0.97	WPH
case	5	0.83	0.88	WPS	measure	2	0.90	0.80	WPS	resistance	3	0.98	0.66	WPS
concentration	4	0.95	0.98	WPH	mechanism	2	0.85	0.76	WPS	response	5	0.89	0.73	WPS
condition	2	0.95	0.96	WPH	model	3	0.96	0.63	WPS	result	4	0.91	0.89	WPH
control	4	0.98	0.97	WPS	pattern	6	0.77	0.81	WP	role	3	0.98	0.99	WPH
culture	2	0.99	0.79	WP	period	3	0.91	0.92	WPS	sequence	2	0.97	0.95	WPS
degree	7	0.76	0.72	WP	point	8	0.92	0.76	WP	set	2	0.98	0.97	WPS
development	5	0.88	0.86	WP	pressure	6	0.79	0.89	WP	site	4	0.96	0.85	WPH
distribution	2	0.96	0.96	WPS	problem	4	0.89	0.67	WP	solution	2	0.99	0.94	WPS
effect	2	0.93	0.75	WPS	process	4	0.85	0.91	WPH	state	4	0.98	0.82	WP
expression	4	0.96	0.81	WPH	product	6	0.95	0.91	WP	strain	3	0.66	0.59	WPS
factor	6	0.96	0.72	WP	profile	3	0.98	0.84	WP	system	4	0.92	0.85	WPS
flow	5	0.83	0.90	WPH	program	5	0.92	0.85	WPH	technique	2	0.91	0.92	WPS
form	4	0.92	0.63	WPS	rate	3	0.95	0.78	WP					

Table 3: Number of semantic types, micro- and macro-averaged accuracy, and the best context setting of 50 target words. W, P, H, S denote word, POS, hard and soft context entity types, respectively.

average, 1428 and 437 nodes were evaluated for each target word and for each target semantic type, respectively. Two annotators evaluated the labels suggested by the MAD algorithm, and the value of Fleiss' Kappa was 0.76, which indicates substantial inter-annotator agreement. Table 3 lists the micro- and macro-averaged accuracy, as well as the best context setting.

4.3 Discussion

Overall, micro-averaged accuracy is very encouraging at 80% or above for 45 target words. A few target words (*degree*, *pattern*, and *pressure*) have higher numbers (6 or 7) of target semantic types. As the number of target semantic types increases for one target word, it becomes harder for the types' contexts to be sufficiently distinct from each other. This phenomenon leads to noisy edge weights in the graph, which in turn leads to poorer classification results. Other target words (*reduction* and *strain*) also have week micro-averaged accuracy despite having fewer (3) target semantic types. In both cases here, the dominant target semantic type is used in such a broad way that a few seed phrases are not sufficient to describe the context. Specifically, a reduction of quantity can be about just anything; and an organism strain can be described at the population, experiment, organism, gene, or molecular level, or can be described via the characteristic effect the strain causes.

Macro-averaged accuracy performs less consistently and varies across target words. The overriding contributing factor here is the skew of the target semantic types' distribution. In our annotations, the most frequent label of one target word constitutes from 23% to 91% of occurrences. When a sparse type is represented by few labeled examples in the graph, naturally there is less generalization power to classify correctly.

In terms of how much context words, POS, and context entity types contribute towards the solution, we are surprised that the use of words and POS alone are sufficient for 28% of the target words to achieve the best experimental setting. While the rest of the target words benefit to varying degrees the hard and soft variants of context entity types, it is worth noting that even a rudimentary estimation of context entity types empowers better context comparisons for the other 72% of target words.

Errors in the classification stem from two main sources. In some cases, the critical cue, be it a word or a context entity type, lies outside of the context window. In other cases, significant expert knowledge is needed to put the puzzle together.

5 Conclusion

In this work, we present a semisupervised method that classifies a word's semantic type in complex noun phrases. With 50 common words, we demonstrate that a small number of labeled seeds can enable a label propagation algorithm to assign both conventional semantic type labels as well as the negative case of uninformative label. We envision that the semantic types of words in a noun phrase make one building block towards more fully utilizing that phrase. In the future, we plan to apply our method to other information extraction modules, and enrich their capability in handling longer phrases that go beyond dictionary entries.

References

Eneko Agirre, Aitor Soroa, and Mark Stevenson. 2010. Graph-based word sense disambiguation of biomedical documents. *Bioinformatics*, 26(22): 2889–2896.

Alan R. Aronson and François-Michel Lang. 2010. An overview of MetaMap: historical perspective and recent advances. *Journal of the American Medical Informatics Association*, 17(3): 229–236.

Shumeet Baluja, Rohan Seth, D. Sivakumar, Yushi Jing, Jay Yagnik, Shankar Kumar, Deepak Ravichandran, and Mohamed Aly. 2008. Video suggestion and discovery for YouTube: taking random walks through the view graph. *Proceedings of WWW*, pp. 895–904.

Weiwei Cheng, Judita Preiss, and Mark Stevenson. 2012. Scaling up WSD with automatically generated examples. *Proceedings of BioNLP*, pp. 231–239.

Jung-Wei Fan and Carol Friedman. 2009. Generating quality word sense disambiguation test sets based on MeSH indexing. *Proceedings of the AMIA Symposium*, pp. 183–187.

Christiane Fellbaum. 1998. WordNet: an electronic lexical database. *MIT Press*.

Wiktoria Golik, Robert Bossy, Zorana Ratkovic, and Claire Nédellec. 2013. Improving term extraction with linguistic analysis in the biomedical domain. *Proceedings of CICLing*, pp. 24–30.

Nathan Harmston, Wendy Filsell, and Michael Stumpf. 2012. Which species is it? Species-driven gene name disambiguation using random walks over a mixture of adjacency matrices. *Bioinformatics*, 28(2): 254–260.

Johannes Hoffart, Mohamed Amir Yosef, Ilaria Bordino, Hagen Fürstenau, Manfred Pinkal, Marc Spaniol, Bilyana Taneva, Stefan Thater, and Gerhard Weikum. 2011. Robust disambiguation of named entities in text. *Proceedings of EMNLP*, pp. 782–792.

Susanne M. Humphrey, Willie J. Rogers, Halil Kilicoglu, Dina Demner-Fushman, and Thomas C. Rindflesch. 2006. Word sense disambiguation by selecting the best semantic type based on Journal Descriptor Indexing: preliminary experiment. *Journal of the American Society for Information Science and Technology*, 57(1): 96–113.

Antonio J. Jimeno-Yepes, Bridget T. McInnes, and Alan R. Aronson. 2011. Exploiting MeSH indexing in MEDLINE to generate a data set for word sense disambiguation. *BMC Bioinformatics*, 12:223.

Claudia Leacock and Martin Chodorow. 1998. Combining local context and WordNet similarity for word sense identification. *WordNet: an electronic lexical database*, 49(2): 265–283.

Bridget T. McInnes. 2008. An unsupervised vector approach to biomedical term disambiguation: integrating UMLS and Medline. *Proceedings of ACL-HLT-SRWS*, pp. 49–54.

Preslav Nakov and Marti A. Hearst. 2013. Semantic interpretation of noun compounds using verbal and other paraphrases. *ACM Transactions on Speech and Language Processing*, 10(3): 13:1–13:51.

Roberto Navigli. 2009. Word sense disambiguation: a survey. *ACM Computing Surveys*, 41(2): 10:1–10:69.

Lev Ratinov, Dan Roth, Doug Downey, and Mike Anderson. 2011. Local and global algorithms for disambiguation to Wikipedia. *Proceedings of ACL-HLT*, pp. 1375–1384.

Ivan A. Sag, Timothy Baldwin, Francis Bond, Ann A. Copestake, and Dan Flickinger. 2002. Word sense disambiguation across two domains: biomedical literature and clinical notes. *Journal of Biomedical Informatics*, 41(6): 1088–1100.

Guergana K. Savova, Anni R. Coden, Igor L. Sominsky, Rie Johnson, Philip V. Ogren, Piet C. de Groen, and Christopher G. Chute. 2008. Multiword expressions: a pain in the neck for NLP. *Proceedings of CICLING*, pp.1–15.

Mark Stevenson, Yikun Guo, and Robert Gaizauskas. 2008. Acquiring sense tagged examples using relevance feedback. *Proceedings of COLING*, pp. 809–816.

Partha P. Talukdar and Koby Crammer. 2009. New regularized algorithms for transductive learning. *Proceedings of ECML PKDD*, part II, pp. 442–457.

Xinglong Wang, Junichi Tsujii and Sophia Ananiadou. 2010. Disambiguating the species of biomedical named entities using natural language parsers. *Bioinformatics*, 26(5): 661–667.

Marc Weeber, James G. Mork, and Alan R. Aronson. 2001. Developing a test collection for biomedical word sense disambiguation. *Proceedings of the AMIA Symposium*, pp. 746–750.

Zhi Zhong and Hwee Tou Ng. 2010. It Makes Sense: a wide-coverage word sense disambiguation system for free text. *Proceedings of ACL*, pp. 78–83.

CoMAGD: Annotation of Gene-Depression Relations

Rize Jin[1] **Jinseon You**[1] **Jin-Woo Chung**[1] **Hee-Jin Lee**[1]
Maria Wolters[2] **Jong C. Park**[1*]

[1]School of Computing
Korea Advanced Institute of Science and Technology
291 Daehak-ro, Daejeon, Republic of Korea
{rizejin, jsyou, jwchung, heejin, park}@nlp.kaist.ac.kr

[2]School of Informatics
University of Edinburgh
Edinburgh, UK
maria.wolters@ed.ac.uk

Abstract

Clinical depression is a mental disorder involving genetics and environmental factors. Although much work studied its genetic causes and numerous candidate genes have consequently been looked into and reported in the biomedical literature, no gene expression changes or mutations regarding depression have yet been adequately collected and analyzed for its full pathophysiology. In this paper, we present a depression-specific annotated corpus for text mining systems that target at providing a concise review of depression-gene relations, as well as capturing complex biological events such as gene expression changes. We describe the annotation scheme and the conducted annotation procedure in detail. We discuss issues regarding proper recognition of depression terms and entity interactions for future approaches to the task. The corpus is available at http://www.biopathway.org/CoMAGD.

1 Introduction

Clinical depression, or major depressive disorder, is a mental disorder of the central nervous system with a pathophysiology involving the neocortex. Genetics and environmental factors are known to contribute to the development of mood disorders (Nestler et al., 2002). Many biomedical research efforts studied the causative factors of genetics in depression, with consequent rapid accumulation of candidate genes (Kao et al., 2011; Piñero et al., 2015). However, the accumulated information is not yet comprehensive enough to explain the role of genes involved in depression.

DisGeNET (Piñero et al., 2015) is a platform for discovering associations of genes and complex diseases including depression, defining gene-depression relations as simple binary relations that consist of *geneId*, *geneSymbol*, *geneName*, *diseaseId*, *diseaseName*, and *score*, where the score is a measure of relevancy based on the supporting evidence. DEPgenes (Kao et al., 2011) gives a prioritizing system that uses combined score to rank candidate genes for depression. Although DEPgenes is a nearly comprehensive candidate gene resource for depression in terms of its volume (5,055 candidate genes), its representation concepts are even simpler than DisGeNET and thus not quite adequate for the full understanding of depression-related phenomena.

In order to fully understand how a particular gene acts in depression, we need detailed information about gene expression changes or mutations and also how the depression level is changed along with the change in the gene. In this regard, we anticipate that text mining systems, which can identify and analyze both genes and depression changes comprehensively from text, would facilitate research on depression much further. Furthermore, if the mined information is annotated and then made available for reuse, key resources would be identified and constructed more effectively (McDonald and Kelly, 2012; Winnenburg

[*] Corresponding author

Proceedings of the 2015 Workshop on Biomedical Natural Language Processing (BioNLP 2015), pages 104–113,
Beijing, China, July 30, 2015.

et al., 2008). Such effort of making relevant corpora has already been made in the studies of genes (Kim et al., 2008; Poux et al., 2014) and of complex diseases such as cancers (Lee et al., 2013; Lee et al., 2014; Pyysalo et al., 2013), but has not yet been applied to depression.

In this paper, we present a depression-specific annotated corpus, CoMAGD, for future text mining systems that target specifically at providing comprehensive information of depression-gene relations as well as capturing complex information such as gene changes and biological events. For this purpose, we follow a multi-faceted annotation scheme for cancers (Lee et al., 2013) while tuning it extensively to depression. In this revised scheme, a piece of annotation is composed of four concepts that together express two events, *gene expression changes* and *depression level or antidepressant effect changes*, and the relationship between these two events. We anticipate that the present corpus and text mined results based on this corpus would contribute meaningfully to the successful exploration of the underlying functional correlation between genes and clinical depression.

The rest of the paper is organized as follows. Section 2 shows the corpus annotation. Section 3 gives details of inter-annotator agreement. Section 4 discusses issues about proper recognition of depression terms and entity interactions for future approaches to the task, before closing the paper in Section 5.

2 Corpus Annotation

2.1 Data collection and pre-processing

We collected PubMed IDs (PMIDs) that contain depression related terms in any of the three fields *title*, *abstract*, and *keyword*, using the query "*depress** OR *dysthymia* OR *cyclothymia*", and randomly selected 500 abstracts among them. The 500 abstracts were then segmented into sentences.

We extracted only the sentences that contain at least one pair of gene and depression/antidepressant related terms. BANNER (Leaman and Gonzalez, 2008) and Moara (Neves et al., 2010) were used to identify and normalize gene names. For depression and antidepressant terms, the system used dictionary-based longest matching. The dictionary consists of 303 entries of depression and antidepressant related terms collected from NCI Thesaurus and other relevant articles. The entries were then edited by a domain expert in mental health.

For the sentences that contain more than one pair, we made their copies, matching the number of depression-gene pairs. We call each of these copies a *co-occurrence*. For example, if there are three gene names and two depression related terms in a sentence, the system makes six co-occurrences for this sentence.

We then tokenized, part-of-speech tagged, and parsed the co-occurrences, using the Charniak-Johnson parser (Charniak and Johnson, 2005) with a biomedical parsing model (McClosky, 2010). The resulting phrase structures were then converted into dependency structures with the Stanford conversion tool (Marneffe et al., 2006). We identified mentions of gene expression changes, using the Turku event extraction system (Björne et al., 2009). Most of the processes above are included in a preprocessed dataset, or EVEX (Landeghem et al., 2012); however, we modified the system and utilized some part of the system separately where necessary.

Finally, we performed manual work to validate automatically identified co-occurrences in order to produce confirmed annotation units, such as manually constructing predicates (i.e., 'depression of [non-human subjects]') to filter out false positives from the dictionary matching outputs of depression-related terms and manually eliminating false relations (hypothesis sentences).

2.2 A multi-faceted annotation scheme

We modify a multi-faceted annotation scheme of (Lee et al., 2013), originally designed to represent ternary relations among genes, cancers and gene changes, in order to address relations not only between depression and genes, but also between antidepressants and genes, so as to provide more details and enable further insights for follow-up studies such as prioritizing depression candidate genes and designing effective treatments and therapy. For example, one may assign a lower weight to a gene if the gene shows expression changes only in antidepressant studies. We also introduce directed causal relations between genes and depression/antidepressants. Identification of the cause and effect not only reflects the methodologies of individual studies, but also provides the facts. While the undirected causality claim usually is interpreted as a necessary and sufficient clause, we find that it could result in false conclusions,

Concept	Value	Definition
Change in Gene Expression (CGE)	increased	Expression level of the gene is increased
	decreased	Expression level of the gene is decreased
Change in Depression Level (CDL) or Change in Antidepressant Effect (CAE)	increased	The depression level/antidepressant effect is increased as CGE
	decreased	The depression level/antidepressant effect is decreased as CGE
	unidentifiable	The information about whether or not CGE accompanies the depression level/antidepressant effect change is not provided
Causality Claim (CC)	none	CGE accompanied by CDL/CAE is reported but the causality between the two is not claimed
	g2x	The causality is claimed as CGE causes CDL/CAE
	x2g	The causality is claimed as CDL/CAE causes CGE

Table 1: Annotation concept values and their definitions

Sentence	CGE	CDL	CC
Example 1. In particular, we found decreased NF-L, PSD95, and SAP102 transcripts in bipolar disorder, and [*decreased*]$_e$ [*SAP102*]$_g$ levels in [*major depression*]$_d$. [PMID: 15054476]	dec.	uni.	non.
Example 2. In conclusion, chronic forced swim stress was a good animal model of [*depression*]$_d$, and it induced depressive-like behavior and [*decreased*]$_e$ [*P-Erk2*]$_g$ in the hippocampus and prefrontal cortex in rats. [PMID: 17050000]	dec.	inc.	x2g
Sentence	**CGE**	**CAE**	**CC**
Example 3. [*Fluoxetine*]$_a$ substantially [*inhibits*]$_e$ [*CYP2D6*]$_g$ and probably CYP2C9/10, moderately inhibits CYP2C19 and mildly inhibits CYP3A3/4. [PMID: 9068931]	dec.	uni.	x2g
Example 4. [*Inhibition*]$_e$ of [*neuronal nitric oxide synthase*]$_g$ in the rat hippocampus induces [*antidepressant-like*]$_a$ effects. [PMID: 9068931]	dec.	inc.	g2x

Gene names, depression related terms, antidepressant related terms, and the keywords for gene expression change are noted in matching square brackets and marked with subscripts 'g', 'd', 'a', and 'e', respectively.

Table 2: Examples of annotated co-occurrences

especially in the studies of depression. For example, depression may decrease the expression level of a particular gene; however, increasing the expression level of that gene may not necessarily reduce the symptom. One reason is that the genetic factor is not the only cause of depression. It is also believed that, compared to oncogenesis, much more genes act together and render a person to become vulnerable to depression (Belmaker and Agam, 2008). As such, a more fine-grained annotation of causal directions will be essential for more complex diseases such as depression. In an answer to these needs, we use a flexible schema for annotating concepts and ever-changing metrics and facts in genetic studies of depression. The flexibility would allow the schema to exploit the location information as well, as studies show that genes may respond differently to the same antidepressant if they are in different parts of a body. More details will be discussed in Section 4.

2.3 Annotation concept

The proposed corpus contains four core annotation concepts: *Change in Gene Expression* (CGE), *Change in Depression Level* (CDL), *Change in Antidepressant Effect* (CAE), and *Causality Claim* (CC). CGE captures whether the expression level of a gene is 'increased' or 'decreased'. CDL/CAE captures the way how the depression level/antidepressant effect changes together with a gene expression level change. If information about such changes is not provided in the sentence,

```
<?xml version="1.0" ?>
<!DOCTYPE gene_depression_corpus [
    <!ELEMENT    gene_depression_corpus (annotation_unit+)>
    <!ELEMENT    annotation_unit (sentence, annotation+)>
    <!ATTLIST    annotation_unit type (depression | antidepressant) #REQUIRED >
    <!ELEMENT    sentence (#PCDATA)>
    <!ATTLIST    sentence pmid CDATA #REQUIRED >
    <!ELEMENT    annotation (gene, expression_change_keyword_1,
                         expression_change_keyword_2, depression_term+, CGE, CDL, CC)>
    <!ATTLIST    annotation id CDATA #REQUIRED>
    <!ELEMENT    gene (#PCDATA)>
    <!ATTLIST    gene offset CDATA #REQUIRED >
    <!ELEMENT    expression_change_keyword_1 (#PCDATA)>
    <!ATTLIST    expression_change_keyword_1 offset CDATA #REQUIRED
                         type (Negative_regulation | Positive_regulation) #REQUIRED>
    <!ELEMENT    expression_change_keyword_2 (#PCDATA)>
    <!ATTLIST    expression_change_keyword_2 offset CDATA #REQUIRED
                         type (None | Gene_expression) #REQUIRED>
    <!ELEMENT    depression_term (#PCDATA)>
    <!ATTLIST    depression_term offset CDATA #REQUIRED>
    <!ELEMENT    CGE EMPTY>
    <!ATTLIST    CGE value (increased | decreased) #REQUIRED>
    <!ELEMENT    CDL EMPTY>
    <!ATTLIST    CDL value (increased | decreased | unidentifiable) #REQUIRED>
    <!ELEMENT    CC EMPTY>
    <!ATTLIST    CC value (x2g | g2x | none) #REQUIRED>
]>
```

Table 3: The XML DTD of the corpus

we assign 'unidentifiable'. CC captures whether the causality between the gene expression change and the CDL/CAE is claimed in the sentence or not, with values 'none', 'x2g', and 'g2x'. Each concept is assigned with one of the pre-specified values to complete a *facet* of annotation. Table 1 shows the pre-specified values and the definitions of the respective values. Three of the four concepts together complete a piece of annotation that express information about a gene's expression level change with a change in depression level or antidepressant effect.

Table 2 shows examples of the annotated sentences and Table 3 shows the DTD schema of the corpus. As mentioned earlier, we collected sentences from PubMed that describe gene expression changes in depression/antidepressants. Each sentence was presented to the annotators as one or more copies with markings for a gene term, keywords for gene expression change, and a depression/antidepressant-related term. The annotators read the sentence with such markings and selected proper values for the annotation concepts. Note that the four annotation concepts are semantically orthogonal, in that the value of a concept can be identified without knowing the values of the other concepts.

2.4 Corpus statistics

The corpus consists of 210 annotation units, where an annotation unit is simply a mention of gene expression change that co-occurs with at least one depression or antidepressant related term in a sentence. These annotation units are derived from 106 different sentences, which in turn are extracted from 73 PubMed abstracts. The corpus contains 82 gene types, 5 depression terms, and 20 antidepressant terms (cf. Table 4).

Tables 5 and 6 show the distribution of annotation concept values and the distribution of the annotated genes, respectively. The values of CGE show a uniform distribution, whereas the others show skewed distributions. In particular, for values of CDL/CAE, 'unidentifiable' is frequently chosen (89% for CDL, 87% for CAE). The value distribution of the concept CC associated with

CAE also exhibits dominance of a single value, or 'x2g'. We compared the genes in our corpus with previous studies: 58% (48) and 95% (79) of our annotated genes (83) are included in DisGeNET and DEPgenes, respectively. Note that DEPgenes only published 169 core genes that exhibit a high chance to be associated with depression from 5,055 candidate genes.

3 Inter-annotator agreement

We annotated the sentence units through two main annotation phases (cf. Table 7) and revised annotation guidelines after each annotation phase. Table 8 shows the IAA values obtained from each annotation phase as well as from the whole corpus. We measured IAAs in three different ways, using simple IAA (the proportion of annotations in common between two annotators over the total number of annotations provided by either annotator), Cohen's kappa, and G-index. IAA values from the final phase show that adequate agreement among the annotators is achieved. The overall IAA values, obtained from the whole corpus, also suggest internal consistency. We resolved all disagreements in the published corpus.

3.1 Disagreements

We identify the following as the major sources for conflicts between the annotators: simple mistakes, subjective readings, the use of reasoning, and the judgements by using prior knowledge. Disagreement rate is greatly reduced in the second annotation phase, as we revised the guidelines after the completion of the first phase.

Simple mistakes are inevitable in manual annotations, contributing a small number of conflicts to all the four annotation concepts. In detail, simple mistakes take up 1% (1 out of 142), 8% (11 out of 142), and 24% (34 out of 142) of the disagreements on CGE, CDL/CAE, and CC values, respectively, in Phase 1, and 9% (6 out of 67), 0% (0 out of 67), and 3% (2 out of 67) in Phase 2.

Disagreements also arise from subjective readings, contributing to most of the disagreements on CC values.

Example 5. [*CRF*]$_g$ is [*increased*]$_e$ during anxiety, [*depression*]$_d$ and pain as well as functional disorders of the pelvic viscera. [PMID: 15538210]

For the annotation unit above, one annotator subjectively interpreted the preposition 'during' as implying a causal relation and assigned 'x2g' to CC, but the other interpreted the word as having its literal meaning and assigned 'none' to CC. After annotator meeting, the annotators agreed to include instructions on such subjectivity issues in the annotation guidelines, and the IAA values on CC show significant improvement in the second annotation phase. Subjective readings induce disagreements on CAE values as well.

	Type	Count
Depress.	Depression	48
	Major depression	17
	Bipolar disorder	14
	Dysthymia	14
	Mood disorder	4
Antidep.	Antidepressant	47
	Fluoxetine	31
	Electroconvulsive therapy	4
	Imipramine	4
	Mirtazapine	4
	Citalopram	3
	Escitalopram	3
	Trazodone	3
	Lithium	2
	SSRI	2
	Carbamazepine	1
	Chlorpromazine	1
	Fluvoxamine	1
	Haloperidol	1
	Papaverine	1
	Perphenazine	1
	Quetiapine	1
	Reboxetine	1
	Sertraline	1
	Venlafaxine	1

Table 4: Statistics of depression/antidepressant related terms

Example 6. BACKGROUND: Indirect evidence suggests that loss of brain-derived neurotrophic factor (BDNF) from forebrain regions contributes to an individual's vulnerability for depression, whereas [*upregulation*]$_e$ of [*BDNF*]$_g$ in these regions is suggested to mediate the therapeutic effect of [*antidepressants*]$_a$. [PMID: 16697351]

For the annotation unit in Example 6, one annotator interpreted the verb 'mediate' as conveying the meaning of 'positive regulation' and as-

	CGE		CDL/CAE			CC		
	Inc.	**Dec.**	**Inc.**	**Dec.**	**Uni.**	**Non.**	**g2x**	**x2g**
Depress.	54(56%)	43(44%)	4(4%)	7(7%)	86(89%)	56(58%)	8(8%)	33(34%)
Antidep.	61(54%)	52(46%)	15(13%)	1(1%)	97(86%)	1(1%)	9(8%)	103(91%)
Total	115(55%)	95(45%)	19(9%)	8(4%)	183(87%)	57(27%)	17(8%)	138(65%)

Table 5: Distribution of the annotation concept values

		Gene	
		inc.	**dec.**
Depress.	**inc.**	PRKCA[d], MAPK3[d], MAPK1[d]	ALB, TNF[d,p], IL2[d], IL1B[d,p], MAPK1[d]
	dec.	MAPK1[d], BDNF[d,p], LEP[d], SLC6A4[d,p]	
	uni.	DLG4, NEFL[d], DLG3, GFAP[d,p], AVP[d], ESR1[d,p], NR3C1[d,p], TRP, CRHR1[d], S100A10[d,p], INS[d], BDNF[d,p], GRM2[d], GRIA3[d], SV2A, IGFBP2[d], PENK, HTR1A[d,p], CD19, CD8[d], GRIN2A[p], GRIN1[p]	PDLIM5[d,p], CRH[d,p], IL6[d,p], CAMK2A[p], CAMK2B, IL1B[d,p], TNF[d,p], IFNA1[d], IL2[d], AVP[d], PDYN[p], FCGR3A, CD4[d], CD8[d], DRD4[d,p], PCNT[d]
Antidep.	**inc.**		TNF[d,p]
	dec.	CHRM1, NOS1[d,p], CYP2D6[dp]	HTR1A[d,p], NR3C1[d,p], BDNF[d,p], PLCG1[d]
	uni.	HTR3A[p], IL1B[d,p], HTR2A[d], TNF[d,p], HTR1A[d,p], FOS[d], FZD3[d], ABCB1[d,p], PLA2G4A[p], IL6[d,p], CACNA1G, CACNA1I, CACNA1H, GSK3A[d], SLC6A3[d,p], SLC6A4[d,p], KCNK2[d], Defa5, VIM, TRA, BRCA1[d], CKB, ACTB, GFAP[d,p], PDE4A[d], CREB1[d,p], CCNA2, CKS1B, BAX	FOS[d], IL6[d,p], HTR2A[d], ALB[d], ADRA2A[d,p], HTR1A[d,p], BDNF[d,p], PDE4A[d], ABCB1[d,p], IGF1[d], S100A10[d,p], HTR1B[d,p], CREB1[d,p], PRL[d], PLA2G4A[p], SYP[d], NCAM1[d], NTRK2[d,p], PLCG1[d], SPR[d], Hspa9, RASEF, PDIA3, SLC6A4[d,p], CDKN1A, CDKN1B, BCL2[d], MAPK1[d]

Genes marked with superscripts d and p are validated with DisGeNET (Piñero et al., 2015) and DEPgenes (Kao et al., 2011), respectively. The reader is referred to the published corpus for more details.

Table 6: Distribution of the annotated genes

signed 'increase' to CAE. However, the other annotator interpreted the word as conveying only the meaning of 'regulation' with no directionality and assigned 'unidentifiable' to CAE. After annotator meeting, the CAE value of the annotation unit above was set to 'increase'.

Example 7. Repeated treatment with antidepressant drugs, $[imipramine]_a$ (Imi) and fluoxetine (Flu), significantly reduced the plasma corticosterone concentration and $[enhanced]_e$ the $[BDNF]_g$ and CREB levels. [PMID: 16519925]

For the annotation unit above, one annotator interpreted the phrase '*repeated treatment*' as conveying the meaning of 'enhance' and assigned 'increase' to CAE. However, the other annotator argued that the nature of the antidepressant drugs did not change and assigned 'unchanged' to CAE.

Another cause of disagreements was the use of reasoning and prior knowledge during annotation.

Example 8. In the current paper, we propose that the rapid $[decrease]_e$ in $[insulin]_g$ level during the postpartum period may be one of the causes of $[postpartum\ mood\ disorders]_d$. [PMID: 16321476]

For the annotation unit in Example 8, one annotator claimed that there is no association between the gene *insulin* and the depression *mood disorders*, as he did not find any explicitly stated piece of information. The other annotator, however, assigned 'decreased' to CGE, as he inferred that the *mood disorders* co-occurs with *insulin* in *postpartum period*. After annotator meeting, the annotators agreed on 'decreased', and added an instruction that allows the inference using logical reasoning to the annotation guidelines.

# Phase	# Units	#Depression	#Antidepressant	#Genes	Data source
Phase 1	142	75	67	47	PubMed abstracts
Phase 2	68	22	46	42	PubMed abstracts
Total/Unique	210/106	97/5	113/20	89/82	PubMed abstracts

Table 7: The annotation phases

	CGE			CDL/CAE			CC		
	Simple	Kappa	G	Simple	Kappa	G	Simple	Kappa	G
Phase 1	1	1	1	0.92	0.69	0.88	0.76	0.47	0.64
Phase 2	0.91	0.81	0.82	1	1	1	0.97	0.93	0.96
Total	0.95	0.91	0.91	0.96	0.85	0.94	0.87	0.7	0.8

Table 8: IAA values

Example 9. All [*antidepressants*]$_a$ [*increased*]$_e$ [*c-fos mRNA*]$_g$ in the central amygdala, as previously shown, while c-fos was also increased in the anterior insular cortex and significantly decreased within the septum. [PMID: 15812568]

One annotator considered the phrase "All antidepressants increased c-fos mRNA" a universal affirmative, and just modified the antidepressant term as the universal quantifier, "All antidepressants". However, the other annotator anchored on the pre-annotated keyword "antidepressants". After annotator meeting, the annotators agreed to specify the quantification type of a term and check the scope of that quantifier.

As we refined annotation guidelines after Phase 1, the disagreements among the annotators were greatly reduced. In Phase 2, almost all the disagreements were found due to simple errors. Compared to the values from Phase 1, IAA values on CDL/CAE and CC from Phase 2 show 13.6% and 50.0% increases in terms of *G index*, respectively.

3.2 Annotation guidelines

The initial annotation guidelines were taken from Lee et al. (2013). After each annotation phase in this work, the annotators held meetings to resolve the disagreements and to revise the guidelines. Table 9 shows the final version of guidelines.

4 Discussion

In this section, we show suggestions to further automating some of the processes described in the

#	Instruction
1	Annotators should annotate the sentences only if the gene exhibits changes in its expression level and this has relations with the depression or anti-depressant related term
2	Annotators can annotate the relations between CGE and CDL/CAE utilizing linguistic clues and textual evidence
3	Annotators can infer omitted fact utilizing reasoning
4	Annotators should interpret the sentences from an 'objective point of view'
5	Annotators need not consider gene expression level changes in healthy people and people with a past history of clinical depression
6	Annotators should not infer information using their prior experience or knowledge about properties of various kinds of depression
7	Annotators should not infer information (i.e., the effects of antidepressants) using their prior knowledge about the functions of genes
8	Annotators should not infer information by using inductive reasoning
9	Annotators need not consider the certainty level of propositions.
10	Annotators need consider universal propositions and particular propositions
11	Annotators should not annotate relations between genes and mania in bipolar disorder

Table 9: Annotation guidelines

previous section, especially those of extracting depression-gene relations.

- ***ML-based event relation recognition***

Example 10. OBJECTIVE: To examine whether the pathogenesis of [*depression*]$_d$ is associated with altered [*activation*]$_e$ and expression of [*Rap-1*]$_g$, as well as expression of Epac, in depressed suicide victims. [PMID: 16754837]

Example 10 shows that there are co-occurrences whose depression and gene name pairs were identified as correct but whose relation was nonetheless incorrect. The present co-occurrence has a relation of study description rather than that of gene expression change event. Besides training to come up with the event relation classifier, we can also build a system that automatically filters out false relations (i.e., hypothesis sentences) based on the previous work such as topic-classified relation recognition (Chun et al., 2006; Kilicoglu and Bergler, 2008) and deep-syntactic parser (Ballesteros et al., 2014; Hara et al., 2005; Masseroli et al., 2006; Skounakis et al., 2008).

- ***Location and contrasting information***

Example 11. Animal studies demonstrate that some antipsychotics and [antidepressants]$_a$ [increase]$_e$ neurogenesis and [*BDNF*]$_g$ expression in the hippocampus, which is reduced in volume in patients with depression or schizophrenia. [PMID: 16652337]

Example 11, and Example 9 too, show that location information turn out to be important in studies of depression and genes may respond differently to the same antidepressant in different parts of a body. Many annotation units do not explicitly provide such location information. However, missing such information will lead to conflicts and even paradoxes among annotated or mined results.

Although the annotation concepts of the presented corpus are originally designed to represent relations between gene changes and depression/antidepressant changes, they must be made to accept other concepts and constantly changing metrics in genetic studies of depression. In this regard, we should extend the annotation scheme to include parts of a body as the location and their hierarchical relationship information.

- ***Pronouns, acronyms, and appositions***

Other difficulties we faced during recognition were in dealing with grammatical constructions such as pronouns, acronyms, and appositions. They may have coped better by using the full resolved forms of pronouns and acronyms for annotation, which in turn require the access of preceding sentences or the whole abstract in the worst case. We also found that text mining tools we used extract both the appositive phrase and the phrase in apposition, but it would be better to utilize only appositives. For example, for the following phrase, we should not annotate the word "*Tricyclic antidepressants*" an antidepressant related term, or annotate "*serotonin reuptake*" a gene.

"*Tricyclic antidepressants, selective serotonin reuptake inhibitors, and serotonin-noradrenaline reuptake inhibitors, as well as the immediate precursor of serotonin*"

Instead, we should identify the three appositives as antidepressant related terms, even if they were not included in the dictionary.

- ***Sense ambiguity of 'depression'***

We also see that using simple dictionary-based matching for detecting depression-related terms produces many ambiguous terms, some of which are not related to the mental disorder at all. In particular, the term 'depression' could also be used in a situation where a certain amount, value, or function is lowered or decreased, among others. We notice that such cases are frequently observed in biomedical texts as exemplified below:

Example 12. Lack of enteral stimulation with PN impairs mucosal immunity and [*reduces*]$_e$ [*IgA*]$_g$ levels through [*depression*]$_d$ of GALT cytokines (IL-4 and IL-10) and GALT specific adhesion molecules. [PMID: 16926565]

Example 13. LTA causes cardiac [*depression*]$_d$ by [*activating*]$_e$ myocardial TNF-alpha synthesis via [*CD14*]$_g$ and induces coronary vascular disturbances by activating Cox-2-dependent TXA2 synthesis. [PMID: 16043646]

In our initial dataset that has 1,251 occurrences of depression-related terms obtained via the simple dictionary-based matching, the term 'depression' is found 730 times, which amounts to more than half of the entire occurrences. Our corpus statistics in Table 4 also show that 'depression' is the most frequent depression-related term. This means that not a few of such terms still have potential sense ambiguities. Although we manually filtered out false positive examples in our corpus, this issue is still important since it could hinder the performance of extracting depression-related terms in a fully pipelined system. Although a few named entity recognizers for biomedical text have been developed (Leaman and Gonzalez, 2008; Campos et al., 2013), none of these tools are capable of recognizing terms referring to depression, especially identifying 'depression' as the mental disorder, to the best of our knowledge.

It is anticipated that the disambiguation of the term 'depression' can be addressed with the conventional methods of word sense disambiguation with various features such as context information or external knowledge resources. Our data analysis suggests that local semantic features would be effective in many cases, among others. In particular, the following three types of syntactic construction could act as strong indicators for false positives: (1) prepositional phrases, (2) prenominal modifiers, and (3) coordinate constructions. First, prenominal modifiers often signal the context where some activity or amount is decreased, such as the physical malfunction ("cardiac depression"), the object or cause of inhibition ("Orx-B-induced depression", "AMPAR depression"), and the degree of decrease ("significant depression", "moderate depression"). Second, prepositional phrases provide information about the location or inhibition of a biological process ("depression in synapses", "depression of synaptic transmission", "depression of gamma interferon"). Last, coordinate constructions allow for exploiting the semantic similarity ("depression and anxiety" vs. "long-term potentiation and depression"). All of these features are highly local; syntactic dependencies do not cross the boundary of noun phrases.

Another possible approach would be to employ the document topic features by assuming that if the abstract of a document discusses the mental disorder, the term 'depression' in the abstract is also likely to refer to the mental disorder. In order to figure out what kind of terms are best indicative of documents that discuss the depressive disorder, we collected a set of 5,000 Medline abstracts that contain unambiguous domain-specific terms in our depression term dictionary such as 'depressive disorder', 'bipolar disorder', and 'antidepressant', and also collected another set of 10,000 abstracts that do not contain any of those terms including 'depression'. The chi-square statistics are employed to measure the discriminative power of terms found in each set of abstracts. Table 10 shows the 10 top-ranked terms for each of two types of term: terms that partially match one of the terms in our depression term dictionary (on the left column) and terms that are not found in the dictionary (on the right column). It is shown that many of the terms in the latter set are used in the context of diagnosis or treatment of depression. One of the possible methods is to use terms of this kind as features for training a binary classifier that determines whether a given document containing 'depression' discusses the mental disorder or not.

Terms in our dictionary		Terms not in our dictionary	
Term	Score	Term	Score
major	3414	treatment	807
antidepressant	2533	reuptake	504
disorder	1957	serotonin	475
depressive	1615	MDD	464
bipolar	986	psychiatric	450
mood	874	rating	356
disorders	695	diagnostic	340
unipolar	523	DSM-IV	312
tricyclic	441	criteria	301
depressed	409	patients	296

Table 10: Discriminative terms for documents related to the depressive disorder

5 Conclusion

In this paper, we presented a depression-specific corpus in support of the development of advanced text mining systems that target specifically at providing a comprehensive information of depression-gene relations. The annotation scheme of current version can express two events, *gene expression changes* and *depression level or antidepressant effect changes*, and the relationship between these two events. The presented corpus shows a high inter-annotator agreement. We also discussed several issues in the domain of depression and made suggestions to extend the annotation scheme further to resolve conflicts and sometimes paradoxes in the acquired knowledge for depression.

Acknowledgements

This work was supported by the National Research Foundation of Korea (NRF) grant funded by the Korea government (MSIP) (No. NRF-2014R1A2A1A11052310).

References

M. Ballesteros, B. Bohnet, S. Mille, L. Wanner. 2014. Deep-Syntactic Parsing. In *Proceedings of the 24th International Conference on Computational Linguistics*. 1402-1413

R. H. Belmaker, G. Agam. 2008. Major depressive disorder. *New England Journal of Medicine*, 358:55-68.

J. Björne, J. Heimonen, F. Ginter, A. Airola, T. Pahikkala, T. Salakoski. 2009. Extracting complex biological events with rich graph-based features sets. In *Proceedings of the BioNLP '09 Shared Task on Event Extraction Association for Computational Linguistics*, 10-18.

D. Campos, S. Matos, J. L. Oliveira. 2013. Gimli: open source and high-performance biomedical name recognition. *BMC Bioinformatics*, 14:54.

E. Charniak, M. Johnson. 2005. Coarse-to-fine n-best parsing and MaxEnt discriminative reranking. In *Proceedings of the 43rd ACL*, 173-180.

H. Chun, Y. Tsuruoka, J. Kim, R. Shiba, N. Nagata, T. Hishiki, J. Tsujii. 2006. Automatic recognition of topic-classified relations between prostate cancer and genes using MEDLINE abstracts. *BMC Bioinformatics*, 7(Suppl 3):S4.

Hee-Jin Lee, Sang-Hyung Shim, Mi-Ryoung Song, Hyunju Lee, Jong C. Park. 2013. CoMAGC: a corpus with multi-faceted annotations of gene-cancer relations. *BMC Bioinformatics*, 14:323.

Hee-Jin Lee, Tien Cuong Dang, Hyunju Lee, Jong C. Park. 2014. OncoSearch: cancer gene search engine with literature evidence. *Nucleic Acids Research*, 42(W1):W416-W421.

T. Hara, Y. Miyao, J. Tsujii. 2005. Adapting a probabilistic disambiguation model of an HPSG parser to a new domain. In *Proceedings of IJCNLP*, 199-210.

C. F. Kao, Y. S. Fang, Z. Zhao, P. H. Kuo. 2011. Prioritization and evaluation of depression candidate genes by combining multidimensional data resources. *PLoS ONE*, 6(4):1-9.

H. Kilicoglu, S. Bergler. 2008. Recognizing speculative language in biomedical research articles: a linguistically motivated perspective. *BMC Bioinformatics*, 2008, 9(Suppl 11):S10.

J. Kim, T. Ohta, J. Tsujii. 2008. Corpus annotation for mining biomedical events from literature. *BMC Bioinformatics*, 9:10.

S. V. Landeghem, K. Hakala, S. Rnnqvist, T. Salakoski, Y. Peer, F. Ginter. 2012. Exploring biomolecular literature with EVEX: connecting genes through events, homology and indirect associations. *Advances in Bioinformatics*, 2012:582765.

R. Leaman, G. Gonzalez. 2008. BANNER: an executable survey of advances in biomedical named entity recognition. In *Proceedings of the Pacific Symposium on Biocomputing*, 652-663.

M. C. D. Marneffe, B. MacCartney, C. D. Manning. 2006. Generating typed dependency parses from phrase structure parses. In *Proceedings of the LREC*, 449-454.

M. Masseroli, H. Kilicoglu, F. Lang, T. Rindflesch. 2006. Argument-predicate distance as a filter for enhancing precision in extracting predications on the genetic etiology of disease. *BMC Bioinformatics*, 7:291.

D. McClosky. 2010. *Any domain parsing: automatic domain adaptation for natural language parsing. PhD Thesis*, Brown University, Department of Computer Science.

D. McDonald, U Kelly. 2012. The value and benefits of text mining. *UK JISC*, [Online. Available: http://www.jisc.ac.uk/reports/value-and-benefits-of-text-mining].

E. J. Nestler, M. Barrot, R. J. DiLeone, A. J. Eisch, S. J. Gold, L. M. Monteggia. 2002. Neurobiology of depression. *Neuron*, 34:13-25.

M. Neves, J. M. Carazo, A. Pascual-Montano. 2005. Moara: a Java library for extracting and normalizing gene and protein mentions. *BMC Bioinformatics*, 11: 157-169.

J. Piñero, N. Queralt-Rosinach, À. Bravo, J. Deu-Pons, A. Bauer-Mehren, M. Baron, F. Sanz, L. I. Furlong. 2015. DisGeNET: a discovery platform for the dynamical exploration of human diseases and their genes. *Database*, 2015:bav028.

S. Poux, M. Magrane, C. N. Arighi, A. Bridge, C. O'Donovan, K. Laiho, The UniProt Consortium. 2014. Expert curation in UniProtKB: a case study on dealing with conflicting and erroneous data. *Database*, 2014:bau016.

S. Pyysalo, T. Ohta, S. Ananiadou. 2013. Overview of the Cancer Genetics (CG) task of BioNLP Shared Task 2013. In *Proceedings of the BioNLP Shared Task 2013 Workshop*, 58-66.

M. Skounakis, M. Craven, S. Ray. 2003. Hierarchical hidden Markov models for information extraction. In *Proceedings of the Eighteenth International Joint Conference on Artificial Intelligence*, 427-433.

R. Winnenburg, T. Wachter, C. Plake, A. Doms, M. Schroeder. 2008. Facts from text: can text mining help to scale-up high-quality manual curation of gene products with ontologies? *Briefings in Bioinformatics*, 9(6):466-78.

Lexical Characteristics Analysis of Chinese Clinical Documents

Meizhi Ju
College of Biomedical Engineering and Instrument Science, Zhejiang University

Haomin Li*
The Children's Hospital of Zhejiang University
The Institute of Translational Medicine, Zhejiang University

Huilong Duan
College of Biomedical Engineering and Instrument Science, Zhejiang University

Abstract

Understanding lexical characteristics of clinical documents is the foundation of sublanguage based Medical Language Processing (MLP) approach. However, there are limited studies focused on the lexical characters of Chinese clinical documents. In this study, a lexical characteristics analysis on both syntactic and semantic levels was conducted in a clinical corpus which contains 3,500 clinical documents generated during daily practices. The analysis was based on the automatic tagging results of a lexicon-based part-of-speech (POS) and semantic tagging method. The medical lexicon contains 237,291 entries annotated with both semantic and syntactic classes. The normalized frequency of different terms, syntactic and semantic classes was calculated and visualized. Major contribution of this paper is providing a wide-coverage Chinese medical semantic lexicon and presenting the lexical characteristics of Chinese clinical documents. Both of these will lay a good foundation for sublanguage based MLP studies in China.

1 Introduction

Clinical documents which contain a tremendous amount of patient information to facilitate inter-provider communication, also the most important part of clinical data for secondary use such as clinical research and administration. Recent advance in MLP technologies (Sager et al., 1987;Sager et al., 1994; Friedman and Hripcsak, 1999; Liu et al., 2012; Irina P. Temnikova et al ,2013; Irina P. Temnikova et al ,2014; Pham et al., 2014), such as de-identification (Meystre et al., 2014; Kayaalp et al., 2014), text classification (Pestian et al., 2007; Vijay, 2012), information retrieval (Uzuner et, al., 2010; Zhu et al., 2013), etc., affords an opportunity to study and analyze clinical documents at an unprecedented scale.

In recent years, Chinese MLP topics have drawn increasing public attention as there are more and more electronic clinical data that major exist in free text format such as clinical documents and reports were accumulated in many hospitals. Some Chinese MLP studies have been reported such as information extraction (Wang et al., 2014), NER(Named Entity Recognition) (Lei et al., 2014). However, systematic studies of lexical characters of Chinese clinical documents, that is the foundation of sublanguage based MLP approach and have been widely studied in other language (Foltz, 1996; Wu and Liu, 2011; Patterson and Hurdle, 2011; Patterson et al., 2010; Friedman et al., 2002), are seldom reported. Lack of accessibility of clinical documents corpus and comprehensive lexical resources for the research community is the major obstacle.

Both syntactic and semantic lexical features are important to understand the medical language structure and grammar (Harris 1968;1991). However, studying lexical features in both syntactic and semantic levels in large scale corpus requires a comprehensive medical lexicon to support the automatic lexical tagging process (Meystre et al., 2008). Unfortunately, such lexical resources in Chinese are not available. In this study, we constructed a 237,291 entries Chinese medical lexicon using computer aided methods at first. Then a lexical analysis which aims to present syntactic and semantic features of Chinese clinical

Proceedings of the 2015 Workshop on Biomedical Natural Language Processing (BioNLP 2015), pages 114–120,
Beijing, China, July 30, 2015.

documents was conducted in a corpus contains 3,500 clinical documents. The lexical features of clinical documents from different departments were reported. The annotated corpus was ready for further utilization such as collection of the co-occurrence patterns (Grishman et al., 1986) and sublanguage grammar.

2 Methods

To understand the lexical characters of language used in a subdomain, a large-scale corpus contains typical language samples from the real word need to be constructed at first. Then this corpus should be annotated manually or automatically with part-of-speech (POS) tags and semantic tags. Then the statistical analysis based on these tagging results will help researchers to understand the features of this type of sublanguage.

2.1 Corpus Collection

The corpus was collected from an EMR system which implemented in a 2000-beded hospital in China. More than 60,000 clinical documents were generated from 2009 to 2011 in total 35 clinical departments. Randomly selected 100 clinical documents from each department were used to construct a corpus for this study. Total 5 document types were included in the 3,500 clinical documents which contain 152,393 sentences and 2,375,909 Chinese characters. In addition, 15 clinical documents were randomly selected and manually annotated as the test set to evaluate the coverage of the lexicon as well as the performance of lexcial tagging methods.

2.2 Lexicon Construction

A general purpose dictionary which used in an open-source Chinese word segmenter Pangu (http://pangusegment.codeplex.com) constituted the basic of this lexicon. While most of the total 146,259 lexemes from this general purpose dictionary are irrelevant to medical concepts. ICD-10, a medication lexicon which was acquired from (http://yao.dxy.com/) using web crawler technology, and a home-grown lexicon were also compiled into this lexicon. Total 237,291 lexemes were included in this lexicon. Learning from the classical Linguistics String Project (LSP) (Grishman et al., 1973), total 24 semantic categories were designed (Listed in Table 1). POS tags were directly inherited from the Pangu systems. Semantic attribute annotations of lexcion were achieved using both statistical method and syntactic rule based method. Medical domain specialty terms such as ICD-10, medication dictionary that with known semantic class will be annotated in batch during their enrollemnts. Some semantic class with obvious morphology was assigned through matching key character of the lexeme. For example, if a character ends with "病" ("disease") with POS attribute "noun", its semantic class will be annotated as "Diagnosis" for further manual review. The ambiguity of semantic classes of many lexemes was resolved based on the most frequently usage in the corpus.

Semantic class	Example	Count
Basic Information	年龄"age"	127
Body Part	脖颈 "neck"	7,411
Nursing Care	常规护理 "nursing routine"	2,212
Chemical	硫酸 "sulfuric acid"	114
Description	交通事故 "traffic accidents"	1,282
Device	呼叫设备 "calling device"	1,618
Diagnosis	肺癌 "lung cancer"	30,209
Document Type	入院记录 "admission notes"	213
Examination	X 射线检查"X-ray examination"	2,066
Expense Name	诊疗费"medical fee"	587
Department	急诊科"emergency department"	155
Irrelevant	法案"law"	146,280
Lab Test	血清总胆固醇测定 "serum total choles-terol determination"	4,544
Medical Entity	医生"doctor"	93
Medication	阿司匹林"aspirin"	20,818
Number	多"more"	55
Organism	血吸虫"schistosome"	959
Phy Function	呼吸"breath"	281
Surgery	骨髓穿刺术"bone marrow puncture"	8,345
Symbol	$,&	303
Symptom	眩晕"dizziness"	4,681
Time	早上"morning"	1,976
Treatment	治疗方案 "therapeutic regimen"	1,340
Unit	pmol/L	236

Table 1: Semantic classes defined in the lexicon.

In addition, semantic class of lexemes with irrelevant POSes such as "Chinese idiom" was tagged as "Irrelevant". Furthermore, lexemes which are not processed with the mentioned approaches were annotated manually.

2.3 Tokenization and Annotation

Supported by the constructed lexicon, the tokenization and annotation of the corpus were conducted in the following steps. Firstly, each clinical document in the corpus with extra space was automatically trimmed in the pre-process. Then a punctuation-driven sentence boundary detection algorithm was applied to obtain sentences and clauses. After that, all clauses were segmented into words or phrases using a Chinese lexical analyzer ICTCLAS (Zhang et al., 2003). Both the semantic and syntactic classes were annotated for each word or phrase based on the lexicon during this process. For words or phrases without semantic attributes in the lexicon will be annotated as "Unknown". To make it simple, all the symbols, Arab numbers and punctuations that without specific meanings were all removed.

2.4 Lexical Characteristics Analysis

A statistical frequencies of different lexical categories in different condition were calculated. As shown in Formula 1, a NF (Normalized Frequency) value was normalized as the count of this type of lexemes in every 10,000 lexemes used in the background. As different categories with significant difference NF values, the logarithm of NF (LoF) will be calculated to plot the diverse values easier (Shown in Formula 2).

$$NF = \frac{N_{Category} * 10000}{N_{Total}} \qquad (1)$$

$$LoF = \begin{cases} \log(NF) & ,\mathrm{NF} \geq 1 \\ 0 & ,\mathrm{NF} < 1 \end{cases} \qquad (2)$$

In Formula 1, the $N_{Category}$ indicated the count of lexemes with specific semantic or syntactic category attribute in corpus or subset of corpus. The N_{Total} represented the total number of lexemes in the same corpus. The LoF value will be set to 0, when there are seldom observation of some category in some subset of corpus.

3 Results

3.1 Evaluation of the Lexicon Coverage and Lexical Tagging Methods

The quality of the lexical characters generated from statistical analysis depends on the coverage and completeness of the lexicon constructed. Comparing with the typical comprehensive medical lexical resources such as UMLS which contains millions of terms, our lexicon scale is relatively small. So we calculate the coverage and completeness of the lexicon during the tokenizing and annotation. Total 13,660 lexemes were unrecognized among all 2,375,909 lexemes in the corpus. The coverage of our lexicon in the corpus was 99.43% calculated by Formula 3. Similarly, the distinct lexemes among the unrecognized lexemes and lexemes in the corpus were 577 and 19,847 respectively. Thus, the completeness of the lexicon was 91.11% calculated by Formula 4.

$$Coverge = \frac{N_{Unrecognized\ lexemes}}{N_{Total}} * 100\% \qquad (3)$$

$$Completeness = \frac{N_{Unrecognized\ distinct\ lexemes}}{N_{Total\ distinct\ lexemes}} * 100\% \qquad (4)$$

Based on the manually annotated test set, we evaluated the accuracy of word segmenter performance and syntax and semantics classification. Word segmentation and annotation regarding POS and semantics were conducted on the test set with the ICTCLAS. As a result, 4,006 lexemes were obtained excluding punctuations and Arabics by the automatic tagging process. Manually checking by one of the authors, the number of error segments caused by ICTCLAS was counted. Meanwhile, the number of lexemes with error POS tag or semantic tag was picked out. The accuracy of word segmentation, POS and semantics was calculated separately by Formula 5 and demonstrated in Table 2.

Evaluate item	Accuracy
Word Segmentation(ICTCLAS)	96.03%
POS	88.09%
Semantics	90.86%

Table 2: The evaluation result of the lexicon.

3.2 Lexical Characters in Chinese Clinical Documents

The semantic class of lexemes usage frequency (NF value) in different clinical departments was plotted in Fig. 1 using heatmap.2 function gplots package in R. It is apparent from the heat map that "body part", "time", "symptom" and "diagnosis" were the top four semantic classes. We can easily distinguish the mental health department from other departments as the "body part" was used in a relatively lower frequency. Some internal medicine department such as rheumatology, hematology and nephrology more interested in the lab test result discussion.

The fluctuation of 22 POS categories in 5 typical document types in Fig. 2.A is basically consistent in general. However, there are observable differences between semantic categories as showed in Fig. 2.B. For example, document type of informed consents has great differences compared with other types of clinical documents.

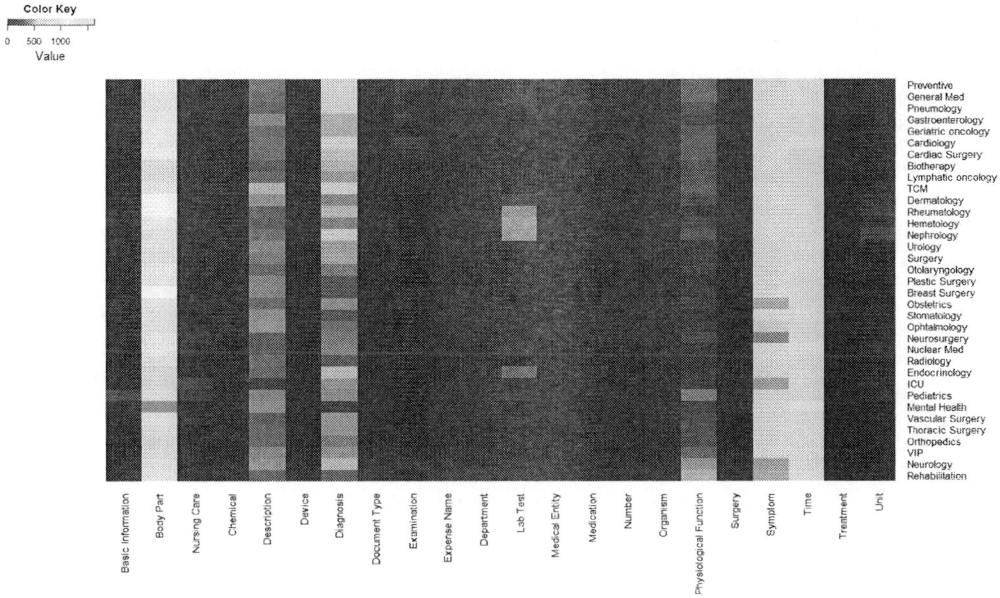

Fig.1. Heat map of original NF value.

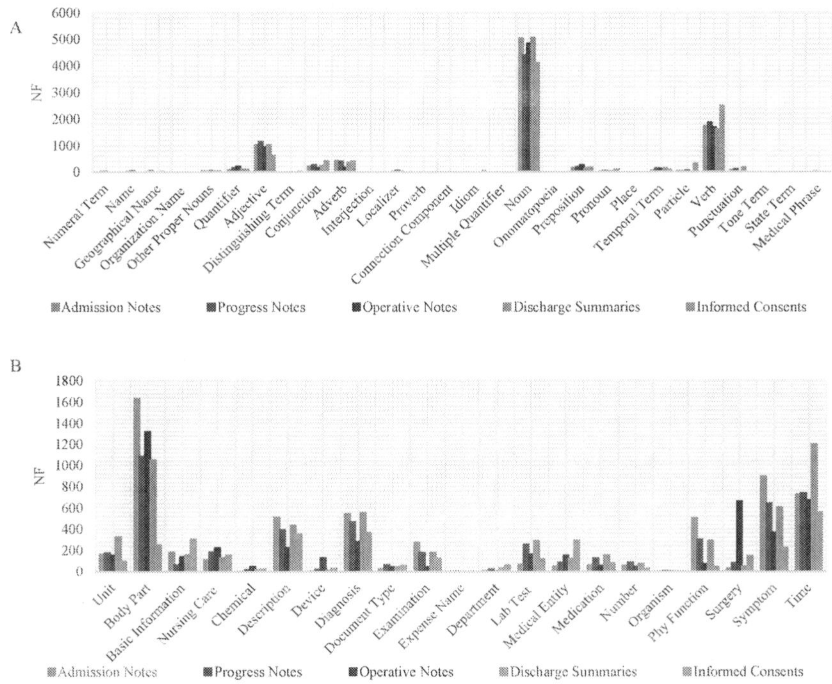

Fig. 2. Sublanguage (A) and POS (B) features of 5 document types in corpus.

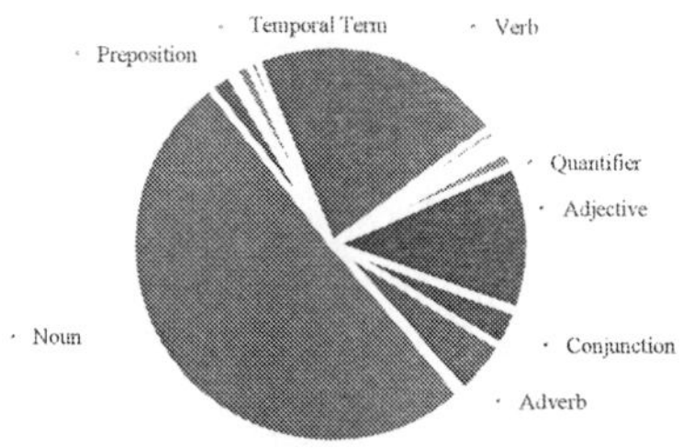

Fig. 3. The POS proportion of Chinese clinical documents.

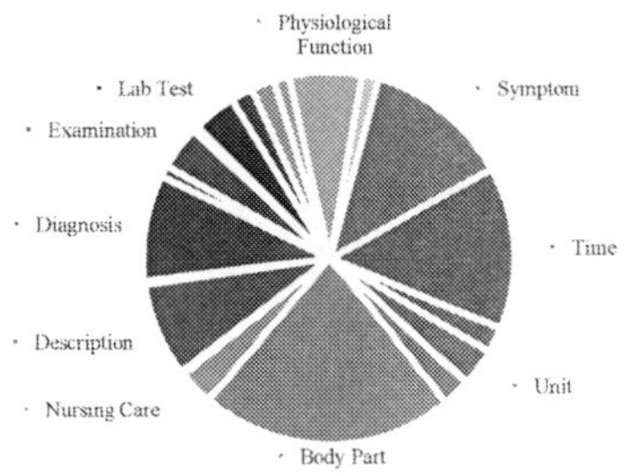

Fig. 4. The sublanguage proportion of Chinese clinical document.

We can also notice that large number of phrases related to "time" were used in discharge summaries, implying that these retrospective documents record many temporal information. Fig. 3 and Fig. 4 show the overall LoF proportion of semantic and POS types in the corpus. All the figures lead us to the conclusion that body part, symptom and diagnosis sublanguages account for the largest portion of Chinese clinical documents.

3.3 Co-occurrence patterns in Chinese Clinical Documents

Furthermore, more than 168,823 nonrepeating clauses were obtained in the corpus including total 565,630 clauses. To count the sematic patterns among these clauses, some frequently used co-occurrence patterns were summarized in Table 3. For each pattern, the example clause was highlighted with different font colors and styles to show corresponding semantic component.

These co-occurrence patterns will lay a foundation to create sublanguage grammars for the Chinese medical language.

Co-occurrence pattern	Sample	Count
Body Part +Irrelevant +Symptom	*心前区*无**隆起**"no **uplift** in *precordium*"，	12,740
Irrelevant +Symptom	为白色粘痰" is white sticky sputum"，	8,588
Body Part +Description	*颈部*对称"*the neck* is symmetrical"，	7,679
Irrelevant +Diagnosis	考虑**脑瘤**"possibly suffer **brain tumor**"，	4,877
Body Part +Diagnosis	*颈椎*肿瘤"*Cervical* **Cancer**"：	4,278
Number +Body Part+Description +Symptom	双*下肢*轻度水肿"two *lower extremities* mild edema"；	3,161

Table 3: Top co-occurrence patterns in the corpus.

4 Discussion and Future Work

In this paper, through constructing a comprehensive medical semantic lexicon, the lexical characteristics of clinical documents both in semantics and syntactic level were analyzed separately. In addition, a number of the most frequent sublanguage co-occurrence patterns of Chinese clinical documents were discovered.

The quality of the lexicon constructed in this study is the major challenge of current analysis. As a mature and high-quality lexical resource such as UMLS will take years and cost millions of dollars to maintain. A Chinese counterpart is urgently needed and its value should be well recognized by governments and funding agencies.

Our future work includes improving the coverage and quality of the lexicon based on the corpus using more computer aided approaches. The accuracy of the automatic tagging process still has plenty of room to improve. Currently most of the errors were caused by ambiguous of semantic type or POS. But the results of this lexical analysis still provide much useful information to Chinese medical language researchers.

Lack accessibility of corpus is one of the obstacles for current Chinese medical language processing studies due to current regulation and privacy concerns. As the automatic de-identification methods already widely accepted in many countries, we will evaluate it in our corpus in the future. After that this annotated corpus will open to the community.

Reference

Anne-Dominique Pham, Aurélie Névéol, Thomas Lavergne, Daisuke Yasunaga, Olivier Clément, Guy Meyer, Rémy Morello and Anita Burgun. 2014. *Natural Language Processing of Radiology Reports for the Detection of Thromboembolic Diseases and Clinically Relevant Incident Findings.* BMC Bioinformatics,15:266.

Carol Friedman and George Hripcsak. 1999. *Natural Language Processing and its Future in Medicine*. Journal of the Association of American Medical Colleges.

Carol Friedman, Pauline Kra and Andrey Rzhetsky. 2002. *Two Biomedical Sublanguages: A Description Based on the Theories of Zellig Harris.* Journal of Biomedical Informatics. 222-235.

Dongqing Zhu, Wu Stephen, Masanz James, Ben Carterette and Hongfang Liu. 2013. *Using Discharge Summaries to Improve Information Retrieval in Clinical Domain.*

Garia, Vijay. 2012. *Kernel Methods and Semantic Techniques for Clinical Text Classification*.

Huaping Zhang, Hongkui Yu, Deyi Xiong and Qun Liu. 2003. *HHMM-based Chinese Lexical Analyzer ICTCLAS.* 2nd SIGHAN workshop affiliated with 41th ACL, Sapporo Japan.

Hui Wang, Weide Zhang, Qiang Zenf, Zuofeng Li, Kaiyan Feng and Lei Liu. *Extracting Important Information from Chinese Operation Notes with Natural Language Processing Methods.* Journal of Biomedical Informatics 48 (2014) 130-136.

Hongfang Liu, Stephen T. Wu, Dingchen Li, Siddhartha Jonnalagadda, Sunghwan Sohn, Kavishwar Wagholikar, Peter J. Hang, Stanley M. Huff and Christopher G Chute. *Towards a Semantic Lexicon for Clinical Natural Language Processing.* AMIA Annual Symposium, 2012.

Irina P. Temnikova, Ivelina Nikolova and William A. Baumgartner Jr. *Closure Properities of Bulgarian Clinical Text.* In Proceedings of RANLP. 2013, 667-675.

Irina P. Temnikova, William A. Baumgartner Jr., Negacy D. Hailu, Ivelina Nikolova, Tony McEnery , Adam Kilgarriff, Galia Angelova and K. Bretonnel Cohen. *Sublanguage Corpus Analysis Toolkit: A Tool for Assessing the Representativeness and Sublanguage Characteristics of Corpora.* In Proceedings of LREC. 2014, 1714-1718.

John P. Pestian, Christopher Brew, Pawel Matykiewicz, Dj J. Hovermale, Neil Johnson, Kevin B. Cohen and Wlodzislaw Duch. 2007. *A Shared Task Involving Multi-label Classification of Clinical Free Text.* Biological, translational, and clinical language processing, pages 97-144, Prague, Association for Computational Linguistics.

Jianbo Lei, Buzhou Tang, Xueqin Lu, Kaihua Gao, Min Jiang and Hua Xu. 2014. *A Comprehensive Study of Named Entity Recognition in Chinese Clinical Text.* J Am Med Inform Assoc, 21:808-814.

Mehmet Kayaalp, Allen C. Browne, Zeyno A. Dodd, Pamela Sagan and Clement J. McDonald. 2014. *De-identification of Address, Date, and Alphanumeric Identifiers in Narrative Clinical Reports.* AMIA Fall Symposium.

Naomi Sager, Carol Friedman and Margaret S. Lyman. 1987. *Medical Language Processing: Computer Management of Narrative Data*.

Naomi Sager, Margaret S. Lyman, Christine Bucknall, Ngo T. Nhan and Leo J. Tick. 1994. *Natural Language Processing and the Representation of Clinical Data. .*

O¨zlem Uzuner, Imre Solti and Eithon Cadag. 2010. *Extracting Medication Information from Clinical Text.* J Am Med Infor Assoc, 17:514-518.

Olga Patterson and John F. Hurdle. 2011. *Document Clustering of Clinical Narratives: A Systematic Study of Clinical Sublanguages.* AMIA Annual Symposium Proceedings, 1099–1107.

Olga Patterson, Sean Igo and John F. Hurdle. 2010. *Automatic Acquisition of Sublanguage Semantic Schema: Towards the Word Sense Disambiguation of Clinical Narratives.* AMIA Annual Symposium proceedings / AMIA Symposium AMIA Symposium, 612–6.

Peter W. Foltz. 1996. *Latent Semantic Analysis for Text-based Research.* Behavior Research Methods, Instruments &Computers, 28(2),197-202.

Ralph Grishman, Lynette Hirschman and Ngo T. Nhan. 1986. *Discovery Procedures for SubLanguage Selectional Patterns: Initial Experiments.* Compusational Lingustics.

Ralph Grishman, Naomi Sager, C. Raze and B. Bookchin. 1973. *The Linguistic String Parser.* Proceedings of national computer conference and exposition p427-434.

Stéphane M. Meystre, Óscar Ferrández, F. Jeffrey Friedlin , Brett R. South, Shuying Shen and Matthew H. Samore. 2014. *Text De-identification for Privacy Protection: A Study of its Impact on Clinical Text Information Content.* Journal of Biomedical Informatics, 50: 142-150.

Stephane M. Meystre, Guergana K. Savova, Karin C. Kipper-Schuler and John F. Hurdle. 2008. *Extracting Information from Textual Documents in the Electronic Health Record: A review of Recent Research.* IMIA

Stephen Wu and Hongfang Liu. 2011. *Semantic Characteristics of NLP-extracted Concepts in Clinical Notes vs.Biomedical Literature.* AMIA Annu Symp Proc. 1550–1558.

Zellig S. Harris. 1968. *Mathematical Structures of Language.* Interscience Publishers.

Zellig S. Harris. 1991. *A Theory of Language and Information: A Mathematical Approach.* Clarendon Press.

Using word embedding for bio-event extraction

Chen Li[1], **Runqing Song**[2], **Maria Liakata**[3],
Andreas Vlachos[4], **Stephanie Seneff**[1], **Xiangrong Zhang**[2,*]
[1] Massachusetts Institute of Technology, United States
[2] Xidian University, China
[3] University of Warwick, United Kingdom
[4] University College London, United Kingdom
* xrzhang@ieee.org

Abstract

Bio-event extraction is an important phase towards the goal of extracting biological networks from the scientific literature. Recent advances in word embedding make computation of word distribution more efficient and possible. In this study, we investigate methods bringing distributional characteristics of words in the text into event extraction by using the latest word embedding methods. By using bag-of-words (BOW) features as the baseline, the result has been improved by the introduction of word-embedding features, and is comparable to the state-of-the-art solution.

1 Introduction

Automated extraction of bio-events from the scientific literature is an important research stage towards extraction of bio-networks, and is the main focus of bio-text-mining [1].

An event represents a biochemical process, e.g. a protein-protein interaction or chemical-protein interaction, within a signalling pathway or a metabolic pathway. An event in text is usually anchored by a word indicating the occurrence of the event, named a trigger, and the other words, which are arguments involved in the reaction. Solutions of extracting events usually begin with detecting trigger words first, and then assemble other detected argument words to a trigger. Some solutions consider event extraction as a structured prediction problem and extract triggers with corresponding arguments at once [2], [3].

BOW is common features of representing tokens when lexcial information is need for prediction, e.g. trigger prediction. However, it has drawbacks of being high dimensional, sparse and discrete. While word embedding is a collective name for a set of language modelling and feature learning techniques, by which words in a vocabulary could by mapped to vectors in a lower dimensional space, which is continuous in and relative to the vocabulary size. It is capable of representing a words distributional characteristics [4]. In this way, word embedding model may capture semantic and sequential information of a word in text. Meanwhile, a word-embedding feature is continuous, since continuous space language models maps integer vector into continuous space via learned parameters. By training a neural network language model, one obtains not just the model itself, but also the learned word embedding.

Due to the size of a dictionary word embedding might involve, computation of word distribution could be expensive. Mikolov et al. proposed two model architectures called CBOW and skip-gram for maing computation of word embedding feasible and efficient [5].

The skip-gram model tries to maximize classification of a word based on another word in the same sentence. Each current word as an input to a log-linear classifier with continuous projection layer, and predict words within a certain range before and after the current word (Figure 2).

Nie et al. utilized word embedding for detecting trigger words [6]. In this paper, we present the experiments using word embedding as token features to extract complete events including triggers and their arguments. The skip-gram model is used to obtain word-embedding features and is compared with a baseline model of using BOW features. The result demonstrates that the introduction of word embedding improves the result, and is comparable to the state-of-the-art solution.

2 Methods and results

2.1 BioNLP GENIA task

A series of efforts has been initiated to evaluate the available solutions and investigate potentials in event extraction technologies. Among them, the

Proceedings of the 2015 Workshop on Biomedical Natural Language Processing (BioNLP 2015), pages 121–126,
Beijing, China, July 30, 2015.

BioNLP Shared Tasks (BioNLP-ST) [7] have been consistently conducted since 2009 and attracted community-wide support. BioNLP-ST GENIA task is a core task and had the third edition in 2013. The task gradually increased its difficulties and complexities, for example, by upgrading from abstract-only text to full-text articles and subsuming co-reference tasks.

In the latest GENIA 2013 task, EVEX achieves the best performance (F-score: 50.97; recall: 45.44; precision: 58.03) [8]. Our system achieves a comparable result with a higher precision (F-score 47.33; recall: 37.14; precision: 65.21).

2.2 Event extraction model

Except binding events, the event extraction process consists of two steps in our system. First, triggers are predicted for each token in a sentence. Then, arguments including themes and causes are predicted to be associated with the triggers. The arguments could be either proteins or other events. The events, which may have other events as arguments, are called recursive events in this paper. During the prediction, this might lead to cyclic referencing. For example, event A is predicted as event Bs argument, while B is also predicted as As. In our model, the candidate events are tested, and the one has lower confidence score given by SVM classifier would be deleted. This method is also extended to bigger number of events, which are referencing each other in a cyclic manner.

For example, in Figure 1, four trigger words indicate four events. After detecting the triggers, the system check proteins one by one to seek the right arguments. The system will start with simple events, the methylation and the gene expression in the example. Then it will check arguments for the triggers of recursive events. This example has two recursive events, a positive regulation and a negative regulation. In the case when a new event is created, the new event has to be tested to see whether it could be an argument of one of the recursive events.

A binding event may have more than one theme. The extraction of binding event consists of three steps. The first two steps are similar to the other event extractions. At the third step, the candidate arguments are constructed with argument in possible combinations. Then, the combinations are tested by an SVM classifier, and the one with the highest confidence score will be kept. In the experiments, we use LibSVM as the implementation of SVM.

2.3 Word embedding for trigger and argument detection

Representing a token in right features is crucial in trigger prediction. BOW is a popular solution. However, it is very high dimensional, sparse and discrete. While word embedding features, which are learnt by a neural-network-based language model called continuous space language model, can represent a words distributional characteristics [4]. This, in a way, may capture semantic and sequential information of a word in text.

One problem of a word embedding model is that the model only represents the distributional characteristics of a word in entire text rather than in a specific context. In another word, the characteristics of an individual word in a sentence cannot be brought into a later prediction model. The lexically same tokens have the same word embedding. This word may indicate different event types in different sentences according to the BioNLP task. Therefore, we also experiment to join word embedding features with BOW features.

Events may have multi-token triggers. For example, mRNA expression is a transcription events trigger in many instances. Meanwhile, expression appears as a gene-expression events trigger in many instances. Biologically, transcription is a more specific process of gene expression. Therefore, for such cases, the system predicts event type as transcription since it is more informative.

In the experiment, training and development data-sets provided in the BioNLP13 are used to obtain word-embedding features in an unsupervised manner. A problem of word embedding method is that it represents a words distributional characteristics in the entire text, however loses the words contextual information in a specific sentence. Thus, during the training, we also consider n-gram features of a token.

After detecting triggers, assembling correct arguments to the triggers is another key link on the chain. As the model described in the section 2.1, the system starts with proteins and then the generated events. If a new event is created, it will be tested against the triggers, which indicate recursive events but have not been constructed as an event yet. The Stanford dependency path is the main feature for argument detection.

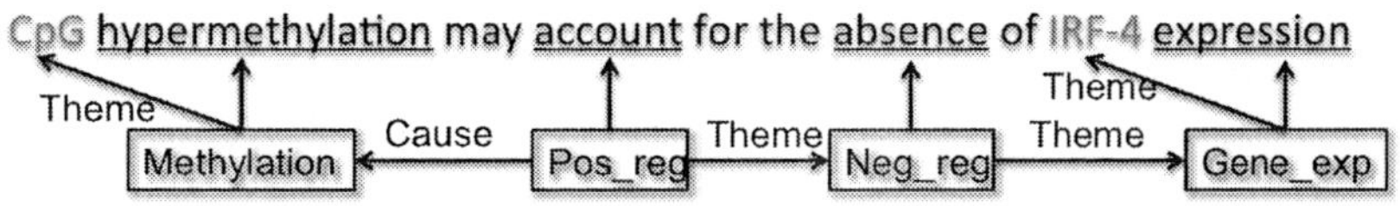

Figure 1: The model of even extraction. The words in orange are the proteins. The underlined words are the triggers.

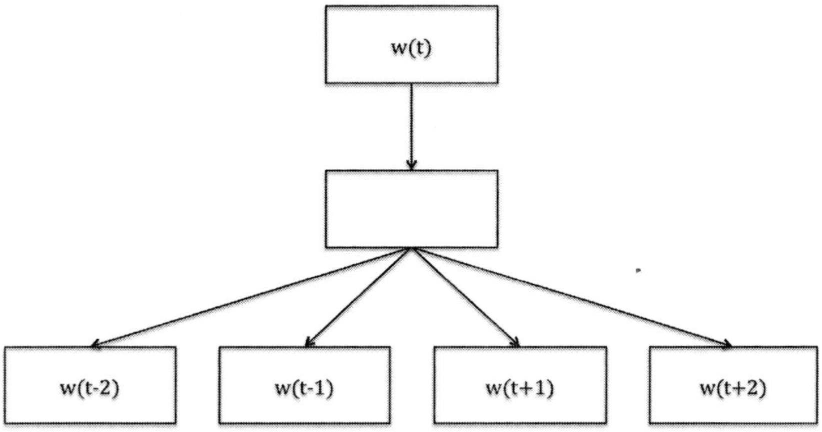

Figure 2: The skip-gram model architecture.

2.4 Results

We evaluate three models on the BioNLP 2013 GENIA test dataset. At the moment, only events described within the boundaries of a sentence are considered.

- BOW + n-gram
- Word embedding
- Word embedding + n-gram

The first model uses BOW and n-gram to represent each token. Then, the model is replaced by another using word embedding only while utilizing the exactly same extraction infrastructure, which is a pipeline converging tokenization, parsing and other pre-processing upon Apache UIMA. At last, we jointly use word embedding with n-gram. In Table 1, it could be observed that the joint model achieves the best performance with 47.33 in F-score. The model only using word embedding achieved the lowest, however, still gets 46.33 in F-score. This is because word embedding loses a word's distributional information in a specific context although the distributional characteristics of words are obtained for the entire text.

Table 2 shows that the detail result of the model performing the best, the joint model. Extraction of simple events achieves an average F-score of 71.98, which is expected, since each simple event contains only one theme and is not recursive. The system achieves 64.00 in F-score for protein modification event. The events are more complicated than simple events since they contain causes besides themes in arguments. The F-score for extracting binding events is 39.85. Regulatory events are the most complex ones because each of them has two arguments and is recursive. Extraction of this type of events achieved 33.97 in F-score.

Since binding is a special event type, which may have unknown number of arguments, we have analysed the extraction of binding events with different extraction strategy. Table 3 is the result with different models of assigning arguments to binding triggers. Single prediction uses one binary classifier to determine the assignment of a candidate argument. Two step prediction firstly check all arguments about whether they could be candidate arguments, then, delete the combinations covered by others. For example, if protein A and protein B are both assigned to a trigger to construct a binding event. Then, the two candidate events with A and B as argument respectively will not be considered. Two steps-confidence scores represents the results that we prune binding events ac-

Event Class	BOW + n-gram	Word embedding	Word embedding + n-gram
Gene expression	76.32	75.91	77.37
Transcription	59.30	46.39	60.24
Protein catabolism	64.00	42.55	64.00
Localization	51.03	58.39	45.33
=[SIMPLE ALL]=	71.66	68.78	71.98
Binding	36.36	35.13	39.85
Protein modification	0.00	0.00	0.00
Phosphorylation	72.66	73.68	70.18
Ubiquitination	12.12	12.12	12.12
Acetylation	0.00	0.00	0.00
Deacetylation	0.00	0.00	0.00
=[PROT-MOD ALL]=	66.25	67.46	64.00
Regulation	16.32	19.78	18.62
Positive regulation	36.01	36.74	35.71
Negative regulation	35.09	38.67	37.50
=[REGULATION ALL]=	33.07	34.45	33.97
==[EVENT TOTAL]==	46.65	46.33	47.33

Table 1: The comparison between the BOW model, the word embedding model and the joint model on the test set of BioNLP 2013. The results are represented in F-scores.

Event Class	Gold (match)	Answer (match)	Recall	Precision	F-score
Gene expression	619 (441)	521 (468)	71.24	84.64	77.37
Transcription	101 (50)	65 (50)	49.50	76.92	60.24
Protein catabolism	14 (8)	11 (8)	57.14	72.73	64.00
Localization	99 (34)	51 (34)	34.34	66.67	45.33
=[SIMPLE ALL]=	833 (533)	648 (533)	63.99	82.25	71.98
Binding	333 (107)	204 (107)	32.13	52.45	39.85
Protein modification	1 (0)	0 (0)	0.00	0.00	0.00
Phosphorylation	160 (102)	131 (102)	63.75	77.86	70.10
Ubiquitination	30 (2)	3 (2)	6.67	66.67	12.12
Acetylation	0 (0)	0 (0)	0.00	0.00	0.00
Deacetylation	0 (0)	0 (0)	0.00	0.00	0.00
=[PROT-MOD ALL]=	191 (104)	134 (104)	54.45	77.61	64.00
Regulation	288 (35)	88 (35)	12.15	39.77	18.62
Positive regulation	1130 (291)	500 (291)	25.75	58.20	35.71
Negative regulation	526 (156)	306 (156)	29.66	50.98	37.50
=[REGULATION ALL]=	1944 (482)	894 (482)	24.79	53.91	33.97
==[EVENT TOTAL]==	3301 (1226)	1880 (1226)	37.14	65.21	47.33

Table 2: The detail result on the BioNLP 2013 GENIA test dataset by using the word-embedding model.

cording to confidence scores (see the section 2.1). Table 3 shows that the performance of dividing Binding events themes extraction in two step is better. Using confidence scores to prune Binding events can improve the performance of Binding events significantly.

3 Conclusion

The paper explores the methods of exploiting distributional characteristics of words in a continuous space into bio-event extraction by using the latest word embedding methods. It is the first system using word embedding to extract complete events from text, and has achieve the result comparable to the state-of-the-art system's.

The system uses the BOW model as the baseline. When the model only using word embedding to represent tokens, the system achieves slightly lower performance than the BOW model's. The model jointly using word-embedding achieves the best performance. This is because n-gram effectively complements the loss of contextual information of words, at the same time when the words' distributional characteristics are introduced by word embedding.

There are various ways we plan to further improve the system. The current experiment uses BioNLP dataset, which is relatively small for achieving word vectors in a continuous space. In the following experiments, we would like to train and obtain the word vectors on a bigger corpus, e.g. a subset containing related articles from Wikipedia. Furthermore, we would like to create a joint model combining the prediction of trigger and arguments [3].

Acknowledgments

The work also benefited from the discussion with Nigel Collier.

Chen Li is sponsored by Quanta Computer Inc., Taiwan.

References

[1] C. Li, M. Liakata, and D. Rebholz-Schuhmann, "Biological network extraction from scientific literature: State of the art and challenges," *Briefings in bioinformatics*, vol. 15, no. 5, pp. 856–877, 2014.

[2] D. McClosky, S. Riedel, M. Surdeanu, A. McCallum, and C. D. Manning, "Combining joint models for biomedical event extraction," *BMC bioinformatics*, vol. 13, no. Suppl 11, S9, 2012.

[3] A. Vlachos and M. Craven, "Biomedical event extraction from abstracts and full papers using search-based structured prediction," *BMC bioinformatics*, vol. 13, no. Suppl 11, S5, 2012.

[4] D. E. Rumelhart, G. E. Hinton, and R. J. Williams, "Learning representations by back-propagating errors," *Cognitive modeling*, vol. 5, 1988.

[5] T. Mikolov, I. Sutskever, K. Chen, G. S. Corrado, and J. Dean, "Distributed representations of words and phrases and their compositionality," in *Advances in Neural Information Processing Systems*, 2013, pp. 3111–3119.

[6] Y. Nie, W. Rong, Y. Zhang, Y. Ouyang, and Z. Xiong, "Embedding assisted prediction architecture for event trigger identification," *Journal of bioinformatics and computational biology*, 2015.

[7] C. Nédellec, R. Bossy, J.-D. Kim, J.-J. Kim, T. Ohta, S. Pyysalo, and P. Zweigenbaum, "Overview of bionlp shared task 2013," in *Proceedings of the BioNLP Shared Task 2013 Workshop*, 2013, pp. 1–7.

[8] J.-D. Kim, Y. Wang, and Y. Yasunori, "The genia event extraction shared task, 2013 edition-overview," in *Proceedings of the BioNLP Shared Task 2013 Workshop*, 2013, pp. 8–15.

Binding event	Gold (match)	Answer (match)	Recall	Precision	F-score
Single prediction	333 (84)	310 (84)	25.23	27.10	26.13
Two-step prediction	333 (64)	148 (64)	19.22	43.24	26.61
Two-step prediction with confidence scores	333 (101)	242 (101)	30.33	41.74	35.13

Table 3: The results of binding event extraction on the test set of BioNLP 2013.

Measuring the readability of medical research journal abstracts

Samuel Severance
University of Colorado
School of Education
Institute of Cognitive Science
Boulder, CO USA
severans@colorado.edu

K. Bretonnel Cohen
University of Colorado
School of Medicine
Biomedical Text Mining Group
Aurora, CO USA
kevin.cohen@gmail.com

Abstract

This study examines whether the readability of medical research journal abstracts changed from 1960 to 2010. Abstracts from medical journals were downloaded from PubMed.org in ten-year batches (1960s, 1970s, etc.). Abstracts in each decade underwent processing via a custom Python script to determine their Coleman-Liau Index (CLI) readability score. Analysis using one-way ANOVA found statistically significant differences between the mean CLI readability scores of each decade ($F(4, 6689135) = 12936.91, p<0.0001$). Post-hoc analysis using Tukey's method also found all pairwise comparisons between decades' mean CLI readability scores to be statistically significant ($p<0.001$). Readability scores increased from decade to decade beginning with a mean CLI score of 16.0813 in the 1960s and ending with a mean CLI score of 16.8617 in the 2000s. These results indicate a 0.7804 grade level increase in the difficulty of reading medical research journal abstracts over time and raises questions about the accessibility of medical research for broader audiences.

1 Introduction

A persistent issue in academic research centers on whether the knowledge published by researchers reaches and is understood by those it could benefit. The medical field takes up this issue in its efforts to translate research into practice, or the idea of "translational research" (Woolf, 2008). Ideally, practitioners can access and thoroughly comprehend research to better ensure new treatments and knowledge reaches patients and that patient care revolves around evidence-based practices (Pravikoff, Tanner, & Pierce, 2005; Woolf, 2008). Beyond seeking to leverage new research among medical practitioners, translational research also focuses on supporting patients in becoming more active and involved in their healthcare (Woolf, 2008). With the advent of the information age, patients and patients' family members have substantial opportunities to research their own medical conditions and their treatment options. Navigating and understanding medical research requires that it proves accessible in terms of its readability.

This study is a diachronic analysis of the readability of medical research. Specifically, this study seeks to answer whether the readability of medical research journal abstracts has changed from the 1960s to the 2000s. Results from this study may have implications for how researchers could communicate their findings to patients and how to address discrepancies between the reading level of medical journals and lay audiences' reading abilities.

2 Relevant Literature

2.1 Readability of health materials in relation to patients

Research on the readability of health materials in relation to patients has a strong presence in the literature. Health literacy researchers have found

Proceedings of the 2015 Workshop on Biomedical Natural Language Processing (BioNLP 2015), pages 127–133,
Beijing, China, July 30, 2015.

that the vast majority of textual information patients typically encounter—from informed consents to patient education materials—surpass the reading ability of patients (Rudd, Moeykens, & Colton, 1999). Such discrepancies may have profound negative influences on patient health outcomes (Paasche-Orlow & Wolf, 2010). Indeed, Baker et al. (1998) found an independent association between low health literacy and increased hospital admission rates where patients with low literacy became hospitalized twice as often as more literate patients. Additionally, patients with high functional health literacy become more involved in their care, including exploring options beyond those presented by a doctor, whereas patients with low functional health literacy tend to limit decisions regarding their care to only those presented to them by doctors (Smith et al. 2009). With implications for personal and community health, a study by Navarra et al. (2014) found that HIV-infected youth with below-grade-level reading skills did not completely adhere to their antiretroviral therapy.

Despite growing evidence of the role of health literacy in patient outcomes, the readability of medical information for patients has not improved over time, even for items intended for patients. The lack of readability of informed consents, in particular, has garnered attention in the literature (Mead & Howser, 1992; Rudd et al., 1999). An examination of the readability of informed consents from 1975 to 1982 at the Veterans Administration Medical Center found them to have a college reading level and that their reading difficulty may have actually increased over the time period examined (Baker & Taub, 1983). Fifteen years later, a study of surgical consents from across the US also showed similarly difficult reading levels with a given consent requiring an average reading level of 12.6 (Hopper et al., 1998). Beyond informed consents, other materials directly aimed at laymen also show readability issues. In an analysis of emergency first-aid instructions, Temnikova (2012a, 2012b) found ten separate categories of readability/complexity problems. Alamoudi and Hong (2015) found the readability of websites related to microtia and aural atresia lacking in terms of facilitating comprehension.

2.2 Identifying and addressing readability issues

A significant body of work focuses on addressing readability issues in health contexts. It makes the significance of the corpus-based study reported here clear: it shows that we can address readability problems, but first *we must know what the readability issues are.*

Elhadad (2006), for instance, shows which terms in a medical journal article a lay reader would likely not understood and presents an application that finds these terms and mines an appropriate definition from the Web. Achieving usable results with a small corpus, Elhadad and Sutaria (2007) presented a parallel-corpus-driven method for finding technical/lay equivalents of medical terms using measures of association. Leroy et al. (2010) pointed out that perceived and the actual difficulty of text influenced the willingness and ability to learn from health information. The researchers manipulated characteristics of health texts and measured perceived and actual difficulty, and found they could improve the perceived difficulty of text. Their technique also uncovered some problems with standard readability formulas. Using lexical and grammatical analysis of a medical corpus to develop a new metric to estimate text difficulty called "term familiarity," Leroy et al. (2012) performed an experiment where individuals showed slightly improved understanding for simplified documents. An evaluation of a writing assistance tool that assists with automated simplification related to term familiarity found that simplified text had strong beneficial effects on both perceived and actual difficulty, with better understanding and more learning after reading simplified text than after reading un-simplified text (Leroy, Kauchak & Mouradi, 2013). In another study, Leroy et al. (2013) examined the effects of lexical simplification and coherence enhancement on readability and showed that they interact in complex ways with both perceived and actual difficulty. Investigating linguistic features, specifically discourse features that correlate with the readability of texts for adults with intellectual disabilities, Fung et al. (2009) presented a tool for rating the readability of texts for these readers. Huenefaurth et al. (2009) compared different methods for evaluating text readability software for adults with intellectual disabilities, finding that multiple-choice questions with illustrations proved more useful than yes/no questions or Likert scales for evaluating simplification programs.

2.3 Work presented in context of relevant literature

Specific research utilizing a diachronic, corpus-based approach to examining the readability of medical journals did not turn up in a review of the literature. However, previous studies taking a diachronic approach to the readability of corpus data do have precedence. Indeed, the inspiration for this study comes from work by Štajner (2011). Štajner performed a diachronic analysis of the Brown "family" of corpora to examine changes in the readability of the English language over time. Similar to this study, Štajner utilized the Coleman-Liau Index as a measure of the readability of the Brown "family" of corpora.

3 Methodology

This study occurred in three main phases in order to answer the research question: How has the readability of medical journal abstracts changed between the 1960s and 2000s?

3.1 Obtaining a medical research corpus

The first phase of this study involved compiling a machine-readable corpus of medical research journal abstracts. PubMed.org contains a large volume of medical research journal abstracts and these provided the basis of a corpus. Abstracts were downloaded in groups by decade (see Table 1).

Table 1. Number of abstracts by decade.

Decade range	Number of abstracts
1960-1969	5324
1970-1979	313053
1980-1989	1049637
1990-1999	2017482
2000-2009	3327954

This study focused solely on journal abstracts dealing with research on human subjects with the assumption that a human-centered research corpus has more meaningful parallels to the potential interests of most patients.

3.2 Measure the readability of abstracts

The Coleman-Liau Index measure of readability (Coleman & Liau, 1975) formula is as follows:

$$CLI = 5.89\frac{c}{w} - 29.5\frac{s}{w} - 15.8$$

In this formula, c is equal to the total number of characters in a given text, w is equal to the total number of words in a given text, and s is equal to the total number of sentences in a given text. The CLI outcome measure is given as a grade-level readability score. For example, a grade of 10.5 would correspond to a text at a reading level of halfway through 10th grade.

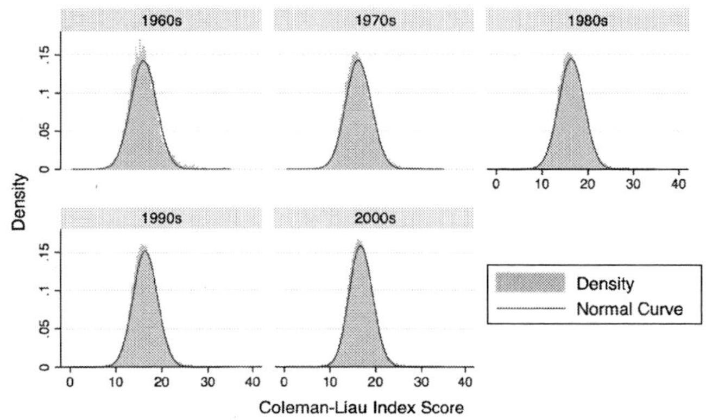

Figure 1: Distribution of CLI Scores by decade.

3.3 Statistical analyses

The next phase of the study involved creating a database and running analyses to determine the mean CLI scores for abstracts in each decade and whether the differences between these mean scores were statistically significant. A statistically significant difference in the mean CLI scores for each decade would indicate changes in the readability of medical journal abstracts over time.

In order to avoid type 1 errors, the analysis did not engage in a series of t-tests to compare the mean CLI scores for each decade. Rather a one-way ANOVA was deemed more appropriate after checking that the data met certain assumptions. Specifically, ANOVA requires that the data have an approximately normal distribution. Evidence for normality includes histograms of each decade's CLI scores with each distribution closely following a normal curve (see Figure 1 above). A Shapiro-Wilk test for normality could not be done because it has an upper limit of 2,000 to 5000 observations (Razali & Wah, 2011), and the data sets in this paper surpass that (see Table 1 above). However, examination of the quantile-quantile plots (Figure 2 below) is consistent with the data being approximately normally distributed in each decade with only a small fraction of overall observations displaying deviations in the tails of some plots.

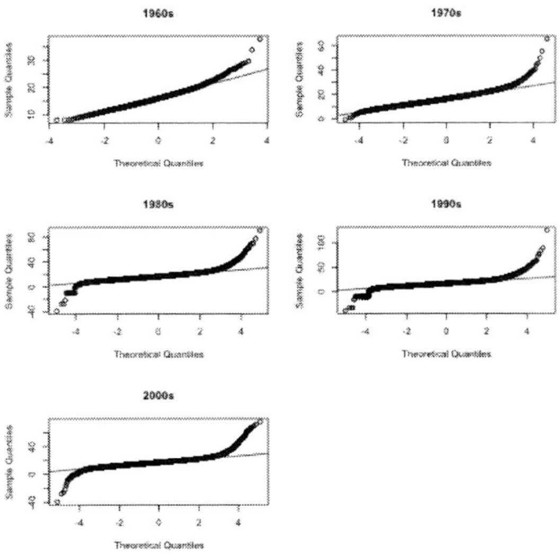

Figure 2: Quantile-quantile plots, by decade.

ANOVA also requires very similar variances for each group. Levene's test for homogeneity of variance, run on subsets created through random sampling of each decade, gave statistically significant values, which means the null hypothesis that the variances were the same could not be rejected. Another assumption for running ANOVA is that the data are independent. Although it is possible that some research articles may have been republished or had text cited in different decades, such instances likely were rare and not significant given the size of the corpus.

With the above assumptions addressed, the one-way ANOVA was carried out. Examination of the output indicated that a statistically significant difference did exist between the mean CLI scores for each decade. A one-way ANOVA, however, is an omnibus test and does not indicate between which groups the statistically significant difference exists, just that a statistically significant difference exists somewhere in the data. To determine between which decades there exists a statistically significant difference in mean CLI scores, a post hoc analysis using Tukey's method was carried out.

4 Results

The mean CLI scores for each decade were calculated (see Table 2.) The 1960s had a mean CLI score of 16.0813 with a 95% confidence interval (CI) of 16.00567 to 16.1569. The 1970s had a mean CLI score of 16.3123 with a 95% CI of 16.3024 to 16.32212. The 1980s had a mean CLI score of 16.3867 with a 95% CI of 16.38137 to 16.39194. The 1990s had a mean CLI score of 16.4302 with a 95% CI of 16.42657 to 16.43385. The 2000s had a mean CLI score of 16.8617 with a 95% CI of 16.85901 to 16.86446. Note that none of the 95% CIs overlap between decades.

Table 2. Mean CLI scores by decade.

Decade	Mean CLI Score	Number of Abstracts
1960s	16.0813	5324
1970s	16.3123	313053
1980s	16.3867	1049637
1990s	16.4302	2017482
2000s	16.8617	3327954

A one-way ANOVA indicated a statistically significant difference between the mean CLI scores for each decade ($F(4, 6689135) = 12936.91$, $p<0.0001$; see table 3). In determining which pairs of mean CLI Scores for each decade had a statistically significant difference, a pairwise comparison of means post hoc analysis using Tukey's method indicated that all possible combinations of CLI Scores for each decade had statistically significant differences ($p<0.001$).

5 Analysis

Having confirmed the statistical significance of the differences between all pairings of the mean CLI scores for each decade, we can consider the mean CLI scores for each decade statistically distinct

Table 3. One-way ANOVA results comparing mean CLI scores by decade.

Source	SS	df	MS	F-statistic	p-value
Between groups	353331.527	4	88332.8818	12936.91	<0.0001
Within groups	45673229.4	6689135	6.82797244		
Total	46026560.9	6689139	6.88079003		

from one another. Given this, we can make higher-level observations based on what patterns the individual means reveal as part of a group. More importantly, we can make assertions that allow us to answer our research question: How has the readability of medical journal abstracts changed between the 1960s and 2000s?

According to the results of this study, the average difficulty in readability of medical research journal abstracts increased over time. Specifically, readability scores increased from decade to decade beginning with a mean CLI score of 16.0813 in the 1960s and ending with a mean CLI score of 16.8617 in the 2000s. The mean CLI score, therefore, increased 0.7804 grade level units within the timespan examined. We should also note the high mean CLI scores for each decade. All scores fell within the level of readability expected for a grade level of 16 or a senior in college.

6 Future work

The work reported here discusses only one readability metric. Fleshing out the data with additional readability metrics would prove useful. Experimental assessment of comprehension by lay readers would be a useful addition to the metrics; for example, by asking them to read abstracts and answer questions. Specific subdomains of the biomedical literature may have their own readability issues, such as formulae and gene names, and identifying these might have implications for approaches to addressing specific readability issues.

7 Conclusion

This study sought to determine whether the readability of medical research journal abstracts changed between the 1960s and 2000s. The results here indicate an increase in difficulty of 0.7804 grade levels during this time period. Medical journal abstracts, we can conclude, have become more and more difficult to read.

For patients attempting to learn more about medical conditions or their treatment options through the reading primary literature, this task has become more difficult to achieve. Importantly, however, the high overall mean CLI scores for each decade indicate that this task likely has always proven difficult for patients. Medical journal abstracts have had readability scores equivalent to grade levels of 16 since the 1960s, well above the average American who reads between a 7th and 8th grade level (NCES, 2003) and certainly above the 9th-grade level considered "difficult" (USDHHS, 2000). This consistent difficulty mirrors other research showing a lack of progress in the readability of medical-related text (Rudd et al., 1999).

From this study's results and the US Department of Health and Human Services recommendations for the reading levels of medical information text, the readability gap between published medical research and the average American patient's reading ability appears equal to 7 grade levels. Bridging this chasm in accessibility will likely require interventions for both the researcher and patient. Shoring up the "health literacy" of Americans would involve a concerted effort to increase the average reading ability of patients. Purposefully addressing health literacy in K-12 education settings and Adult Basic Education settings may prove beneficial (Nielsen-Bohlman et al., 2004; Rudd et al., 1999). Such efforts, however, will likely not bridge the 7 grade level gap entirely. Instead, the medical research community should consider taking steps—for example, developing reading guides or parallel publications aimed at lay readers—to increase the readability of their research given patients' information needs and to support patient self-advocacy.

Despite a desire by patients to access and comprehend research that would increase their involvement in their own care, members of the medical research and publishing community continue to place a premium on complex writing skills putting such research out of the reach of most patients. Lakoff (1992) makes a strong case for academics in general being rewarded for difficult writing, and perhaps even being published for incomprehensible writing. With typical reading levels of almost 17, most scientific writing is now beyond the reading level of not only the average patient but also most health professionals who typically have a bachelor's degree equivalent to a grade level of 16.

Acknowledgments

We thank the participants in LING 5200, Computational Corpus Linguistics, in Fall 2014 for their input into this project. Noemie Elhadad and Gondy Leroy provided helpful comments on a late draft of the work.

References

Alamoudi, U., and Hong, P. (2015) Readability and quality assessment of websites related to microtia and aural atresia. *International Journal of Pediatric Otorhinolaryngology* 79(2):151-156.

Baker, M. T., & Taub, H. A. (1983). Readability of informed consent forms for research in a Veterans Administration medical center. *Jama*, *250*(19), 2646-2648.

Baker, D. W., Parker, R. M., Williams, M. V., & Clark, W. S. (1998). Health literacy and the risk of hospital admission. *Journal of general internal medicine*,*13*(12), 791-798.

Coleman, M., & Liau, T. L. (1975). A computer readability formula designed for machine scoring. *Journal of Applied Psychology*, *60*(2), 283.

Elhadad, N. (2006) Comprehending technical texts: predicting and defining unfamiliar terms. *American Medical Informatics Association Symposium Proceedings*, pp. 239-243.

Elhadad, N., and Sutaria, K. Mining a lexicon of technical terms and lay equivalents. *BioNLP 2007*, pp. 49-56.

Feng, L., Elhadad, N., and Huenefauth, M. (2009) Cognitively motivated features for readability assessment. *EACL 2009*, pp. 229-237.

Hartley, J. (2004). Current findings from research on structured abstracts.*Journal of the Medical Library Association*, *92*(3), 368.

Hopper, K. D., TenHave, T. R., Tully, D. A., & Hall, T. E. (1998). The readability of currently used surgical/procedure consent forms in the United States.*Surgery*, *123*(5), 496-503.

Huenefauth, M., Feng, L, and Elhadad, N. (2009) Comparing evaluation techniques for text readability software for adults with intellectual disabilities. *ACM SIGACCESS conference on computers and accessibility*, pp. 3-10.

Joint Commission. (2007) What did the doctor say? Improving health literacy to protect patient safety. *Health Care at the Crossroads* series. http://www.jointcommission.org/nr/rdonlyres/d5248b2e-e7e6-4121-887499c7b4888301/0/improving_health_literacy.pdf

Lakoff, R.T. (1992) *Talking power: the politics of language*. Basic Books.

Leroy, G., Endicott, J. E., Kauchak, D., Mouradi, O., & Just, M. (2013). User evaluation of the effects of a text simplification algorithm using term familiarity on perception, understanding, learning, and information retention. *Journal of medical Internet research*, *15*(7).

Leroy, G., Endicott, J. E., Mouradi, O., Kauchak, D., & Just, M. L. (2012). Improving perceived and actual text difficulty for health information consumers using semi-automated methods. In *AMIA Annual Symposium Proceedings* (Vol. 2012, p. 522). American Medical Informatics Association.

Leroy, G., Helmreich, S., & Cowie, J. R. (2010). The influence of text characteristics on perceived and actual difficulty of health information. *International journal of medical informatics*, *79*(6), 438-449.

Leroy, G., Kauchak, D., & Mouradi, O. (2013). A user-study measuring the effects of lexical simplification and coherence enhancement on perceived and actual text difficulty. *International journal of medical informatics*, *82*(8), 717-730.

Meade, C. D., & Howser, D. M. (1991, December). Consent forms: how to determine and improve their readability. In *Oncology nursing forum* (Vol. 19, No. 10, pp. 1523-1528).

Navarra, A.M., Neu, N., Toussi, S., Nelson, J., & Larson, E.L. (2014) Health literacy and adherence to antiretroviral therapy among HIV-infected youth. *J. Assoc. Nurses AIDS Care.* 25(3):203-213.

National Center for Education Statistics (2003). National Assessment of Adult Literacy (NAAL). http://nces.ed.gov/naal.

Pravikoff, D. S., Tanner, A. B., & Pierce, S. T. (2005). Readiness of US nurses for evidence-based practice: many don't understand or value research and have had little or no training to help them find evidence on which to base their practice. *AJN The American Journal of Nursing*, *105*(9), 40-51.

Razali, N. M., & Wah, Y. B. (2011). Power comparisons of Shapiro-Wilk , Kolmogorov-Smirnov , Lilliefors and Anderson-Darling tests. *Journal of Statistical Modeling and Analytics*, *2*(1), 21–33.

Roter, D. L., Rudd, R. E., Keogh, J., & Robinson, B. (1986). Worker produced health education material for the construction trades. *International Quarterly of Community Health Education*, *7*(2), 109-121.

Rudd, R. E., Moeykens, B. A., & Colton, T. C. (1999). Health and literacy: a review of medical and public health literature. *Office of Educational Research and Improvement.*

Smith, S. K., Dixon, A., Trevena, L., Nutbeam, D., & McCaffery, K. J. (2009). Exploring patient involvement in healthcare decision making across different education and functional health literacy groups. *Social science & medicine*,*69*(12), 1805-1812.

Štajner, S. (2011, September). Towards a Better Exploitation of the Brown 'Family' Corpora in Diachronic Studies of British and American English Language Varieties. In *RANLP Student Research Workshop* (pp. 17-24).

Temnikova, I. (2012a) Improving emergency instructions. *Communicator* (pp. 48-53).

Temnikova, I. (2012b) *Text complexity and text simplification in the crisis management domain.* University of Wolverhampton doctoral thesis.

United States Department of Health and Human Services. (2000) Saying it clearly. http://www.talkingquality.gov/docs/section3/3_4.htm.

Weiss, B. D., Coyne, C., Michielutte, R., Davis, T. C., Meade, C. D., Doak, L. G., ... & Furnas, S. (1998). Communicating with patients who have limited literacy skills-Report of the National Work Group on Literacy and Health.*Journal of Family Practice*, *46*(2), 168-176.

Woolf, S. H. (2008). The meaning of translational research and why it matters. *JAMA*, *299*(2), 211-213.

Translating Electronic Health Record Notes from English to Spanish: A Preliminary Study

Weisong Liu, PhD
University of Massachusetts Medical School,
Worcester, MA
Weisong.Liu@umassmed.edu

Shu Cai
Information Sciences Institute
Marina del Rey, CA
shucai@isi.edu

Balaji P Ramesh, PhD
University of Massachusetts Medical School,
Worcester, MA
balaji288@gmail.com

Germán Chiriboga, MPH
University of Massachusetts Medical School,
Worcester, MA
German.Chiriboga@umassmed.edu

Kevin Knight, PhD
Information Sciences Institute
Marina del Rey, CA
knight@isi.edu

Hong Yu, PhD
University of Massachusetts Medical School,
Worcester, MA
Hong.Yu@umassmed.edu

Abstract

The Centers for Medicare & Medicaid Services Incentive Programs promote meaningful use of electronic health records (EHRs), which, among many benefits, allow patients to receive electronic copies of their EHRs and thereby empower them to take a more active role in their health. In the United States, however, 17% population is Hispanic, of which 50% has limited English language skills. To help this population take advantage of their EHRs, we are developing English-Spanish machine translation (MT) systems for EHRs. In this study, we first built an English-Spanish parallel corpus and trained $\text{NoteAid}_{\text{Spanish}}$, a statistical MT (SMT) system. Google Translator and Microsoft Bing Translator are two baseline MT systems. In addition, we evaluated hybrid MT systems that first replace medical jargon in EHR notes with lay terms and then translate the notes with SMT systems. Evaluation on a small set of EHR notes, our results show that Google Translator outperformed $\text{NoteAid}_{\text{Spanish}}$. The hybrid SMT systems first map medical jargon to lay language. This step improved the translation. A fully implemented hybrid MT system is available at http://www.clinicalnotesaid.org. The English-Spanish parallel-aligned MedlinePlus corpus is available upon request.

1 Introduction

The Centers for Medicare & Medicaid Services Incentive Programs promote meaningful use of electronic health records (EHRs), which, among many benefits, allow patients to receive electronic copies of their health records and thereby empower them to take a more active role in their health. EHRs present a new and *personalized* communication channel that has the potential to increase patient involvement in care and improve communication between physicians and patients and their caregivers. In particular, allowing patients access to their physicians' notes has the potential to enhance patients' understanding of their conditions and disease and improve medication adherence and self-managed care.

However, most EHRs are written in English. In the United States, 17% population is Hispanic, of which 50% has limited English language skills. Many general-purpose MT systems are available. For example, Google Translate is a free service that has been used by health professions. Like most general-purpose MT systems, it is based on SMT, looking for patterns in hundreds of millions of WWW documents. In contrast, EHRs contain medical terms, shortened forms, complex disease and medication names, and other domain-specific jargon that do not typically appear in WWW documents, and therefore Google Translate may not perform well for EHRs, as was found in a prior study that evaluated general-purpose MT systems (Zeng-Treitler et al., 2010). Furthermore, the Health Insurance Portability and Accountability Act of 1996 protects the privacy and security of individually identifiable health information, so a secure MT system may be needed for US hospitals.

Therefore we are developing an EHR domain–specific English-Spanish MT system called *$\textit{NoteAid}_{\textit{Spanish}}$*, which may help over 37 million Spanish speaking US residents to meaningfully use their EHRs.

Proceedings of the 2015 Workshop on Biomedical Natural Language Processing (BioNLP 2015), pages 134–140,
Beijing, China, July 30, 2015.

2 Background

MT has been an active research field for the past 60–70 years. Early systems mainly applied bilingual dictionaries and manually crafted rules. However, since the 1990s, research has turned to SMT (Brown et al., 1990). The best SMT systems are built from translation patterns that are learned automatically from parallel, human-translated text corpora (Koehn, 2010). Translation patterns include phrase translations that translate input text by translating sequences of words at a time (Koehn et al., 2003; Och, 2002), re-ordering tendencies allowing swapping of words or phrases (Tillmann, 2004), hierarchical phrase translations with variables (Chiang, 2007), and syntax-based transformations (Galley et al., 2004). Automatic learning enables systems to imitate human translation behavior and adapt to particular domains. The bulk of current MT research is tested on domains such as news and politics. The BLEU (Papineni et al., 2002) score is a standard evaluation metric for MT. It measures n-gram overlap with human translations and has shown correlation with human judgment.

Comparatively few MT systems have been developed in the medical domain. Early work focused on knowledge-based approaches for phrase translation (Eck et al., 2004; Humphrey et al., 1998; Liu et al., 2006; Merabti et al., 2011). Several research groups built parallel corpora, then trained SMT systems (Wu et al., 2011; Yepes et al., 2013). The Shared Task of Medical Translation provided both parallel aligned and monolingual corpora (Bojar et al., 2014). Eight teams participated the shared task and most of the systems were based on the Mose phrase-based toolkit with in-domain and out-of-domain language models.

Zeng-Treitler et al (Zeng-Treitler et al., 2010) evaluated a general-purpose MT tool called Babel Fish to translate 213 EHR note sentences from English into Spanish, Chinese, Russian, and Korean and then evaluated the comprehensibility and accuracy of the translation. They found, however, the majority of the translations were incomprehensible and/or incorrect.

3 Methods

We first built a domain-specific English-Spanish parallel aligned corpus and then developed and evaluated SMT and hybrid machine translation (HMT) systems for translating EHR notes from English to Spanish. This study was approved by the Institutional Review Board of University of Massachusetts Medical School. All EHR notes have been deidentified.

3.1 English-Spanish Parallel Aligned Bio-medical Corpora

The MedlinePlus ($ESPAC_{MedlinePlus}$)

Source: The NIH's MedlinePlus ((U.S.),) web site hosts web pages of medical articles of different health topics. Most of the articles in English have a corresponding Spanish version translated by human. 2,999 articles have Spanish translations, which we crawled to build the parallel aligned corpus. We conducted data cleaning and sentence alignment. We split $ESPAC_{MedlinePlus}$ into a training set (60%), a tuning set (20%) and a testing set (20%) by interleaving sentence by sentence. Table 1 shows the statistics of the data. Unknown words or word types on the English side are 4,580 and 3,308 for tuning and 4,558 and 3,309 for testing.

	Sentence Pairs	Word tokens (English)	Sent. Length (English)	Word tokens (Spanish)	Sent. Length (Spanish)
Training	85,540	1,005,342	11.7	1,135,080	13.27
Tuning	29,299	341,821	11.7	386,754	13.20
Testing	29,258	338,431	11.6	382,239	13.06

Table 1. Statistics of $ESPAC_{MedlinePlus}$

The EHR Corpus ($ESPAC_{EHR}$)

The UMass Amherst Translation Center translated three de-identified EHR notes (108 sentences, 13.4 word tokens per sentence, and a total of 1,445 words) from English to Spanish.

3.2 MT Systems

Phrase-Based SMT

Using $ESPAC_{MedlinePlus}$, we trained an initial phrase-based Moses (Koehn et al., 2007) system. The training aligns the words in sentence pairs and extracts phrase pairs consistent with those alignments. We set the maximum phrase pair length to 7 words. We trained a 3-gram language model on the Spanish side using SRILM (Stolcke, 2002; Stolcke et al., 2011). We first used the default feature weights in Moses, then adjusted these feature weights using MERT (Och, 2003).

HMT Systems

EHR notes contain medical jargon that differs significantly from the consumer-oriented medical corpora most MT systems are trained on. We therefore speculate that if we replace medical jargon with lay terms and then feed the transformed EHR note to a SMT system, we may improve the MT performance. In our HMT system, we first applied the Metemap (Aronson, 2001) to map free text to UMLS concepts. For those mapped concepts, we replace the medical jargon with lay terms. A concept is clinically relevant if it belongs to one of the 18 UMLS semantic types, as described in the *NoteAid* system (Ramesh et al., 2013). A term is a lay term if it appears in the Consumer Health Vocabulary of the UMLS. A term is also a lay term if it appears in MedlinePlus. We also identify abbreviations and replace them with their expanded full terms. The second component of the HMT systems is an SMT system. We explored two state-of-the-art SMT systems, Google Translate and Microsoft Bing Translator, resulting in two HMT systems, *NoteAid-Google*$_{Spanish}$ and *NoteAid-Bing*$_{Spanish}$.

Baseline MT Systems

The baseline systems are the state-of-the-art general purpose Google and Bing MT systems in which EHR notes are directly fed into the systems without any medical jargon replacement.

3.3 Evaluation Metrics and Procedure

All the MT systems were evaluated by single-reference, case-insensitive BLEU score using the Moses package. We also asked a bilingual domain expert to manually evaluate the five MT system outputs of the three EHR notes.

4 Results

4.1 Automatic Evaluation

The BLEU score of *NoteAid-Moses*$_{Spanish}$ on the tuning and testing medical parallel data are 41.8 (1.097) and 41.2 (1.104) before MERT and 50.4 (0.99) and 49.8 (0.99) after MERT. The BLEU score of Google Translate, which was 49.9 (0.99). Table 2 shows the performance (macro-average) of the translation systems on the three de-identified EHR notes. We found that 17.9% of all terms in EHR notes do not appear in the MedlinePlus corpus,

	BLEU score (ave. ± SD)	Sentence length ratio (ave. ± SD)
Bing	21.33 ± 7.38	1.02 ± 0.07
***NoteAid-Bing*$_{Spanish}$**	18.17 ± 7.38	1.03 ± 0.07
Google	14.05 ± 6.30	1.24 ± 0.04
***NoteAid-Google*$_{Spanish}$**	11.05 ± 5.63	1.23 ± 0.04
***NoteAid-Moses*$_{Spanish}$**	5.82 ± 1.95	1.10 ± 0.02

Table 2. MT systems on $ESPAC_{EHR}$

4.2 Evaluation by a Domain Expert

A bilingual human expert performed a blind review of the outputs of all five MT systems on the three EHR notes (a total of 15 Spanish outputs). He ranked all five MT systems. In addition, he marked up the errors by each MT system.

The expert judged that each MT system had a few translation omissions. For example, "symptomatically," was omitted by all the MT systems. Of the three EHR notes, *Google Translate* performed the best for two. *NoteAid-Google*$_{Spanish}$ and *NoteAid-Bing*$_{Spanish}$ were second on three. Bing Translator was the best for one. *NoteAid-Moses*$_{Spanish}$ was the last.

The expert also performed a blind comparison of *Google Translate* versus *NoteAid-Google*$_{Spanish}$. He found that the hybrid system simplified the medical jargon and translated well. However, it introduced inconsistencies a few times. Therefore, the rating for *Google translation* is slightly better on two out of the three EHR notes.

5 Discussion

There are a number of challenges for translating EHR notes from English to Spanish. Spanish translation frequently increases token length. In addition, rhetoric styles differ, which can considerably affect text length in cases where the medical note is more of a narrative than a sequence of facts and isolated sentences (Valero-Garces, 1996). Finally, it is expensive to create English-Spanish parallel aligned EHR corpora.

Both *NoteAid-Moses*$_{Spanish}$ and Google Translate achieved a competitive performance for *ESPAC*$_{MedlinePlus}$. Several factors could have contributed to the excellent MT performance. Since 25% of our data is redundant, during the training process the decoder memorized those sentences. This combined with the fact that the total percentage of unknown words and sentences were small (~16%) may have contributed to the good results. In addition, we found that 37% of the sentences in the tuning and testing sets had less than seven words, and about half of those sentences overlapped with the training set. These sentences were memorized as phrases during training, although their contribution to the overall performance was less significant than longer sentences. Finally, translating sentences with one word is easier than translating sentences with multiple words because one-word sentences do not have a re-ordering problem, which is one of the challenges in MT.

The evaluation of MT systems on EHR notes (Table 2) showed much reduced performance. The results are not surprising since 17.9% terms in EHR notes do not appear in the MedlinePlus.

In addition, all HMT systems performed worse than their SMT counterparts. The lower performance of HMT systems can be attributed to the lack of gold standards that exactly match the source text of hybrid systems. The gold standard consists of original English notes translated to Spanish by human translators. But, the HMT systems modify the original notes by replacing the medical jargon with lay terms and then translate the notes to Spanish. Since, the BLEU score calculates what percentage of the n-grams or phrases from the translations also appear in the gold standard and the HMT systems modify the original text before translation, it is expected to yield a lower performance.

We also found that sentences in EHR notes were not always grammatically well formed. Whereas, when humans translated the text, they inferred the context from the note and formed coherent and logical sentences by inserting the missing verb or conjunction. The translation systems translated the original ill-formed sentences into Spanish word for word. This resulted in a lower BLEU score performance for MT systems.

Our manual analyses show that the baseline and the HMT systems perform well and make very few mistakes on EHR notes. The mistakes include:

- Translation omission when they encounter typos in the source language. For example, the MT systems failed to translate typos like "possily" and "phychological."
- Failure to take context into consideration when translating the text. For example, in "we are redrawing blood cultures," the MT systems failed to recognize that "redraw" refers to removing blood cultures, and translated it as "redibujando" or "rediseñando," meaning redrawing or redesigning something.
- Incorrect grammatical gender assignment although the translation is correct. For example, "Skin: Warm and dry" is translated as "Piel: Cálido y seco" ignoring the fact that the grammatical gender context of "Piel"/Skin is feminine.
- Errors in verb conjugation. For example, "to drain" is translated as "para drenar" instead of "á drenar."

We select and describe three examples of errors by MT systems, as shown below.

In the example below, all the five MT systems fail to accurately translate the sentence and change the meaning when translated back to English. We also observed that human translators often translate the text using different words while maintaining the semantic sense of the sentence.

Source: Acute renal failure with neutropenia likely medication induced
Human Translation: fallo renal grave con neutropenia probablemente debido a medicamento.
Human Back Translation: severe renal failure with neutropenia probably due to medication
Google Translate: La insuficiencia renal aguda con neutropenia probable medicación inducida
Human Back Translation: acute renal failure with neutropenia probably induced medication
Bing Translate: Insuficiencia renal aguda con medicación probable neutropenia inducida
Human Back Translation: acute renal failure with medication, probably induced neutropenia
NoteAid-Moses$_{Spanish}$: insuficiencia renal aguda con la neutropenia probable *Medicines induced*
Human Back Translation: Probable medication induced acute renal failure with neutropenia
NoteAid-Google$_{Spanish}$: insuficiencia renal aguda con neutropenia probables Medicamentos inducidos
Human Back Translation: acute renal failure with neutropenia, probable induced medications
NoteAid-Bing$_{Spanish}$: la insuficiencia renal aguda con neutropenia indujeron probables medicamentos
Human Back Translation: acute renal failure with neutropenia induced probable medications

In this example, *NoteAid-Moses$_{Spanish}$* conserves only some of the source text's context and format but omits translation of several words, including medical jargon. The *NoteAid-Bing$_{Spanish}$* omits only one word but the remaining MT systems do not omit any word. The Google translate and both the hybrid systems make a grammatical mistake by assigning incorrect gender to the patient in Spanish.

Source: ASSESSMENT AND PLAN: The patient was scheduled for a kidney biopsy today, but she was informed by the Renal Transplant Service that they were going to delay this since there was some improvement in her creatinine (today's creatinine is not yet available).

Human Translation: EVALUACION Y PLAN: Se proyectaba que la paciente tuviese una biopsia del riñón hoy, pero el Servicio de Trasplante Renal le informó que iban a retrasarla pues ha habido una mejoría en su creatinina (la creatinina de hoy todavía no está disponible).

Google Translate: EVALUACIÓN Y PLAN: **El** paciente fue programado para una biopsia de riñón hoy, pero fue informado por el Servicio de Trasplante Renal de que iban a demorar esto, ya hubo alguna mejora en su creatinina (creatinina de hoy todavía no está disponible).

Bing Translate: EVALUACIÓN y PLAN: La paciente estaba programada para una biopsia de riñón hoy, pero fue informada por el servicio de Trasplante Renal que iban a retrasar esto ya que hubo cierta mejoría en la creatinina sérica (creatinina de hoy aún no está disponible).

NoteAid-Moses$_{Spanish}$: *ASSESSMENT AND PLAN: The* paciente se programado para una biopsia del riñón *today*, pero que ella estaba informado por la *Renal Transplant Service* que fueron de irse a retrasar este dado que no hubo alguna mejoría en su creatinina (*today's* creatinina aún no se *available*).

NoteAid-Google$_{Spanish}$: EVALUACIÓN Y PLAN: **El** paciente fue programado para una biopsia de riñón hoy, pero fue informado por el Servicio de Trasplante Renal de que iban a demorar esto, ya hubo alguna mejora en sus creatininas (hoy creatininas aún no está disponible).

NoteAid-Bing$_{Spanish}$: EVALUACIÓN y PLAN: **El** paciente fue programado para una biopsia de riñón hoy, pero fue informada por el servicio de trasplante de riñón que iban a retrasar esto ya que hubo cierta mejoría en sus *creatinines* (hoy *creatinines* aún no está disponible).

Unlike the narrative in the previous examples, the non-narrative context consists of a list of medical jargon, the format of which is quite common in clinical notes. While the baseline and hybrid MT translation systems perform well, *NoteAid-Moses$_{Spanish}$* omits many of the terms.

Source: Lungs: bilateral: reduced breath sounds, expiratory wheezes present (in the upper lobe, in the lower lobe), at or near baseline? , no rhonchi present, . Skin: lesion, . b/l venous stasis changes on distal anterior tibias trace tibial edema

Human Translation: Pulmones: bilateral: sonidos respiratorios reducidos, presencia de resoplido espiratorio (en el lóbulo superior, en el lóbulo inferior), ¿en o cerca de la línea de base? , no hay presencia de estertor roncus, . Piel: lesión, . cambios bilaterales de estasis venosa en tibias anteriores distales indicio edema tribial

Google Translate: Pulmones: bilateral : ruidos respiratorios reducida , sibilancias espiratorias presentes (en el lóbulo superior , en el lóbulo inferior) , en o cerca de la línea de base ? , No hay roncus presentes , . Piel: lesión , . b / l estasis venosa cambios en la tibia distal tibial anterior traza edema

Bing Translate: Pulmones: bilateral: sonidos respiratorios reducida, resuellos espiratorios presentes (en el lóbulo superior, en el lóbulo inferior), en o cerca de línea de base. , presentes, no roncus. Piel: lesión. b/l estasis venosa cambios en tibias anteriores distales rastrear el edema tibial

NoteAid-Moses$_{Spanish}$: *Lungs*: bilateral: reducción de aliento *sounds*, *wheezes* presente (in dispositivos de la parte superior del *lobe*, en la parte inferior de *lobe*), , en o cerca de *baseline*? , no *present*, estertores . *Skin*: *lesion*, . b/l de estasis venosa cambios en el edema tibial anterior distal tibias trace

NoteAid-Google$_{Spanish}$: Pulmones: bilateral: reducción de sonidos pulmonares espiratorio sibilancias presentes (en el lóbulo superior , en el lóbulo inferior) , en o cerca de la línea de base la visión ? , No hay rhonchis presentes , . Piel: lesión . cambios b / l venostasis en distal rastro tibias anterior tibial Edema

NoteAid-Bing$_{Spanish}$: Pulmones: bilateral: reducido sibilancias espiratorio de sonidos pulmonares presentes (en el lóbulo superior, en el lóbulo inferior),, en o cerca de base de la visión? , no rhonchis presente,. Piel: lesión. cambios b/l lindo tibias anteriores distales rastrear el Edema tibial

NoteAid-Moses$_{Spanish}$ performed poorly on the EHR notes, suggesting that the system needs to be trained on bigger data sets, or be trained directly on the EHR notes. We found that some errors by *NoteAid-Google$_{Spanish}$* were due to engineering errors, which can be fixed.

6 Limitations, Conclusion and Future Work

This pilot study has limitations. The SMT system was built on the limited MedlinePlus data. We plan to incorporate other biomedical corpora (e.g., Medline and ClinicalTrial.gov). The corpus size of EHR notes for evaluation is small and we plan to build such a corpus.

The BLEU score does not provide a measurement in terms of whether the semantic content is correctly translated. In the future work we may explore other domain-specific evaluation metrics (Castilla et al., 2005).

In this application, we have experimented with simple MT approaches. In the future we may explore other MT approaches, including incorporating biomedical knowledge resources (e.g., the UMLS), domain adaptation, semantic role labelling and abstract meaning represenation.

Acknowledgement: The authors thank the anonymous reviewers for invaluable comments.

References

A. R Aronson. 2001. Effective mapping of biomedical text to the UMLS Metathesaurus: the MetaMap program. *Proc AMIA Symp*:17–21.

M Baldry, C Cheal, B Fisher, M Gillett, and V Huet. 1986. Giving patients their own records in general practice: experience of patients and staff. *British Medical Journal (Clinical research ed.)*, 292(6520):596–598, March. PMID: 3081187PMCID: PMC1339574.

Ondrej Bojar, Christian Buck, Christian Federmann, Barry Haddow, Philipp Koehn, Johannes Leveling, Christof Monz, Pavel Pecina, Matt Post, Herve Saint-Amand, and others. 2014. Findings of the 2014 workshop on statistical machine translation. In *Proceedings of the Ninth Workshop on Statistical Machine Translation*, pages 12–58. Association for Computational Linguistics Baltimore, MD, USA.

Peter F. Brown, John Cocke, Stephen A. Della Pietra, Vincent J. Della Pietra, Fredrick Jelinek, John D. Lafferty, Robert L. Mercer, and Paul S. Roossin. 1990. A statistical approach to machine translation. *Computational linguistics*, 16(2):79–85.

A. C. Castilla, A. S. Bacic, and S. S. Furuie. 2005. Machine Translation on the Medical Domain: The Role of BLEU/NIST and METEOR in a Controlled Vocabulary Setting. *Proceedings of the Tenth Machine Translation Summit. Phuket, Thailand*:47.

David Chiang. 2007. Hierarchical phrase-based translation. *computational linguistics*, 33(2):201–228.

James J Cimino, Vimla L Patel, and Andre W Kushniruk. 2002. The patient clinical information system (PatCIS): technical solutions for and experience with giving patients access to their electronic medical records. *International journal of medical informatics*, 68(1-3):113–127, December. PMID: 12467796.

Tom Delbanco, Jan Walker, Sigall K. Bell, Jonathan D. Darer, Joann G. Elmore, Nadine Farag, Henry J. Feldman, Roanne Mejilla, Long Ngo, James D. Ralston, Stephen E. Ross, Neha Trivedi, Elisabeth Vodicka, and Suzanne G. Leveille. 2012. Inviting Patients to Read Their Doctors' Notes: A Quasi-experimental Study and a Look Ahead. *Annals of Internal Medicine*, 157(7):461–470, October.

Darren A DeWalt, Robert M Malone, Mary E Bryant, Margaret C Kosnar, Kelly E Corr, Russell L Rothman, Carla A Sueta, and Michael P Pignone. 2006. A heart failure self-management program for patients of all literacy levels: a randomized, controlled trial [ISRCTN11535170]. *BMC health services research*, 6:30. PMID: 16533388.

Matthias Eck, Stephan Vogel, and Alex Waibel. 2004. Improving statistical machine translation in the medical domain using the Unified Medical Language System. In *Proceedings of the 20th international conference on Computational Linguistics*, page 792.

Michel Galley, Mark Hopkins, Kevin Knight, and Daniel Marcu. 2004. What's in a Translation Rule? In *HLT-NAACL*, pages 273–280.

B. Humphrey, D. A. B. Lindberg, H. M. Schoolman, and G. O. Barnett. 1998. The Unified Medical Language System: An Informatics Research Collaboration. *Journal of the American Medical Association*, 5:1–11.

Philipp Koehn. 2010. *Statistical Machine Translation*. Cambridge University Press, 1 edition edition, January.

Philipp Koehn, Franz Josef Och, and Daniel Marcu. 2003. Statistical phrase-based translation. In *Proceedings of the 2003 Conference of the North American Chapter of the Association for Computational Linguistics on Human Language Technology-Volume 1*, pages 48–54.

P. Koehn, H. Hoang, A. Birch, C. Callison-Burch, M. Federico, N. Bertoldi, B. Cowan, W. Shen, C. Moran, R. Zens, and others. 2007. Moses: Open source toolkit for statistical machine translation. In *Proceedings of the 45th Annual Meeting of the ACL on Interactive Poster and Demonstration Sessions*, pages 177–180.

Fang Liu, Michael Ackerman, and Paul Fontelo. 2006. BabelMeSH: development of a cross-language tool for MEDLINE/PubMed. *AMIA ... Annual Symposium proceedings / AMIA Symposium. AMIA Symposium*:1012. PMID: 17238631 PMCID: PMC1839504.

Bradley M Mathers, Louisa Degenhardt, Hammad Ali, Lucas Wiessing, Matthew Hickman, Richard P Mattick, Bronwyn Myers, Atul Ambekar, and Steffanie A Strathdee. 20. HIV prevention, treatment, and care services for people who inject drugs: a systematic review of global, regional, and national coverage. *The Lancet*, 375(9719):1014–1028.

Michael Meltsner. 2012. A Patient's View of OpenNotes. *Annals of Internal Medicine*, 157(7):523–524, October.

Tayeb Merabti, Lina F. Soualmia, Julien Grosjean, Olivier Palombi, Jean-Michel Müller, and Stéfan J. Darmoni. 2011. Translating the Foundational Model of Anatomy into French using knowledge-based and lexical methods. *BMC medical informatics and decision making*, 11(1):65.

F.J. Och. 2003. Minimum error rate training in statistical machine translation. In *Proceedings of the 41st Annual Meeting on Association for Computational Linguistics-Volume 1*, pages 160–167.

Franz Josef Och. 2002. *Statistical machine translation: from single-word models to alignment templates*. Ph.D. thesis, Bibliothek der RWTH Aachen.

Kishore Papineni, Salim Roukos, Todd Ward, and Wei-Jing Zhu. 2002. BLEU: a method for automatic evaluation of machine translation. In *Proceedings of the 40th Annual Meeting on Association for Computational Linguistics*, pages 311–318, Stroudsburg, PA, USA. Association for Computational Linguistics.

Balaji Ramesh, Thomas Houston, Cynthia Brandt, Julia Fang, and Hong Yu. 2013. Improving Patients' Electronic Health Record Comprehension with NoteAid. *The 14th World Congress on Medical and Health Informatics. Best Student Paper.*

Dean Schillinger, Margaret Handley, Frances Wang, and Hali Hammer. 2009. Effects of self-management support on structure, process, and outcomes among vulnerable patients with diabetes: a three-arm practical clinical trial. *Diabetes care*, 32(4):559–566, April. PMID: 19131469.

J F Seitz, A Ward, and W H Dobbs. 1978. Granting patients access to records: the impact of the Privacy Act at a federal hospital. *Hospital & community psychiatry*, 29(5):288–289, May. PMID: 640590.

Andreas Stolcke. 2002. SRILM—An extensible language modeling toolkit. In *In Proceedings of the 7th International Conference on Spoken Language Processing (ICSLP 2002*, pages 901–904.

Andreas Stolcke, Jing Zheng, Wen Wang, and Victor Abrash. 2011. SRILM at sixteen: Update and outlook. In

Christoph Tillmann. 2004. A unigram orientation model for statistical machine translation. In *Proceedings of HLT-NAACL 2004: Short Papers*, pages 101–104.

National Library of Medicine (U.S.). Fact SheetMedlinePlus®.

C. Valero-Garces. 1996. Contrastive ESP Rhetoric: Metatext in Spanish-English Economics Texts. *English for Specific Purposes*, 15(4):279–294, November.

Warren J. Winkelman, Kevin J. Leonard, and Peter G. Rossos. 2005. Patient-Perceived Usefulness of Online Electronic Medical Records: Employing Grounded Theory in the Development of Information and Communication Technologies for Use by Patients Living with Chronic Illness. *Journal of the American Medical Informatics Association*, 12(3):306–314, May.

Cuijun Wu, Fei Xia, Louise Deleger, and Imre Solti. 2011. Statistical Machine Translation for Biomedical Text: Are We There Yet? *AMIA Annual Symposium Proceedings*, 2011:1290–1299. PMID: 22195190PMCID: PMC3243244.

Antonio Jimeno Yepes, Élise Prieur-Gaston, and Aurélie Névéol. 2013. Combining MEDLINE and publisher data to create parallel corpora for the automatic translation of biomedical text. *BMC Bioinformatics*, 14(1):146, April. PMID: 23631733.

Qing Zeng-Treitler, Hyeoneui Kim, Graciela Rosemblat, and Alla Keselman. 2010. Can multilingual machine translation help make medical record content more comprehensible to patients? *Studies in health technology and informatics*, 160(Pt 1):73–77. PMID: 20841653.

2014. ACL 2014 Ninth Workshop on Statistical Machine Translation Shared Task: Medical Translation. Technical report.

Automatic Detection of Answers to Research Questions from Medline Abstracts

Abdulaziz Alamri and Mark Stevenson
Department of Computer Science
The University of Sheffield
Sheffield, UK
adalamri1@sheffield.ac.uk; mark.stevenson@sheffield.ac.uk

Abstract

Given a set of abstracts retrieved from a search engine such as *Pubmed*, we aim to automatically identify the claim zone in each abstract and then select the best sentence(s) from that zone that can serve as an answer to a given query. The system can provide a fast access mechanism to the most informative sentence(s) in abstracts with respect to the given query.

1 Introduction

The large amount of medical literature hinders professionals from analyzing all the relevant knowledge to particular medical questions. Search engines are increasingly used to access such information. However, such systems retrieve documents based on the appearance of the query terms in the text despite the fact that they may describe another problem.

The search engine Pubmed® for example is a well known IR system to access more than 24 million abstracts for the biomedical literature including Medline® (Wheeler et al., 2008). The engine takes a query from user and returns a list of abstracts that can be relevant or partially irrelevant to the query, which requires from the user to go through each abstract for further analysis and evaluation.

Researchers who conduct a systematic review (Gough et al., 2012) tend to use the same approach to collect the studies of interest; however, they are found to spend significant effort identifying the studies that are relevant to the research question. Relevancy is usually measured by scanning the result and conclusion sections to identify authors claim and then comparing the claim with the review question; where a claim can be defined as the summary of the main points presented in a research argument.

Incorporating a middle tier system between the search engine and the user will be useful to minimize the effort required to filter the results. This research presents a system that aids those searching for studies that discuss a particular research question. The system acts as a mediator between the search engine and the user. It interprets the search engine results and returns the most informative sentence(s) from the claim zone of each abstract that are potential answers to the research question. The system reduces the cognitive loads on the user by assisting their identification of relevant claims within abstracts

The system comprises two components. The first component identifies the claim zone in each abstract using the rhetorical moves principle (Teufel and Moens, 2002), and the second component uses the sentences in the claim zone to predict the most informative sentence(s) from each abstract to the given query.

This paper makes three contributions: presenting a new set of features to build a classifier to identify the structure role of sentences in an abstract that is at least shows similar performance to the current systems; building a classifier to detect the best sentence(s) (lexically) that can be an answer to a given query; and introducing a new feature (*Z-score*) for this task.

2 Related Work

We are not aware of any work that has explicitly discussed the detection of claim sentence most related to a predefined question, however, studies have discussed related research.

Ruch et al. (2007) for example used the rhetorical moves approach to identify the conclusion sentences in abstracts. Their system was based on a Bayesian classifier, and normalized n-grams and relative position features. The main objective of that research was to identify sentences that belong to the conclusion sections of abstracts; they re-

Proceedings of the 2015 Workshop on Biomedical Natural Language Processing (BioNLP 2015), pages 141–146,
Beijing, China, July 30, 2015.

garded such information as *key* information to determine the research topic. Our research is similar to that work since we use the conclusion section to identify the key information in an abstract with respect to a query, but we also include the result sections.

Hirohata et al. (2008) showed a similar system using CRFs to classify the abstract sentences into four categories: objective, methods, results, and conclusions. That classifier takes into account the neighbouring features in sentence S_n such as the n-grams of the previous sentence S_{n-1} and the next sentence S_{n+1}.

Agarwal et al. (2009) described a system that automatically classifies sentences appear in full biomedical articles into one of four rhetorical categories: introduction, methods, results and discussions. The best system was achieved using Multinominal Naive Bayes. They reported that their system outperformed their baseline system which was a rule-based.

Recently, Yepes et al. (2013) described a system to index Gene Reference Into Function (GeneRIF) sentences that show novel functionality of genes mentioned in Medline. The goal of that work was to choose the most likely sentences to be selected for GeneRIF indexing. The best system was achieved using Naive Bayes classifier and various features including the discourse annotations (the NLM category labels) for the abstracts sentences.

Our research is close to Hirohata et al. (2008) system since we use the same algorithm, but use a different set of features to build the model. Moreover, it similar to Yepes et al.(2013) system since we use the value of the *nlmCategory* attribute rather than the labels provided by the authors to learn the role of sentences.

3 Method

3.1 Claim Zoning Component

This component is based on the hypothesis that the contribution of a research paper tend to be found within the result or conclusion sections of its abstract (Lin et al., 2009). Identifying these sections manually especially in unstructured abstracts is a tedious task. Medical abstracts tend to have logical structure (Orasan, 2001) in which each section represent a different role.

Unfortunately, about 70% of Medline abstracts are unstructured (have no section labels). Structured abstracts use a variety of these labels. The National Library of Medicine (NLM) have reported that 2,779 headings have been used to label abstracts sections in Medline (Ripple et al., 2012).

Relying on the labels provided by the abstracts authors to identify the roles of the sentences could be useful for research purpose; but in practice this means all Medline abstracts need to be reannotated even the structured abstracts to guarantee that they are labelled with the same set of annotations to understand their roles. This is not efficient especially when we consider the huge volume of the Medline repository.

To accommodate that problem, we use the NLM category value assigned to each section in the XML abstract (*nlmCategory* attribute). The NLM assigns five possible values (categories): *Objective, Background, Methods, Results and Conclusions*. This research uses these categories as an alternative way to learn the roles of abstracts sentences. This resolves two problems: first, the roles of sentences in structured abstracts can be automatically learned from the the value of the *nlmCategory* attribute without any further processing, consequently, the roles of sentences in 30% of the Medline abstracts can be accurately identified; second, those labels can be used to build a machine learning classifier to predict the role sentences of the unstructured abstracts in Medline.

The claim zoning component regards identifying the roles of sentences as a sequence labelling problem. This requires an algorithm that takes into account the neighbouring observations rather than only current observation as in other ordinary classifiers e.g. SVM and Naive bayes. Conditional Random Fields (CRF) algorithm have been used successfully for such task (Hirohata et al., 2008; Lin et al., 2009). Therefore, we use the CRF algorithm along with lexical, structural and sequential features to build a classifier model to identify the claim zones in abstracts. The classifier is implemented using the CRFsuite library (Okazaki, 2007) using L-BFGS method. Note that we modify the NLM five categories to become four where the *Background* and *Objective* categories are merged into a new category called *Introduction*. That is because the background and objectives sections in Medline tend to overlap with each other (Lin et al., 2009). Moreover, these sections usually appear sequentially and merging them together is sensible to avoid the overlapping problem. Therefore, this component identifies the sentences roles in abstracts by labelling them with one of the four possible categories: *Introduction, Methods, Results and Conclusions*.

3.1.1 Data

The claim zoning component is built using a dataset consisting of 10,000 structured abstracts collected from Medline using the query "*cardiovascular disease*".

3.1.2 Features

The claim zoning component employs various features:

N-grams: N-grams are lexical features that have been reported as useful to capture the general context of text (Turney, 2002; Yu and Hatzivassiloglou, 2003). For every sentence, uni-grams and bi-grams are extracted from the abstract's title, the current sentence S_n, the previous sentence S_{n-1}, and the next sentence S_{n+1}.

Sentence-Title similarity (*st-sim*): This feature is the cosine similarity score $sim(s, T)$ between each sentence in an abstract and its title. This feature has been previously found useful for summarization tasks (Teufel and Moens, 2002). Achieving an accurate similarity score between the sentences and the title in an abstract is not a straightforward task. Many abstracts in the medical domain use multiple forms (i.e abbreviation and its expansions) to describe the same medical concept e.g. ACE and angiotensin-converting enzyme.

Such variation may cause inaccurate scores particularly when computing the similarity between an abbreviation and its expansion. Fortunately, the pattern of using abbreviations and their expansions in medical research can be predicted using an algorithm developed by Schwartz and Hearst (2003). We automatically replace all long-forms concepts with their abbreviations to unified their appearance within an abstract. Similarity scores are binned into 11 values starting from 0 to 10.

Relative Sentence location: The relative location of a sentence is important to identify its role within the abstract. The introduction sentences for instance tend to occur at the beginning of an abstract and the conclusion sentences occur at the end. Rather than using the original position of the sentence, we adjusted the all sentences positions to have the same scale from 1 to 10.

Tense feature: The tense of verbs used in sentences often correlates with its rhetorical moves (Teufel and Moens, 2002). For example, some authors use the present perfect tense in the introduction section and past simple in the conclusion section. For each sentence in an abstract, the main verb tense (ROOT-0, verb) is extracted using the dependency tree generated from the Stanford parser (de Marneffe and Manning, 2008).

3.2 Answers Detection Component

This component uses the sentences that belong to the result or conclusion sections of abstracts (claim zone) to identify the most informative sentence(s) to a given query. It relies on three assumptions, two from the literature (Lin et al., 2008; Ruch et al., 2007; Lin et al., 2009; Otani and Tomiura, 2014) and the last one that is conventional: the first assumption is that any sentence in abstract that shares many words with the title tends to express important information about the topic. The second is that any sentence that applies the first assumption within certain threshold and exist in the result and conclusion sections is considered as a key sentence concerning the research topic. The third assumption is that any sentence that applies the previous two assumptions and has a high lexical similarity score with the query is considered an informative sentence with respect to the query.

The component classifier is built using a decision tree algorithm (Quinlan, 1993). The decision tree algorithm builds a tree-like model that can be converted into rules which can be easily interpreted and analysed by human. We use the open source implementation of decision tree (J48) in Weka (Hall et al., 2009) to build the model.

3.2.1 Data

This component uses two subsets (corpus-2 and corpus-3) of a corpus that was originally developed to recognize contradictory claims in medical abstracts. That corpus consists of abstracts that were collected from the studies used in systematic reviews that discuss various problems about cardiovascular diseases. Note that each systematic review attempts to answer one question. Two independent annotators were asked to identify the best claim sentence from each abstract that answers the review question e.g. (1). In this research the most informative information with respect to the research question is considered to be the claim.

1. In patients with dilated cardiomyopathy, are HLA genes associated with development of Dilated Cardiomyopathy? **[Question]**

2. In the IDC group, the frequency of human leukocyte antigen DR4 was similar to that reported in the normal population. **[PMID#9220309][ANSWER]**

The classifier of answer detection is trained and evaluated using corpus-2 (structured abstracts). That corpus consists of 183 sentences annotated as *answers* and 987 sentences annotated as *non-answers* to 24 review questions. Note that it is possible for more than one sentence to answer a review question, however, only the most informative sentence was annotated as answer.

Corpus-3 (unstructured abstracts) consists of 69 abstracts (69 *answer* sentences and 357 *non-answer* sentences) which answer 15 review questions. It is used to evaluate the system resulted from the integration of the claim zoning component and the answer detection component.

3.2.2 Features

This component uses four features which are extracted from the result and conclusions sentences:

Sentence Structure Role (*role-label*): This feature comes from annotating the abstract sentences using the claim zoning component if the abstract is unstructured, otherwise the value of *nlm-Category* is extracted and used as a feature.

Sentence-Title Similarity (*st-sim*): This feature is similar to *st-sim* feature used in the claim zoning component. The scores are normalized to a scale of 0 to 50 since this was shown to improve performance.

Sentence-Query Similarity (*sq-sim*): This feature captures the relationship between the research question and sentences in the abstract. Those with a high lexical similarity to the question are more likely to be answers to it than others. Similar to *st-sim* feature, the cosine similarity score between sentences and their related questions are computed and the scores are normalized to a scale of 0 to 50.

Z-score Value: This feature is used to exploit assumption (2) described in section 3.2. This feature identifies the position of the similarity score of a sentence with respect the distribution of the similarity scores of the other sentences within an abstract. It assumes that the similarities of the sentences in the result and conclusion sections are normally distributed. The goal of using this feature is to enable the classifier to learn a similarity threshold score that can be used to identify the potential answer sentences.

The *Z-score* value is a standard score that shows the number of standard deviations (σ) above the mean (μ) (Wonnacott and Wonnacott, 1990). This value is identified for each sentence by exploring all possible Z values using equation (1) that makes the similarity *st-sim* of that sentence is just equal or above the score X.

$$X = \mu + Z\sigma \qquad (1)$$

4 Result and Discussion

Table (1) describes the performance of the claim zoning component using corpus-1. Table (2) describes the performance of Hirohata et al. (2008) system using the same corpus. Although, the difference was not significant, our system showed an alternative set of features that can achieve at least similar performance to the state of the art systems.

	Precision	Recall	F1-score
Introduction	0.96	**0.95**	**0.96**
Method	**0.83**	0.82	0.83
Results	0.87	**0.89**	**0.88**
Conclusions	**0.93**	**0.92**	**0.92**
Overall	**0.89**	**0.89**	**0.89**

Table 1: Claim zoning performance

	Precision	Recall	F1-score
Introduction	0.96	0.94	0.95
Method	0.81	**0.84**	0.83
Results	**0.88**	0.86	0.87
Conclusions	0.91	0.91	0.91
Overall	0.88	0.88	0.88

Table 2: Hirohata et al. (2008) system performance.

The output of the first component, particularly the sentences in the results and conclusions sections were then used as input in the answer detection component. That component was trained and evaluated on *corpus-2* using 10-folds cross validation. Table (3) shows the component's performance using five different combinations of features as follows:

- feature-set 1: *st-sim, sq-sim*
- feature-set 2: *Z-score, sq-sim*
- feature-set 3: *st-sim, role-label*

- feature-set 4: *Z-score, role-label*
- feature-set 5: *st-sim, sq-sim, role-label*
- feature-set 6: *Z-score, sq-sim, role-label*

The goal of trying different features combinations was to measure the effect of the *Z-score* feature on enhancing the overall performance of the component. The component achieved F1-score of 45% using set 1 compared to 56% using set 2. At this stage it was clear that the *Z-score* feature outperformed the *st-sim* feature.

Next, the *sq-sim* feature was replaced with the *role-label* as in set 3 and 4; however the results showed that using set 3 enhanced the F1-score by 22% compared to using set 1; and 19% using set 4 compared to set 2. This suggested that combining the *st-sim* feature with *sq-sim* was better than combining the *Z-score* and *sq-sim*.

The experiment was repeated using set 5 and set 6 which included the *sq-sim* feature in set 3 and 4; and the results were consistent with the results of using set 3 and 4. The component using set 5 outperformed set 6 due to the recall score (85%) in set 5. However, the precision score using set 6 was higher than using set 5 (73% vs 70%). This result was consistent with the component performance using set 3 and 4.

The above experiments showed a comparison between the *st-sim* and the *Z-score* features. The results suggest that using the *Z-score* feature contributes more than the *st-sim* feature with respect to the precision score, but less with respect to the recall score.

	Precision	Recall	F1-score
features-set(1)	0.68	0.34	0.45
features-set(2)	0.67	0.48	0.56
features-set(3)	0.70	**0.85**	**0.77**
features-set(4)	**0.73**	0.78	0.75
features-set(5)	0.70	0.83	0.76
features-set(6)	**0.73**	0.75	0.74

Table 3: The performance of the answer detection component using different combinations of features

Table (4) shows the performance of integrating the two components (the claim zoning and answer detection) using *corpus-3*. Note that the corpus only consists of unstructured abstracts (see *section (3.2.1)*. The integrated system was able to achieve precision of 56%, recall of 57% and F1-score of 56%. The main reason for the reduction in the performance score was due to the number of the answers examples used in the corpus being relatively small (69 answers). Another reason was the errors generated from the claim zoning component, which may have influenced the decisions made by the answer detection component.

	Precision	Recall	F1-score
Answer	0.56	0.57	0.56
Non-answer	0.92	0.92	0.92
Overall	0.86	0.86	0.86

Table 4: Answer detection performance using both components

5 Conclusion

This paper explored the problem of identifying the sentence(s) in an abstract that are the most informative information for a given query. It described a system for automatically identifying these sentences that consisted of two components: claim zone detection and answers detection. The system used the attribute value of *nlmCategory* to learn the sentences roles, which was found useful. Moreover, the component used different set of features that achieved at least similar performance to other systems for similar task. Finally, the research examined a new feature (*Z-score*) that was extracted from the same information used in *(st-sim)* feature. The *Z-score* feature was found more useful to enhance the precision score of the system compared with the *st-sim*.

References

Shashank Agarwal and Hong Yu. 2009. Automatically classifying sentences in full-text biomedical articles into introduction, methods, results and discussion. *Bioinformatics*, 25(23):3174–3180, Dec.

Marie-Catherine de Marneffe and Christopher D. Manning. 2008. The stanford typed dependencies representation. In *Coling 2008: Proceedings of the workshop on Cross-Framework and Cross-Domain Parser Evaluation*, CrossParser '08, pages 1–8, Stroudsburg, PA, USA. Association for Computational Linguistics.

David Gough, Sandy Oliver, and James Thomas. 2012. *An introduction to systematic reviews*. Sage Publications.

Mark Hall, Eibe Frank, Geoffrey Holmes, Bernhard Pfahringer, Peter Reutemann, and Ian H. Witten. 2009. The weka data mining software: an update. *SIGKDD Explor. Newsl.*, 11(1):10–18.

Kenji Hirohata, Naoaki Okazaki, Sophia Ananiadou, and Mitsuru Ishizuka. 2008. Identifying sections in scientific abstracts using conditional random fields. In *In Proc. of the IJCNLP 2008*.

Antonio J. Jimeno-Yepes, J Caitlin Sticco, James G. Mork, and Alan R. Aronson. 2013. Generif indexing: sentence selection based on machine learning. *BMC Bioinformatics*, 14:171.

Ryan T.K. Lin, Hong-Jei Dai, Yue-Yang Bow, Min-Yuh Day, Richard Tzong-Han Tsai, and Wen-Lian Hsu. 2008. Result identification for biomedical abstracts using conditional random fields. In *Information Reuse and Integration, 2008. IRI 2008. IEEE International Conference on*, pages 122–126.

Ryan T. K. Lin, Hong-Jie Dai, Yue-Yang Bow, Justing Lian-Te Chiu, and Richardg Tzon-Han Tsai. 2009. Using conditional random fields for result identification in biomedical abstracts. *Integr. Comput.-Aided Eng.*, 16(4):339–352, December.

Naoaki Okazaki. 2007. Crfsuite: a fast implementation of conditional random fields (crfs). http://www.chokkan.org/software/crfsuite/.

Constantin Orasan. 2001. Patterns in scientific abstracts. In *In Proceedings of Corpus Linguistics 2001 Conference*, pages 433–443. Lancaster University.

S. Otani and Y. Tomiura. 2014. Extraction of key expressions indicating the important sentence from article abstracts. In *Advanced Applied Informatics (IIAIAAI), 2014 IIAI 3rd International Conference on*, pages 216–219.

Ross Quinlan. 1993. *C4.5: Programs for Machine Learning*. Morgan Kaufmann Publishers, San Mateo, CA.

Anna M. Ripple, James G. Mork, John M. Rozier, and Lou S. Knecht. 2012. Structured abstracts in medline: Twenty-five years later.

Patrick Ruch, Clia Boyer, Christine Chichester, Imad Tbahriti, Antoine Geissbhler, Paul Fabry, Julien Gobeill, Violaine Pillet, Dietrich Rebholz-Schuhmann, Christian Lovis, and Anne-Lise Veuthey. 2007. Using argumentation to extract key sentences from biomedical abstracts. *I. J. Medical Informatics*, 76(2-3):195–200.

A.S. Schwartz and M.A. Hearst. 2003. A simple algorithm for identifying abbreviation definitions in biomedical text. In *In Proceedings of Pacic Symposium on Biocomputing*, volume 4, pages 451–462, November.

Simone Teufel and Marc Moens. 2002. Summarizing scientific articles: Experiments with relevance and rhetorical status. *Comput. Linguist.*, 28(4):409–445.

Peter D. Turney. 2002. Thumbs up or thumbs down?: Semantic orientation applied to unsupervised classification of reviews. In *Proceedings of the 40th Annual Meeting on Association for Computational Linguistics*, ACL '02, pages 417–424, Stroudsburg, PA, USA. Association for Computational Linguistics.

David L. Wheeler, Tanya Barrett, Dennis A. Benson, Stephen H. Bryant, Kathi Canese, Vyacheslav Chetvernin, Deanna M. Church, Michael Dicuccio, Ron Edgar, Scott Federhen, Lewis Y. Geer, Yuri Kapustin, Oleg Khovayko, David L, David J. Lipman, Thomas L. Madden, Donna R. Maglott, James Ostell, Vadim Miller, Kim D. Pruitt, Gregory D. Schuler, Edwin Sequeira, Steven T. Sherry, Karl Sirotkin, Re Souvorov, Grigory Starchenko, Roman L. Tatusov, Tatiana A. Tatusova, Lukas Wagner, and Eugene Yaschenko. 2008. Database resources of the national center for biotechnology information. *Nucleic Acids Res*, pages 13–21.

Thomas H. Wonnacott and Ronald J. Wonnacott. 1990. *Introductory Statistics*. John Wiley and Sons, fifth edition edition.

Hong Yu and Vasileios Hatzivassiloglou. 2003. Towards answering opinion questions: Separating facts from opinions and identifying the polarity of opinion sentences. In *Proceedings of the Conference on Empirical Methods in Natural Language Processing (EMNLP)*.

A preliminary study on automatic identification of patient smoking status in unstructured electronic health records

Jitendra Jonnagaddala
School of Public Health and Community Medicine, University of New South Wales, Australia
z3339253@unsw.edu.au

Hong-Jie Dai*
Department of Computer Science and Information Engineering, National Taitung University, Taiwan
hjdai@nttu.edu.tw

Pradeep Ray
Asia-Pacific Ubiquitous Healthcare Research Centre, University of New South Wales, Australia
p.ray@unsw.edu.au

Siaw-Teng Liaw*
School of Public Health and Community Medicine ,
University of New South Wales,
Australia
siaw@unsw.edu.au

Abstract

Identifying smoking status of patients is vital for assessing their risk for a disease. With the rapid adoption of electronic health records (EHRs), patient information is scattered across various systems in the form of structured and unstructured data. In this study, we aimed to develop a hybrid system using rule-based, unsupervised and supervised machine learning techniques to automatically identify the smoking status of patients in unstructured EHRs. In addition to traditional features, we used per-document topic model distribution weights as features in our system. We also discuss the performance of our hybrid system using different feature sets. Our preliminary results demonstrated that combining per-document topic model distribution weights with traditional features improve the overall performance of the system.

1 Introduction

Electronic health records (EHRs) carry vital patient information. EHRs generally store information such as medical history, procedures and tests, medications, admissions data and social history. Social history includes details on a patient's smoking habits, alcohol and drug usage. However, most of the information stored in EHRs are in the free-text form as clinical narratives. Natural language processing (NLP) and text mining can be used to extract this valuable information from unstructured EHRs. The extracted information in turn can be used to build a number of applications such as clinical decision support, medical coding, cohort selection and registry systems (Jensen, Jensen, & Brunak, 2012; Jonnagaddala, Dai, Ray, & Liaw, 2015).

Smoking is known to be one of the major risk factors in the development of coronary artery disease, cardiovascular disease, chronic kidney disease and cancer. Thus, identifying smoking status automatically from unstructured EHRs is crucial for preventive medicine. Smoking status can be used to assess risk for a particular disease and provide interventions based on clinical guidelines (Jonnagaddala, Liaw , et al., 2015). Identifying smoking status automatically in unstructured EHRs is not straightforward and often complex. Clinicians usually report smoking information in various formats. For example, few clinicians report in packs per day and others simply classify patient as just smoker or non-smoker.

Previous studies have reported success in using support vector machines (SVMs) to automatically identify smoking status in unstructured EHRs (Clark et al., 2008; Cohen, 2008; Khor et al., 2013; Savova et al., 2010; Savova, Ogren, Duffy,

Proceedings of the 2015 Workshop on Biomedical Natural Language Processing (BioNLP 2015), pages 147–151,
Beijing, China, July 30, 2015.

Buntrock, & Chute, 2008). Similarly, Bui et al developed a system using SVMs by automatically learning regular expressions from two different datasets (Bui & Zeng-Treitler, 2014). However, most of these studies developed their automated systems using traditional features like unigrams, bigrams and POS tags in combination with few rules (Uzuner, Goldstein, Luo, & Kohane, 2008). In this study, we developed a hybrid system using topic modelling and SVMs to automatically identify patients smoking status in unstructured EHRs. Per-document topic distribution weights obtained from unsupervised topic modelling technique are used as features together with traditional features. For the purpose of this study we combined two different datasets to form one large dataset. The system classifies patients into five categories depending on their smoking history using rule-based and machine learning techniques.

2 Materials and Methods

2.1 Dataset

The dataset used in the study is generated by merging datasets from the 2006 and 2014 NLP challenges set forth by the information for integrating biology to the bedside (i2b2) project(Amber Stubbs, Kotfila, Xu, & Uzuner, 2015; A. Stubbs & Uzuner, 2015; Uzuner et al., 2008). The 2006 i2b2 dataset has one document per patient. The 2014 has multiple documents (from multiple encounters) per patient. In this study, we aim to identify the smoking status of a given document irrespective of the fact that one patient might have multiple documents with varying smoking status. In other words, we aimed to develop an automated system to identify smoking status at document level. The final merged dataset consisted of documents classified into one of the five possible smoking categories listed below:

Current Smoker: A current smoker class is assigned to a document when it explicitly state that the patient was a smoker within the past year. If the document mentions, patient has quit smoking within the past one year, the document is still classified as current smoker.

Past Smoker: A past smoker is when a document explicitly state that the patient used to smoke more than a year ago.

Past or Current Smoker: A past or current smoker is assigned when a document mentions that patient smokes, but not possible to determine the status either as past or current.

Non-Smoker: A non-smoker is when documents explicitly states that they never smoked.

Unknown: An unknown status is assigned to a document if there is no mention of smoking.

2.2 Baseline System

The smoking status classifier of nttmuClinical.NET (Chang, Dai, Jonnagaddala, Chen, & Hsu, 2015) was used as the baseline system in this study. For the detection of smoking status, a list of smoking-related keywords, such as "smoking" and "cigarette", was matched with the given document by the classifier. If no match was found, the document was automatically assigned with the UNKNOWN class. Otherwise, the line containing the listed terms was regarded as a

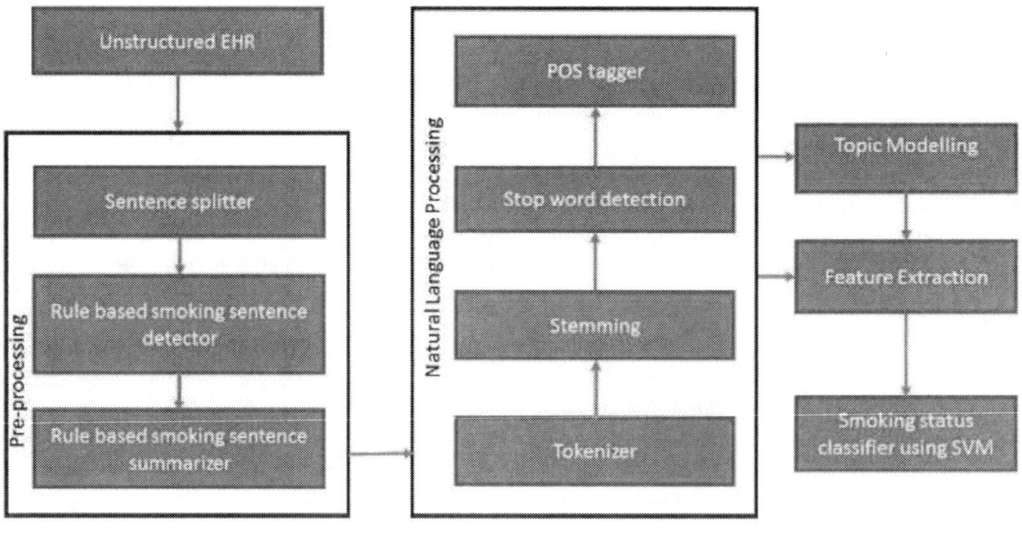

Figure 1: Overview of components in smoking identification pipeline

context that could provide more information for detecting the smoking status of a patient in the document. The context-aware algorithm with several weighted rules developed by leveraging document creation information for different smoking statuses was then applied on the document to determine the smoking status. The algorithm starts by checking the current context. If the context did not provide sufficient temporal information to determine the smoking status, the classifier extends the current context to include more sentences and re-apply the developed rules until either the status was determined or no further updated context was available.

2.3 Hybrid system for smoking classification

Our smoking status identification system takes advantage of the fact that, most of the documents had smoking related information present in a particular section of the document. Thus, instead of using the whole document for classification, we first extracted the smoking related sentences and then used those sentences to identify smoking status. Our system comprised of the following components (Figure1).

Sentence Splitter: To split the documents into individual sentences we used sentence segmentation available in Stanford coreNLP (Manning et al., 2014). The tool was modified to preserve the section headers like "Family History" and "Social History".

Smoking sentence detector: This component was developed to extract the smoking related sentences from the documents. The component identified the smoking status related terms and extracted surrounding sentences.

Smoking sentence summarizer: As some of the documents had multiple smoking related instances a rule-based component was developed to summarize these sentences. The rules were created based on the headers, like Social History, Habits etc.

NLP Component: Once the smoking related sentences were identified and summarized, they were processed further using multiple core NLP components - tokenizer, stemming, stop words removal and POS tagging to generate features.

Feature Extraction: After NLP was done, multiple feature sets were developed including unigrams, bigrams, POS bigrams, word POS pairs and topic models. We generated ten topics using Latent dirichlet allocation (LDA) and Gibbs sampling (Blei, Ng, & Jordan, 2003). The per-document distribution weights of the topics were later incorporated into the feature sets used to train smoking status classifier.

SVM Classifier: Linear SVM classifier was used to classify the documents into one of the five classes discussed above. The cost parameter was optimized to 0.01 for better performance. The SVM classifier was developed using training set and evaluated on test set. The performance of the developed system is presented in the form of precision (P), recall (R) and F1 score (F1) in micro and macro averaged settings.

3 Results

We observed that the 2006 and 2014 i2b2 NLP smoking datasets are not identical in structure and smoking classification classes. We implemented few changes to standardize the smoking status in the merged dataset. Similarly, we also manually annotated documents where smoking status was missing, even though available in documents. Where the smoking status cannot be determined we labeled them as unknown. The summary of number of documents available in final merged dataset (training and test) with the class distribution is presented in Table 1.

Smoking classification classes	Training	Test
Current Smoker	100	46
Past Smoker	185	124
Non-Smoker	251	136
Past Or Current Smoker	29	6
Unknown	623	306
Total no. of documents	1188	618

Table 1 Document level class distribution of dataset

The training set was processed through our hybrid system to generate features and train linear SVM classifier to perform multi class classification. Initially the training set generated model was evaluated using tenfold cross validation on same. This evaluation allowed us to tweak the parameters of our components for better performance. We also used grid search to identify best parameters for linear SVM. The results on the test set with best performing parameters are reported in Table 2. The feature set which incorporated topic modelling based features performed better than baseline and traditional feature set. The topic modelling based feature set trained SVM classifier achieved F1 measure of 83.66% whereas the traditional feature set achieved F1 measure of 82.69% and baseline system 81.85%.

Feature set	Micro averaged		
	P	R	F1
Baseline	0.8185	0.8185	0.8185
Unigrams, Bigrams, POS bigrams, Word POS pairs	0.8269	0.8269	0.8269
Unigrams, Bigrams, POS bigrams, Word POS pairs, Topic models	0.8366	0.8366	0.8366

Table 2: Micro averaged results on test set

4 Discussion

Linear SVMs were used in this study and during the development stage it was observed that the linear kernel performs better than non-linear kernels like radial basis function (RBF). The reason behind the better performance of the linear kernel may be attributed to the presence of a large number of features. It is also believed that when the number of features is much greater than the number of instances then mapping the feature space to a higher dimension like in RBF adds no improvement to the performance of the system.
We also noticed that adding topic models as features did increase the performance of the classifier. However, we believe that the overall performance of classifier can be further increased by optimizing the number of topics to be extracted. The high number of topics we chose to extract using LDA algorithm in current setting are creating sparse features for SVM classifier. Further investigation into choosing optimal number of topics is required.

Both training and test sets in the merged dataset included almost half of documents with unknown class. SVMs in general tend to be biased towards majority classes giving less priority to minority classes. This resulted in significant gap between micro and macro averaged scores. This problem can be solved by taking a multi layered classification approach. As the system is detecting smoking related sentences first, one of the ways to classify is to mark all the instances with no smoking reference as unknown and then classify the remaining into two groups smoker and non-smoker followed by past and current smoker. Another option to address this imbalance problem is by assigning weights to the SVM classifier (Chew, Bogner, & Lim, 2001). Our system also failed to classify current smoker and past smoker efficiently mainly due to negation. The performance of our system can be further improved by implementing a negation component in conjunction with temporal component which can leverage discharge/admission dates and document generated dates as demonstrated in the baseline system. During our error analysis we also noticed that few documents included smoking related administration data in the form of billing and medication codes. We can also use this information to improve the performance of our system (Wiley, Shah, Xu, & Bush, 2013).

5 Conclusion

In summary, we presented the results of a preliminary study in automatically identifying smoking status in unstructured EHRs using SVMs and topic models. Our approach encompassed usage of per-document topic distribution weights generated from topic modelling as features in conjunction with several other traditional features extracted from NLP pipeline. We compared the results of our system using various feature sets against a baseline system. The results demonstrated that topic modelling is useful in identifying smoking status, however, proper topic sampling strategies should be employed. Also, the need for the inclusion of negation and temporal information recognition components in smoking identification is highlighted. In future, we would like to improve our system performance by employing negation and temporal related features. We also would like to explore optimal topic size for smoking identification from relevant smoking related sentences and compare the performance of our system against various smoking identification systems available like Apache cTAKES (Savova et al., 2010).

Acknowledgements

The authors would like to thank the organizers of 2014 and 2006 i2b2/UTHealth Shared-Tasks. De-identified health records used in this research were provided by the i2b2 National Center for Biomedical Computing funded by grants 2U54LM008748 and 1R13LM01141101 from National Institute of health (NIH). The authors would like to thank the organizers of 2014 i2b2/UTHealth Shared-Tasks. De-identified health records used in this research were provided by the i2b2 National Center for Biomedical Computing funded by grants 2U54LM008748 and 1R13LM01141101 from National Institute of health (NIH). This study was conducted as part of the electronic Practice Based Research Network (ePBRN) and Translational Cancer research network (TCRN) research programs. ePBRN was/is funded in part by the School of Public Health & Community Medicine, Ingham Institute for Applied Medical Research, UNSW Medicine and South West Sydney Local Health District. TCRN is funded by Cancer Institute of New South Wales and Prince of Wales Clinical School, UNSW Medicine. The content of this publication is solely the responsibility of the authors and does not necessarily reflect the official views of the funding bodies.

References

Blei, D. M., Ng, A. Y., & Jordan, M. I. (2003). Latent dirichlet allocation. *the Journal of machine Learning research, 3*, 993-1022.

Bui, D. D. A., & Zeng-Treitler, Q. (2014). Learning regular expressions for clinical text classification. *Journal of the American Medical Informatics Association, 21*(5), 850-857.

Chang, N.-W., Dai, H.-J., Jonnagaddala, J., Chen, C.-W., & Hsu, W.-L. (2015). A Context-Aware Approach for Progression Tracking of Medical Concepts in Electronic Medical Records. Manuscript submitted.

Chew, H.-G., Bogner, R. E., & Lim, C.-C. (2001). *Dual ν-support vector machine with error rate and training size biasing.* Paper presented at the 2001 IEEE International Conference on Acoustics, Speech, and Signal Processing, 2001. Proceedings.(ICASSP'01).

Clark, C., Good, K., Jezierny, L., Macpherson, M., Wilson, B., & Chajewska, U. (2008). Identifying smokers with a medical extraction system. *Journal of the American Medical Informatics Association, 15*(1), 36-39.

Cohen, A. M. (2008). Five-way smoking status classification using text hot-spot identification and error-correcting output codes. *Journal of the American Medical Informatics Association, 15*(1), 32-35.

Jensen, P. B., Jensen, L. J., & Brunak, S. (2012). Mining electronic health records: towards better research applications and clinical care. *Nature Reviews Genetics, 13*(6), 395-405.

Jonnagaddala, J., Dai, H.-J., Ray, P., & Liaw, S.-T. (in press). Mining electronic health records to guide and support good clinical decision support systems. In J. Moon & M. P. Galea (Eds.), *Improving Health Management through Clinical Decision Support Systems*: IGI-Global.

Jonnagaddala, J., Liaw , S.-T., Ray , P., Kumar , M., Chang, N.-W., & Dai, H.-J. (2015). Coronary artery disease risk assessment from unstructured electronic health records using text mining. *Journal of Biomedical Informatics*.

Khor, R., Yip, W.-K., Bressel, M., Rose, W., Duchesne, G., & Foroudi, F. (2013). Practical implementation of an existing smoking detection pipeline and reduced support vector machine training corpus requirements. *Journal of the American Medical Informatics Association*, amiajnl-2013-002090.

Manning, C. D., Surdeanu, M., Bauer, J., Finkel, J., Bethard, S. J., & McClosky, D. (2014). *The Stanford CoreNLP natural language processing toolkit.* Paper presented at the Proceedings of 52nd Annual Meeting of the Association for Computational Linguistics: System Demonstrations.

Savova, G. K., Masanz, J. J., Ogren, P. V., Zheng, J., Sohn, S., Kipper-Schuler, K. C., & Chute, C. G. (2010). Mayo clinical Text Analysis and Knowledge Extraction System (cTAKES): architecture, component evaluation and applications. *Journal of the American Medical Informatics Association, 17*(5), 507-513.

Savova, G. K., Ogren, P. V., Duffy, P. H., Buntrock, J. D., & Chute, C. G. (2008). Mayo clinic NLP system for patient smoking status identification. *Journal of the American Medical Informatics Association, 15*(1), 25-28.

Stubbs, A., Kotfila, C., Xu, H., & Uzuner, O. (2015). Practical applications for NLP in Clinical Research: the 2014 i2b2/UTHealth shared tasks.

Stubbs, A., & Uzuner, O. (2015). Annotating Risk Factors for Heart Disease in Clinical Narratives for Diabetic Patients. *J Biomed Inform*. doi: 10.1016/j.jbi.2015.05.009

Uzuner, Ö., Goldstein, I., Luo, Y., & Kohane, I. (2008). Identifying patient smoking status from medical discharge records. *Journal of the American Medical Informatics Association, 15*(1), 14-24.

Wiley, L. K., Shah, A., Xu, H., & Bush, W. S. (2013). ICD-9 tobacco use codes are effective identifiers of smoking status. *Journal of the American Medical Informatics Association, 20*(4), 652-658.

Restoring the intended structure of Hungarian ophthalmology documents

Borbála Siklósi[2] **and Attila Novák**[1,2]
[1]MTA-PPKE Hungarian Language Technology Research Group,
[2]Pázmány Péter Catholic University
Faculty of Information Technology and Bionics
50/a Práter street, 1083 Budapest, Hungary
{siklosi.borbala, novak.attila}@itk.ppke.hu

Abstract

Clinical documents have been an emerging target of natural language applications. Information stored in documents created at clinical settings can be very useful for doctors or medical experts. However, the way these documents are created and stored is often a hindrance to accessing their content. In this paper, an automatic method for restoring the intended structure of Hungarian ophthalmology documents is described. The statements in these documents in their original form appeared under various subheadings. We successfully applied our method for reassigning the correct heading for each line based on its content. The results show that the categorization was correct for 81.99% of the statements in our testset, compared to a human categorization.

1 Introduction

Documents created in clinical settings contain a large amount of practical information characteristic of the local community. Collecting and processing such documents may provide doctors and medical experts a valuable source of information (Meystre et al., 2008; Sager et al., 1994; Friedman et al., 1995).

In a broad sense, there are two sources of clinical documents regarding the nature of these textual data. First, they might be produced through an EHR (Electronic Health Records) system. In this case, practitioners or assistants type the information into a predefined template, resulting in structured documents. The granularity of this structure might depend on the actual system and the habit of its users. The second possibility is that the production of these clinical records follows the nature of traditional hand-written documents, i.e. even though they are stored in a computer, it is only used as a typewriter, resulting in raw text, having some clues of the structure only in the manual formatting. These are the two extremes, and the production of such records is usually somewhere in between, depending on institutional regulations, personal habits and the actual clinical domain.

In this paper, an automatic method is described that is able to assign labels of structural units to statements in Hungarian ophthalmology documents. In Hungarian hospitals, the usage of EHR systems is far behind expectations. Assistants or doctors are provided with some documentation templates, but most of them complain about the complexity and inflexibility of these systems. This results in keeping their own habit of documentation, filling most of the information into a single field and manually copying patient history.

Moreover, ophthalmology has been reported to be a suboptimal target of application of EHR systems in several surveys carried out in the US (Chiang et al., 2013; Redd et al., 2014; Elliott et al., 2012). The special requirements of documenting a mixture of various measurements, some of them resulting in tabular data, while others in single values or textual descriptions make the design of a usable system for storing ophthalmology reports in a structured and validated form very hard.

2 The corpus of Hungarian ophthalmology notes

We were provided with anonymized clinical records from the ophthalmology department of a Hungarian clinic. Due to the lack of a sophisticated clinical documentation system, the structure of the raw documents can only be inferred from the formatting or by understanding the actual content. Besides basic separations – that are not even uniform through documents – there were no other clues for determining structural units. Moreover, a significant portion of the records were redundant:

Proceedings of the 2015 Workshop on Biomedical Natural Language Processing (BioNLP 2015), pages 152–157,
Beijing, China, July 30, 2015. ©2015 Association for Computational Linguistics

medical history of a patient is sometimes copied to later documents at least partially, making subsequent documents longer without additional information regarding the content itself. Moreover, the language of these documents contains a high ratio of word forms not commonly used: such as Latin medical terminology, abbreviations and drug names. Many of the authors of these texts are not aware of the standard orthography of this terminology. Figure 1 shows a document after processing, but the original format is kept in the example and the English translation is provided.

The documents of ophthalmology investigated in this research were especially characterized by nontextual information interspersed with sections containing texts. These (originally tabular) data behave as noise in such a context. Non-textual information inserted into free-word descriptions includes laboratory test results, numerical values, delimiting character series and longer chains of abbreviations and special characters. Moreover, these statements do not follow any standard patterns even by themselves and they further vary from document to document according to the style of the doctor or assistant.

Regarding the textual parts of the documents, these are also quite different from general Hungarian. Consult Siklósi et al. (2014) and Siklósi and Novák (2014) for a detailed comparison of Hungarian and the medical sublanguage.

3 Structuring and categorizing lines

First, a preprocessing chain adapted to these special characteristics was applied to the documents, which included tokenization (Orosz et al., 2013), spelling correction (Siklósi et al., 2014), and part-of-speech tagging (Orosz et al., 2014). Thus, an enriched representation of the corpus was achieved. This provided the basis for structuring and categorizing the content of each document. This was performed in two steps. First, formatting clues were recognized and labelled. Second, each line was classified into a content unit defined on statistical observations from the corpus.

3.1 Structuring

Even though the documentation system used when creating these documents did provide a basic template for labelling each section of the document to be created, these were very rarely followed by the administrative personnel. However, some of these system generated labels were printed into the final documents, which we could consider as 'clues' of the intended structure. These system generated labels followed a consistent pattern, and as such, could easily be recognized based on features such as the amount of white space at the beginning of the line, capitalization, and the recurring text of the headline. Thus, such structural units were identified and labelled with a `PART` tag.

Similarly, tables of codes were also printed by the system in a predefined format. These tables contain the BNO-codes (the Hungarian system of ICD coding) of diagnoses and the applied treatments. Such tables, though printed as raw text, could also be recognized by the spacing used in them and were labelled with an `SPART` tag.

3.2 Detecting patient history

We found it very often that findings about a patient recorded in documents of earlier visits were copied to the actual record, and in some cases minor adjustments were also introduced during the replication. Thus, although these partial recurrences contain only redundant information, they could not be recognized by simply looking for exact matches. Moreover, the short and dense statements of findings are often formatted the same way in the case of different patients or even doctors. In order to filter these copied sections, first we detect all date stamps in each document. Date stamps may occur in the headers, in the notation of some examinations, in the tables of codings or might be inserted manually at any point in the documents. The dates were labelled with a `DATE` tag. Then, the contents between these tags were ordered in increasing order and partial matches were found by comparing the `md5` coded form of each part. Those sections that had a matching under an earlier date stamp, were labelled with a `COPY` tag. Furthermore, these `DATE` tags were used to partition each document corresponding to separate visits. Thus, patient history could be retrieved by referring to the same ID and each date. All the information that was originally in a single document can thus be retrieved in order.

3.3 Categorizing statements

Even though the `PART` tags have labelled each part according to the documentation template of the system, the title of these fields is rarely in accordance with the content. For example, the status field is frequently used to include all the information, be it originally anamnesis, treatment, therapy, or any other comments. Thus, it was necessary to categorize each statement in each part of the documents. Table 1 shows the categories and their description used for classification. It should be noted, that these categories are defined directly for the ophthalmology domain. For other specialties, the tagset should be redefined.

Prior to categorization, units of statements had to be declared. The documents were exported from the original system in a way that kept the fixed width of the original input fields. Thus, linebreaks were inserted to the text at certain positions corresponding to this width. In order to restore the original units intended to be single lines, these linebreaks were deleted from the end of a line which could be continued by the next one. That is, if the second line does not start with capital letter, does not start with whitespace and if the length of the actual line plus the length of the first word of the second line is larger than the fixed width (hyphenation was not implemented in the system, thus if a word would pass the right margin, then the whole word is transmitted to a new line). Moreover, lines containing tabular data were also recognized during this processing step. The units of categorization were these concatenated lines and since these lines were either short or contained usually only one type of information, each received one tag. Longer sections of neighboring lines falling into the same category could be merged after labelling each line. The categorization was done in three steps.

First, using the preprocessed version of the texts, some patterns were identified based on part-of-speech tags and the semantic concept categories assigned to the most frequent entities. For example, due to the rare use of verbs, if a past tense verb was recognized in a sentence, it was a good indicator of being part of the anamnesis or the complaints of the patient (Siklósi, 2015).

Second, some indicator words were extracted from the documents. At the first place, these were those line initial words and short phrases that started with capital letter and were followed by a colon and some more content. These phrases were then ordered by their occurrence frequencies. Then, they were manually assigned a category label referring to the type of the statement that the phrase could be an indicator of. For example the phrase, *korábbi betegségek* 'previous illnesses' was given the label `Ana` referring to anamnesis. Table 2 shows some more examples of tags and phrases labelled by them. After having all the phrases occurring at least 10 times in the whole corpus labelled, they were matched against the lines of each document that were found in `PART` sections and were not recognized as tabular data. If the line started with a phrase or any of its variations (case variations, misspellings, punctuation marks and white spaces were allowed differences), then the line was labelled with the tag the phrase belonged to. These first two steps were able to categorize 34% of the concatenated lines in the documents.

tag	phrase	English translation
`Ana`	egyéb betegség	other illness
	panasz	complaint
	család	family
	korábbi	earlier
	hypertonia	hypertonia
	anamnézis	anamnesis
`T`	eredmény	result
	ultrahang	ultrasound
	Topo	Topo
	Schirmer	Schirmer
`RL`	réslámpa	slit lamp
	macula	macula
	fundus	fundus
	rl	sl (for slit lamp)
	lencse	lens
`Ther`	th	th (for therapy)
	szemcsepp	eyedrop
	terápia	therapy
	rendelés	prescription
	javasolt	recommended

Table 2: Examples of tags and some of the phrases labelled by the tag.

In the third step, the rest of the lines were given a label. In order to do this, all lines labelled in the first two steps were collected for each tag (they will be referred to as tag collections). Then, for each line, the most similar tag collection was determined and the tag of this collection was assigned to the actual line. The similarity measure applied was the tf-idf weighted cosine similarity between a line (l) and a tag collection (c) defined by Formula 1.

tag	meaning	description
Tens	Tension	Measurements of the tension of the eye
V/Refr	Refraction	Refraction data
Ana	Anamnesis	Complains of the patient, other/past diseases, family history, etc.
Dg	Diagnosis	The actual diagnoses
Beav	Treatment	Applied treatments, except operations and medication
Vél	Opinion	Opinion of the doctor, except diagnoses and treatments
St	Status	Actual status of the patient
Ther	Therapy	Prescribed/applied medication
BNO	BNO (ICD)	Statements used with their BNO codes
T	Test	Tests, other than those in the Rl category
V	Visus	Visus data
Rl	Slit lamp	Tests carried out using the slit lamp (most of the tests are done with it)
Kontr	Control	Information about when the patient should return to the doctor
Műtét	Operation	Operations applied or prescribed
XXX	-	Other statements that can not be categorized

Table 1: The tags used in categorizing statements

$$sim(\vec{l}, \vec{c}) = \frac{\sum\limits_{w \in l,c} tf_{w,l} tf_{w,c} (idf_w)^2}{\sqrt{\sum\limits_{l_i \in l} (tf_{l_i,l} idf_{l_i})^2} \times \sqrt{\sum\limits_{c_i \in c} (tf_{c_i,c} idf_{c_i})^2}} \quad (1)$$

, where $\vec{l}$ contained the normalized set of words in line l, and $\vec{c}$ the normalized set of words contained in the tag collection c. During normalization, stop-words and punctuation marks were removed and numbers were replaced by the character x, so that the actual numerical values do not mislead the representation. As a result, all lines within PART sections were labelled by a tag. Finally, tabular lines were assigned the tag Vis, since these contained the detailed information about the visual acuity of the patient.

4 Results

The labels of 1000 lines were checked manually. This testset was selected randomly only from PART sections, since the categorization was applied only to these portions of the documents. However, the label XXX was also allowed in the system when it was not able to assign any meaningful labels. The rest of the lines were assigned one of the 15 labels. Figure 1 shows the processed state of a document. In the example, the format and whitespaces of the original document is kept. Tags are shown at the beginning of the lines. Tags starting with a number of # symbols are used for the separation of structural units. Categorization is applied to lines in a Part section, here Part:Státusz. Lines ending with an @ symbol were concatenated with the next line. tags regarding structural units and classification of statements. The English translation for the meaningful parts are inserted between the lines.

In the evaluation setup, the labels were considered either as correct, non-correct or undecidable. Lines of this latter category either did not include enough information referring to the content, or it was too difficult even for the human evaluator to decide what category the line belonged to. The label XXX was accepted as correct, if the line did not belong to any category (e.g. a single date). Out of the 1000 lines in the test set, its 7.8% could not be categorized by the human expert. For the rest of the lines, 81.99% of these lines were assigned the correct label and only 18.01% the incorrect one. Regarding the errors, most of them were due to the lack of contextual information for the algorithm. For example, if the anamnesis of a patient included some surgery, then the label for surgery was assigned to it, which is correct at the level of standalone statements, but incorrect in the context of the whole document. The other main source of the errors was that some longer lines included more than one types of statements and the system was unable to choose a correct one. In these cases, the human annotation assigned the "more relevant" tag as correct. Thus, a significant part of these errors could be eliminated by a more ac-

```
###DOCTYPE:AMBULÁNS KEZELŐLAP
`T                               A M B U L Á N S   K E Z E L Ő L A P

###PART:Státusz //Status
St       Státusz

##DOCDATE##
##DATE-TIME##
XXX `T       2010.10.19 12:28   Székelyhidi/Füst

Beav `C        Olvasó szemüveget szeretne. Néha könnyeznek a szemei.
                    //S/he would like reading glasses, eyes are sometimes watering.
V          V:0,7+0,75Dsph=1,0
V            1,0 +0,5 Dsph élesebb

V\Refr         +2.0 Dsph mko Cs IV

St          St.o.u: halvány kh, ép cornea, csarnok kp mély tiszta, iris ép békés, pupilla@
                    // St.o.u: blanch conj, intact cornea, chamber deep clean, iris intact, calm, papil
    `C   rekciók rendben, lencse tiszta, jó vvf.
                    //reactions allright, clean lens, good rbl.
Ther          Átfecskendezés mko sikerült.
                    //Successful squishing at both side
V\Refr `C        Olvasó szemüveg javasolt: +2.0 Dsph mko.
                    //Reading glasses are suggested: +2.0 Dsph both side
Vél `C         Éjszakánként műkönnygél ha szükséges.
                    //Artificial tears can be used at night if necessary
Kontr          Kontroll: panasz esetén
                    //Contorl: in case of further complaints


###SPART:Diagnózis  //Diagnoses
      Diagnózis
        DIAGNÓZISOK megnevezése                          Kód    Dátum      Év   K V T
##DOCDATE##
        Látászavar, k.m.n.                               H5390  2010.10.19          3

###SPART:Beavatkozások  //Treatments
      Beavatkozások
         Kód      Megnevezés                                   Menny.  Pont
         11041    Vizsgálat                                         1   750
```

Figure 1: The processed state of a document.

curate segmentation for separating each statement and by the incorporation of contextual features to the categorization process, which are among our future plans.

5 Conclusion

A method for structuring Hungarian ophthalmology notes has been described. The original form of these records created at clinical settings contains a large amount of noise and lacks almost any structure. Thus, in order to be able to use these documents either as the input of information retrieval algorithms or as a searchable database for medical experts, their intended structure had to be restored by assigning medical headings to each statement. This categorization was achieved by our method in three steps, relying on (1) the formatting clues of the original documents, (2) domain-specific keywords derived from the ophthalmology notes and (3) a statistical classification approach. Compared to a manually created gold standard, the results showed relatively high accuracy.

References

Michael F. Chiang, Sarah Read-Brown, Daniel C. Tu, Dongseok Choi, David S. Sanders, Thomas S. Hwang, Steven Bailey, Daniel J. Karr, Elizabeth Cottle, John C. Morrison, David J. Wilson, and Thomas R. Yackel. 2013. Evaluation of electronic health record implementation in ophthalmology at an academic medical center (an american ophthalmological society thesis). *Trans Am Ophthalmol Soc*, 111:70–92, Sep.

Amanda Elliott, Arthur Davidson, Flora Lum, Michael Chiang, Jinan B. Saaddine, Xinzhi Zhang, John E. Crews, and Chiu-Fang Chou. 2012. Use of electronic health records and administrative data for public health surveillance of eye health and vision-related conditions. *Am J Ophthalmol*, 154(6 0):S63–S70, Dec.

C. Friedman, S.B. Johnson, B Forman, and J Starren. 1995. Architectural requirements for a multipurpose natural language processor in the clinical environment. *Proc Annu Symp Comput Appl Med Care*, pages 347–51.

S.M. Meystre, G.K. Savova, K.C. Kipper-Schuler, and JF Hurdle. 2008. Extracting information from textual documents in the electronic health record: a review of recent research. *Yearb Med Inform*, 35:128–44.

György Orosz, Attila Novák, and Gábor Prószéky, 2013. *Hybrid text segmentation for Hungarian clinical records*, volume 8265 of *Lecture Notes in Artificial Intelligence*. Springer-Verlag, Heidelberg.

György Orosz, Attila Novák, and Gábor Prószéky. 2014. Lessons learned from tagging clinical Hungarian. *International Journal of Computational Linguistics and Applications*, 5(1):159–176.

Travis K. Redd, Sarah Read-Brown, Dongseok Choi, Thomas R. Yackel, Daniel C. Tu, and Michael F. Chiang. 2014. Electronic health record impact on productivity and efficiency in an academic pediatric ophthalmology practice. *Journal of AAPOS*, 18(6):584–589.

Naomi Sager, Margaret Lyman, Christine Bucknall, Ngo Nhan, and Leo J. Tick. 1994. Natural language processing and the representation of clinical data. *Journal of the American Medical Informatics Association*, 1(2), Mar/Apr.

Borbála Siklósi and Attila Novák. 2014. A magyar beteg. *X. Magyar Számítógépes Nyelvészeti Konferencia*, pages 188–198.

Borbála Siklósi, Attila Novák, and Gábor Prószéky. 2014. Context-aware correction of spelling errors in Hungarian medical documents. *Computer Speech and Language*, in press(0):–.

Borbála Siklósi. 2015. Clustering relevant terms and identifying types of statements in clinical records. In Alexander Gelbukh, editor, *Computational Linguistics and Intelligent Text Processing*, volume 9042 of *Lecture Notes in Computer Science*, pages 619–630. Springer International Publishing.

Evaluating distributed word representations for capturing semantics of biomedical concepts

Muneeb T H[1], **Sunil Kumar Sahu**[1] and **Ashish Anand**[1]

[1]Department of Computer Science and Engineering,
Indian Institute of Technology Guwahati, Assam - 781039, India
{muneeb, sunil.sahu, anand.ashish}@iitg.ac.in

Abstract

Recently there is a surge in interest in learning vector representations of words using huge corpus in unsupervised manner. Such word vector representations, also known as word embedding, have been shown to improve the performance of machine learning models in several NLP tasks. However efficiency of such representation has not been systematically evaluated in biomedical domain. In this work our aim is to compare the performance of two state-of-the-art word embedding methods, namely *word2vec* and *GloVe* on a basic task of reflecting semantic similarity and relatedness of biomedical concepts. For this, vector representations of all unique words in the corpus of more than 1 million full-length research articles in biomedical domain are obtained from the two methods. These word vectors are evaluated for their ability to reflect semantic similarity and semantic relatedness of word-pairs in a *benchmark data set* of manually curated semantic similar and related words available at `http://rxinformatics.umn.edu`. We observe that parameters of these models do affect their ability to capture lexico-semantic properties and word2vec with particular language modeling seems to perform better than others.

1 Introduction

One of the crucial step in machine learning (ML) based NLP models is how we represent word as an input to our model. Most of earlier works were treating word as atomic symbol and were assigning one hot vector to each word. Length of the vector in this representation was equal to the size of the vocabulary and the element at the word index is 1 while the other elements are 0s. Two major drawbacks with this representation are: first, length of the vector is huge and the second, there is no notion of similarity between words. The inability of one-hot vector representation to embody lexico-semantic properties prompted researchers to develop methods which are based on the notion that the "similar words appear in similar contexts". These methods can broadly be classified into two categories (Turian et al., 2010), namely, *distributional representation* and *distributed representation*. Both group of methods works in unsupervised manner with huge corpus. Distributional representations are mainly based on co-occurrence matrix O of words in the vocabulary and their contexts. Here, among other possibilities, contexts can be documents or words within a particular window. Each entry O_{ij} in the matrix may indicate either frequency of word i in the context j or simply whether the word i has appeared in the context j at least once. Co-occurrence matrix can be designed in variety of ways (Turney and Pantel, 2010). The major issue with such methods is size of the matrix O and reducing its size generally tends to be computationally very expensive. Nevertheless, the requirement of constructing and storing the matrix O are always there. The second group of methods are mainly based on language modeling (Bengio et al., 2003). We discuss more about these methods in the section 3.

Outside the biomedical domain, this kind of representation has shown significant improvement in the performance of many NLP tasks. For example, Turian et al. (2010) have improved the performance of chunking and named entity recognition by using word embedding also as one of the features in their CRF model. In one study, Collobert et al. (2011) have formulated the NLP tasks of parts of speech tagging, chunking, named entity recognition and semantic role labeling as multi-

Proceedings of the 2015 Workshop on Biomedical Natural Language Processing (BioNLP 2015), pages 158–163,
Beijing, China, July 30, 2015. ©2015 Association for Computational Linguistics

task learning problem. They have shown improvement in the performance when word vectors are learned together with other NLP tasks. Socher et al. (2012) improved the performance of sentiment analysis task and semantic relation classification task using recursive neural network. One common step among these models is: learning of word embedding from huge unannotated corpus like Wikipedia, and later use them as features.

Motivated by the above results, we evaluate performance of the two word embedding models, word2vec (Mikolov et al., 2013a; Mikolov et al., 2013b) and GloVe (Pennington et al., 2014) for their ability to capture syntactic as well as semantic properties of words in biomedical domain. We have used full-length articles obtained from PubMed Central (PMC) open access subset[1] as our corpus for learning word embedding. For evaluation we have used publicly available validated reference dataset (Pakhomov et al., 2010; Pedersen et al., 2007) containing semantic similarity and relatedness scores of around 500 word-pairs. Our results indicate that the word2vec word embedding is capturing semantic similarity between words better than the GloVe word embedding in the biomedical domain, whereas for the task of semantic relatedness, there does not seem to be any statistical significant difference among different word-embeddings.

2 Related Work

In a recent study, Miñarro-Giménez et al. (2015) have evaluated the efficiency of word2vec in finding clinical relationships such as "may treat", "has physiological effect" etc. For this, they have selected the manually curated information from the National Drug File - Reference Terminology (NDF-RT) ontology as reference data. They have used several corpora for learning word-vector representation and compared these different vectors. The word-vectors obtained from the largest corpus gave the best result for finding the "may treat" relationship with accuracy of 38.78%. The relatively poor result obtained for finding different clinical relationships indicates the need for more careful construction of corpus, design of experiment and finding better ways to include domain knowledge.

In another recent study, Nikfarjam et al. (2015) have described an automatic way to find adverse drug reaction mention in social media such as twitter. Authors have shown that including word embedding based features has improved the performance of their classifier.

Faruqui and Dyer (2014) have developed an online suit to analyze and compare different word vector representation models on a variety of tasks. These tasks include syntactic and semantic relations, sentence completion and sentiment analysis. In another recent work, Levy et al. (2015) have done extensive study on the effect of hyperparameters of word representation models and have shown their influence on the performance on word similarity and analogy tasks. However in both the studies (Faruqui and Dyer, 2014; Levy et al., 2015) the benchmark datasets available for NLP tasks are not suitable for analyzing vector representations of clinical and biomedical terms.

3 Word Embedding

As discussed earlier, word embedding or distributed representation is a technique of learning vector representation for all words present in the given corpus. The learned vector representation is generally dense, real-valued and of low-dimension. As contrast to one-hot vector representation each dimension of the word-vector is supposed to represent a latent feature of lexico-semantic properties of the word. In our work we considered two state of the art word embedding techniques, namely, *word2vec* and *GloVe*. Although in literature there exists several word-embedding techniques (Hinton et al., 1986; Bengio et al., 2003; Bengio, 2008; Mnih and Hinton, 2009; Collobert et al., 2011), the selected two word embedding techniques are very much computationally efficient and are considered as state-of-the art. We have summarized the basic principles of the two methods in subsequent sections.

3.1 word2vec Model

word2vec generates word vector by two different schemes of language modeling: continuous bag of words (CBOW) and skip-gram (Mikolov et al., 2013a; Mikolov et al., 2013b). In the CBOW method, the goal is to predict a word given the surrounding words, whereas in skip-gram, given a single word, window or context of words are predicted. We can say skip-gram model is opposite of CBOW model. Both models are neural network based language model and take huge corpus as an input and learn vector representation for

[1] http://www.ncbi.nlm.nih.gov/pmc/tools/ftp/

each words in the corpus. We used freely available *word2vec*[2] tool for our purpose. Apart from the choice of architecture skip-gram or CBOW, word2vec has several parameters including *size of context window*, *dimension of vector*, which effect the speed and quality of training.

3.2 GloVe Model

GloVe (Pennington et al., 2014) stands for Global Vectors. In some sense, GloVe can be seen as a hybrid approach, where it considers global context (by considering co-occurrence matrix) as well as local context (such as skip-gram model) of words. *GloVe* try to learn vector for words w_x and w_y such that their dot product is proportional to their co-occurrence count. We used freely available *glove*[3] tool for all analysis.

4 Materials and Methods

4.1 Corpus Data and Preprocessing

PubMed Central® (PMC) is a repository of biomedical and life sciences journal literature at the U.S. National Institutes of Health's National Library of Medicine (NIH/NLM). We have downloaded the gzipped archived files of full length texts of all articles in the open access subset [4] on 19th April, 2015. This corpus contains around 1.25 million articles having around 400 million tokens altogether.

In pre-processing step of the corpus, we mainly perform following two operations-

- we put all numbers in different groups based on number of digits in them. For example, all single digit numbers are replaced by the token "number1", all double digit numbers by the token "number2" and so on.
- each punctuation mark is considered as separate token.

4.2 Reference Dataset

Pakhomov et al. (2010) have constructed a reference dataset of semantically similar and related word-pairs. These words are clinical and biomedical terms obtained from control vocabularies maintained in the Unified Medical Language System(UMLS). This reference dataset contains 566 pairs of UMLS concepts which were manually rated for their semantic similarity and 587 pairs of UMLS concepts for semantic relatedness. We removed all pairs in which at least one word has less than 10 occurrences in the entire corpus as such words are removed while building vocabulary from the corpus. After removing less frequent words in both reference sets, we obtain 462 pairs for semantic similarity having 278 unique words, and 465 pairs for semantic relatedness having 285 unique words. In both cases, each concept pair is given a score in the range of $0 - 1600$, with higher score implies similar or more related judgments of manual annotators. The semantic relatedness score span the four relatedness categories: completely unrelated, somewhat unrelated, somewhat related, closely related.

4.3 Experiment Setup

We generate the word vectors using the two word embedding techniques under different settings of their parameters and compare their performance in semantic similarity and relatedness tasks. Dimension of word-vector is varied under the two different language models, CBOW and SKIP-GRAM, for word2vec word embedding. For GloVe, only dimension of word vector is changed. For each model, word vectors of 25, 50, 100, and 200 dimensions are generated. Due to limited computing power, we could not go for higher dimensions. For window size, we did not perform any experiment and simply considered 9 as window size for all models.

4.4 Evaluation

As discussed earlier, both reference data have provided a score for each word-pair in them. We calculate cosine similarity between the two words of each pair present in the reference data using learned word vectors. Now, each word pair has two scores: one given in the dataset and the other cosine similarity based on learned word vectors. We calculate Pearson's correlation between these two scores.

Further we visualize a limited number of manually selected words for qualitative evaluation. For this we use the t-SNE (van der Maaten and Hinton, 2008) tool to project our high dimensional word vectors into two-dimensional subspace. t-SNE is being widely used for this purpose.

[2]https://code.google.com/p/word2vec/

[3]http://nlp.stanford.edu/projects/glove/

[4]http://www.ncbi.nlm.nih.gov/pmc/tools/ftp/

Dimension	Semantic Similarity			Semantic Relatedness		
	CBOW	**Skip**	**GloVe**	**CBOW**	**Skip**	**GloVe**
25	0.32	0.39	0.28	0.30	0.34	0.27
50	0.36	0.44	0.34	0.33	0.38	0.36
100	0.42	0.48	0.41	0.39	0.43	0.41
200	0.46	0.52	0.42	0.41	0.45	0.42

Table 1: Correlation between cosine similarity and the score provided in the benchmark dataset.

5 Results and Discussion

Table 1 shows the correlation values in all cases. We observe that increasing the dimension of word vectors improve their ability to capture semantic properties of words. The above results indicate that less than $d = 200$ dimension will likely to be a bad choice for any NLP tasks. Due to the limited computing power, we could not complete our experiments with 500 and 1000 dimensional vector representations. We have also calculated the Spearman and Kendall-Tau's correlation in each case and have observed similar trends in all cases.

Skip-gram model seems to be better than both CBOW and GloVe models in the semantic similarity task for all dimensions. However this does not seem to be the case with the relatedness task. So we perform the statistical significance test to check whether correlation corresponding to word2vec skip-gram model is significantly higher than correlation corresponding to other two models. In the statistical test, we evaluate the null-hypothesis " correlation corresponding to alternate model (CBOW or GloVe) is equal to that corresponding to the skip-gram model" at significance level $\alpha = 0.05$. We use cocor (Diedenhofen and Musch, 2015) package for statistical comparison of dependent correlations.

It turns out that for the semantic similarity task, word2vec skip-gram model is significantly better (i.e., correlation is higher corresponding to skip-gram word vectors) than word2vec-CBOW (p-value: 0.01) and GloVe (p-value: 0.0007) models. On the other hand correlation in skip-gram model is not found significantly higher than the correlations in the other two models for the semantic relatedness task. The above observation is made for the 200 dimensional vectors. But we can not say the same for results obtained by lower dimensional vectors. For example, in case of 25-dimensional vectors, correlation obtained by skip-gram model is significantly higher than that obtained by GloVe model for both tasks. However similar observation was made in case of comparison between CBOW and skip-gram as in 200 dimensional case.

We further look at nearest neighbors of some manually selected words. If word-vectors truly represent latent features of lexical-semantic properties of words, then their nearest neighbors must be related words. We tested this hypothesis on a small set of manually selected seed-words and their nearest neighbors. We selected 8 seed-words representing *disease*, *disorder*, *organ* and *treatment*: *eye (organ)*, *liver (internal organ)*, *fever (disorder/symptom)*, *tumour (disease/disorder)*, *thyroid (gland)*, *cough (symptom)*, *surgery (procedure/treatment)*, *leg (external organ)*, *aids (disease)*. Table 2 shows the 10 nearest neighbors of some of the seed-words (similar results are observed for other seed-words) as picked by the three methods. As it can be seen from the table that the nearest neighbors are very much related to the seed-words. Not only words like "coughs", "coughing", but also words like "wheezing", "dyspnea" are within the top-10 nearest neighbors of "cough". The first set of examples indicates ability of the learned word-vectors to capture lexical properties of words, whereas the later set of words shows vectors' ability to capture semantic properties as well.

Next we visualize (Figure 1) the 4 seed-words (shown in Table 2) and their 25 nearest neighbors using t-SNE. Here we have shown the result obtained from the word2vec skip gram model (dimension = 200) only. Due to space constraints we have not shown the results of other methods but similar observation was made for the other methods. t-SNE projects high-dimensional vectors into $\mathbf{R}^2$ by preserving the local structure of high-dimensional space.

Figure 1 clearly shows the ability of the learned word-vectors to automatically group similar words together. This again provides another evidence of the vectors' ability to represent semantic properties.

seed word	CBOW	Skip	GloVe
eye	eye, eyes, eyeball, hemifield, hemibody, forelimb, eyebrow, midline, head, face	eye, eyes, face, head, ocular, mouth, pupillary, fovea, angle, Eye	eye, eyes, SEFsupplementary, ocular, visual, vision, cornea, optic, retina, ear
cough	cough, coughing, breathlessness, Cough, dyspnea, wheezing, wheeze, hemoptysis, coughs, haemoptysis	cough, breathlessness, expectoration, coughing, wheezing, dyspnea, phlegm, shortness, haemoptysis, sore	cough, coughing, shortness, breathlessness, TDITransition, dyspnea, wheezing, sore, bronchitis, expectoration
surgery	surgery, operation, decompression, dissection, resection, parathyroidectomy, stenting, surgeries, esophagectomy, resections	surgery, surgical, operation, procedure, esophagectomy, surgeries, laparoscopic, elective, reintervention, postoperative	surgery, surgical, BCSBreast-conserving, surgeries, operative, eBack, postoperative, PSMPositive, operation, resection
tumour	tumour, tumor, tumoral, tumoural, glioma, melanoma, PDAC, HNSCC, tumors, neoplastic	tumour, tumor, tumors, tumours, malignant, metastatic, metastasis, metastases, tumoral, melanoma	tumour, tumor, Tprimary, tumors, VHLVon-Hippel-Lindau, tumours, metastatic, metastasis, malignant, EHSEngelbreth-Holm-Swarm

Table 2: 10 Nearest neighbors of selected seed-words.

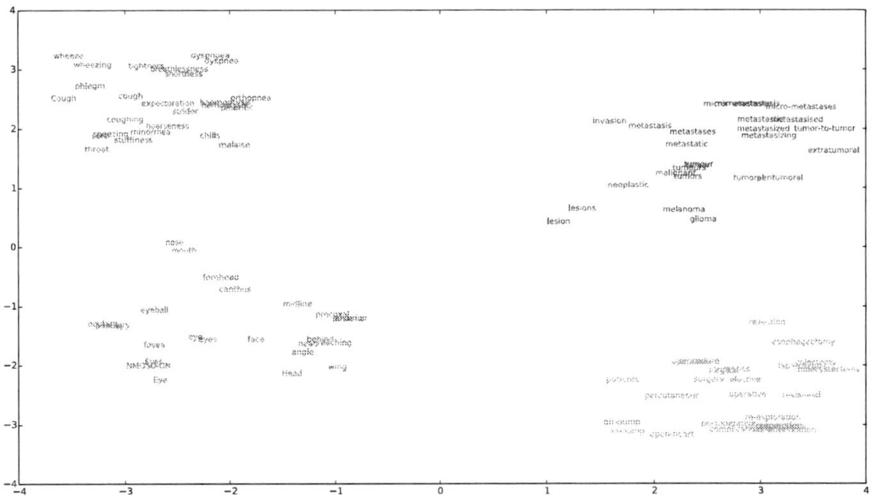

Figure 1: t-SNE projection of 100 biomedical words after applying word2vec skip-gram model. These words are nearest neighbors of the 4 seed-words 'eye', 'cough', 'surgery', and 'tumour'. All nearest neighbors of a particular seed-word are in closer proximity of each other than the nearest-neighbors of other seed-words.

6 Conclusion and Future Work

In this study, we have shown that while *word2vec* with skip-gram model gave the best performance compared to other models in the semantic similarity task, none of the model significantly outperformed others in the semantic relatedness task. Our results indicate that word-vectors should be at least of dimension 200, irrespective of the embedding model. However, further systematic evaluation of all models on more complex NLP tasks, such as medical concept and relation extraction, is required to find out which model will work best.

References

Yoshua Bengio, Réjean Ducharme, Pascal Vincent, and Christian Janvin. 2003. A neural probabilistic language model. *J. Mach. Learn. Res.*, 3:1137–1155, March.

Yoshua Bengio. 2008. Neural net language models. *Scholarpedia*, 3(1):3881.

Ronan Collobert, Jason Weston, Léon Bottou, Michael Karlen, Koray Kavukcuoglu, and Pavel Kuksa. 2011. Natural language processing (almost) from scratch. *J. Mach. Learn. Res.*, 12:2493–2537, November.

Birk Diedenhofen and Jochen Musch. 2015. cocor: A comprehensive solution for the statistical comparison of correlations. *PLoS ONE*, 10(4):e0121945, 04.

Manaal Faruqui and Chris Dyer. 2014. Community evaluation and exchange of word vectors at wordvectors.org. In *Proceedings of ACL: System Demonstrations*.

Geoffrey E Hinton, James L McClelland, and David E Rumelhart. 1986. Distributed representations. In *Parallel distributed processing: Explorations in the microstructure of cognition. Volume 1: Foundations*, pages 77–109. MIT Press.

Omer Levy, Yoav Goldberg, and Ido Dagan. 2015. Improving distributional similarity with lessons learned from word embeddings. *Transactions of the Association for Computational Linguistics*, 3:211–225.

Tomas Mikolov, Kai Chen, Greg Corrado, and Jeffrey Dean. 2013a. Efficient estimation of word representations in vector space. *arXiv preprint arXiv:1301.3781*.

Tomas Mikolov, Ilya Sutskever, Kai Chen, Greg S Corrado, and Jeff Dean. 2013b. Distributed representations of words and phrases and their compositionality. In *Advances in Neural Information Processing Systems*, pages 3111–3119.

José Antonio Miñarro-Giménez, Oscar Marín-Alonso, and Matthias Samwald. 2015. Applying deep learning techniques on medical corpora from the world wide web: a prototypical system and evaluation. *CoRR*, abs/1502.03682.

Andriy Mnih and Geoffrey E Hinton. 2009. A scalable hierarchical distributed language model. In *Advances in neural information processing systems*, pages 1081–1088.

Azadeh Nikfarjam, Abeed Sarker, Karen OConnor, Rachel Ginn, and Graciela Gonzalez. 2015. Pharmacovigilance from social media: mining adverse drug reaction mentions using sequence labeling with word embedding cluster features. *Journal of the American Medical Informatics Association*, page ocu041.

Serguei Pakhomov, Bridget McInnes, Terrence Adam, Ying Liu, Ted Pedersen, and Genevieve B Melton. 2010. Semantic similarity and relatedness between clinical terms: an experimental study. *AMIA annual symposium proceedings*, 2010:572.

Ted Pedersen, Serguei V.S. Pakhomov, Siddharth Patwardhan, and Christopher G. Chute. 2007. Measures of semantic similarity and relatedness in the biomedical domain. *Journal of Biomedical Informatics*, 40(3):288 – 299.

Jeffrey Pennington, Richard Socher, and Christopher D. Manning. 2014. Glove: Global vectors for word representation. In *Proceedings of EMNLP*.

Richard Socher, Brody Huval, Christopher D. Manning, and Andrew Y. Ng. 2012. Semantic Compositionality Through Recursive Matrix-Vector Spaces. In *Proceedings of the 2012 Conference on Empirical Methods in Natural Language Processing (EMNLP)*.

Joseph Turian, Lev Ratinov, and Yoshua Bengio. 2010. Word representations: A simple and general method for semi-supervised learning. In *Proceedings of the 48th Annual Meeting of the Association for Computational Linguistics*, ACL '10, pages 384–394, Stroudsburg, PA, USA. Association for Computational Linguistics.

Peter D. Turney and Patrick Pantel. 2010. From frequency to meaning: Vector space models of semantics. *J. Artif. Int. Res.*, 37(1):141–188, January.

L.J.P. van der Maaten and G.E. Hinton. 2008. Visualizing high-dimensional data using t-sne.

Investigating Public Health Surveillance using Twitter

Antonio Jimeno Yepes[◇♠], **Andrew MacKinlay**[◇♠], **Bo Han**[◇]
[◇] IBM Research Australia, Melbourne, VIC, Australia
[♠] Dept. of Computing and Information Systems, University of Melbourne, Australia
{antonio.jimeno,admackin,bohan.ibm}@au1.ibm.com

Abstract

Microblog services such as Twitter are an attractive source of data for public health surveillance, as they avoid the legal and technical obstacles to accessing the more obvious and targeted sources of health information. Only a tiny fraction of tweets may contain useful public health information but in Twitter this is offset by the sheer volume of tweets posted. We present a system which can identify medical named entities in a real-time stream of Twitter posts and determine their geographic locations, as well as preliminary experiments in using this information for health surveillance purposes.

1 Introduction

Public health surveillance (Nsubuga et al., 2006) is the systematic collection, analysis and monitoring of population health for the public good using a variety of tools. For instance, syndromic surveillance (monitoring for symptoms as signatures of diseases) can be used for tracking and early detection of infectious diseases to flag potential outbreaks, assist in disease modelling, or detect cases of biological terrorism. Meanwhile, pharmacovigilance (WHO and others, 2002) can be used to detect adverse effects associated with pharmaceutical products, while statistics on population health and wellbeing can inform governmental health policy. However, to be effective, these applications require large volumes of real-world data on health statistics (such as from hospital records), which are in most cases difficult to access because of privacy regulations and technical challenges.

The proliferation of social media might enable legitimate large scale collection of health information. Users of forums (e.g., PatientsLikeMe) and microblogs (e.g., Twitter), which we focus on here, post health-related messages with varying levels of frequency. These might cover diseases they have, symptoms they have experienced or drugs they have taken. Twitter may have a large enough volume of data to partially make up for its lack of a health-specific focus. Some judiciously-used data is better than no data at all which is often all that can be obtained from health-specific sources. Such information can be leveraged in analytics to provide insights on public health, e.g., for drug safety (Sarker et al., 2015). However, it is still unclear how large a contribution social media could make to population health surveillance.

In this paper, we perform analysis of health related Twitter data for public health surveillance. The large volume of data in Twitter (approximately 5000 posts per second) is the reason it is useful for such tasks, but each of these posts must be examined (in real-time for practical applications) to determine whether is it relevant, and if so, stored for subsequent analysis. Here, we consider a relevant post to be one containing medical named entities, as identified by an in-domain named-entity tagger (Jimeno Yepes et al., 2015) which we run over our entire data-set after applying some pre-filtering heuristics. A second challenge with Twitter is that location information is scarce, with only around 2% of messages containing reliable geographic coordinates (Cheng et al., 2010). Location information is needed, for instance, in syndromic surveillance to identify the possible location of an outbreak. We handle this by adapting and tuning an existing geotagger to augment the tweets with automatically-determined geographic information (Han et al., 2013). We

Proceedings of the 2015 Workshop on Biomedical Natural Language Processing (BioNLP 2015), pages 164–170,
Beijing, China, July 30, 2015.

then analyse the data, by examining the trend of geolocated medical entities in different regions, presenting commonly discussed medical entities in different categories, and identifying salient medical entities and common topics for a given medical entity. Our results show promising outcomes of utilising Twitter data in health surveillance applications and also raise some limitations of using this data. Overall, the contributions of this paper are twofold: (1) it helps us to understand to what extent Twitter data supports public health surveillance and (2) it provides pilot results that indicate future directions to explore when utilising Twitter data for public health.

2 Related Work

Several sources of data have been previously considered for public health surveillance. Biosurveillance has been usually achieved by monitoring emergency department notes (Espino et al., 2004). The data is reliably sourced, however, there are severe issues in processing time and data aggregations when the data is collected from several departments in various forms and with different time latencies. In addition, access to these sensitive electronic health records is also restricted by privacy issues.

Search engine query logs are an abundant source of data for the organisations which own the search engines, and have been exploited in the health realm. Google[1] (Carneiro and Mylonakis, 2009) finds a spatio-temporal correlation between flu-related queries and data from the United States Centers for Disease Control (CDC). Similarly, Yom-Tov and Gabrilovich (2013) have used Yahoo search data to identify adverse-drug reactions. However, since the search logs are not publicly accessible, these methods are only viable for the companies which own the search log data.

An alternative approach is to monitor information from news data. Collier et al. (2008) identified health rumours and compared them to CDC data, however this might be less successful for real time monitoring and less public disease outbreaks, because only large outbreaks of diseases are newsworthy, and they will have some time lag. For health information of individuals, it is more likely to appear in search logs or medical forums (Segura-Bedmar et al., 2014; Metke-Jimenez et al., 2014; Cameron et al., 2013).

Twitter data has also been considered to identify trends in the 2009 swine flu outbreak in the UK that correlated with official data (Lampos and Cristianini, 2010) and to track alcohol consumption (Kershaw et al., 2014) using geolocated tweet data. Some initial work on exploring health topics in Twitter has been previously done (Paul and Dredze, 2011; Paul and Dredze, 2012; Prier et al., 2011; Signorini et al., 2011), showing the presence of health-related information. These systems typically rely on the Twitter API data with location information.

While there has been some work on medical text mining in social media (e.g., identification of relevant tweets for adverse drug events (Nikfarjam et al., 2015)), a critical assessment of performance of current text mining technology has not been performed. In this work, we have taken a closer look into Twitter data for public health surveillance.

3 Methods

Our pipeline for processing and analysing the Twitter stream is represented in Figure 1. Medical named entities are identified in tweets and those tweets are then geotagged if they do not contain accurate GPS labels. From the large volume of source Twitter data, this yields a much smaller number of tweets containing of medical named entities along with geographical information. This smaller data set is then stored in a MongoDB[2] document database for querying and filtering.

3.1 Micromed: medical NER for Twitter

We have developed a medical named entity recogniser, named `Micromed` (Jimeno Yepes et al., 2015), which uses supervised learning to recognise three types of entities: diseases, symptoms and pharmacological substances.[3] It uses a linear-chain CRF (condi-

[1]Google Flu Trends: `http://www.google.org/flutrends`

[2]`https://www.mongodb.org`

[3]For performance reasons the CRF implementation used here was different to the original system and no POS-based features were used, resulting in a roughly

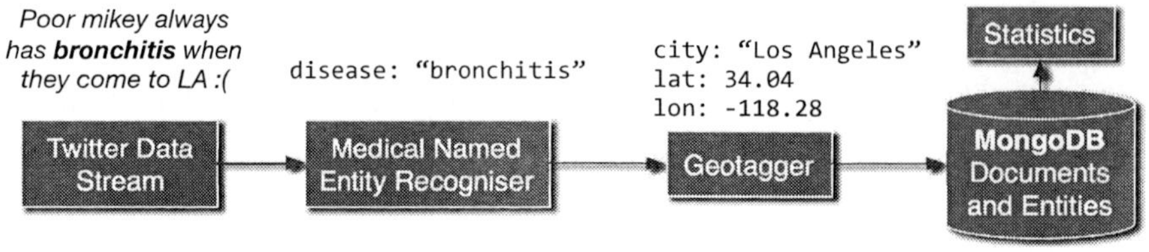

Figure 1: Annotation pipeline

tional random field) (Lafferty et al., 2001), and is trained on a publicly available[4] set of 1300 tweets which have been manually annotated with relevant medical entities. The three entity types are correspond with entries in the Unified Medical Language System (UMLS) (Bodenreider, 2004) Semantic types – specifically *T047 (Diseases or Syndrome)* for diseases, *T184 (Sign or Symptom)* for symptoms and *T121 (Pharmacologic Substance)* for pharmacologic substances. Table 1 shows the performance of `Micromed` on our annotated set for exact matching of the boundaries of the entities, which outperforms systems like MetaMap (Aronson and Lang, 2010) or Stanford NER (Finkel et al., 2005). A comparison is available in (Jimeno Yepes et al., 2015).

Entity Type	Precision	Recall	F1
Disease	0.7987	0.5020	0.6165
Pharm.Subs.	0.8142	0.3948	0.5318
Symptom	0.7193	0.6028	0.6559

Table 1: `Micromed` performance evaluated using 13-fold cross-validation

3.2 Geotagger

To obtain geolocation information for the vast majority of tweets, we adapted and tuned an off-the-shelf geotagger `LIW-META` (Han et al., 2013). `LIW-META` leverages location indicative words to infer geolocations for tweets which lack GPS labels. It applies various feature selection methods to extract words associated with particular locations. Both explicit gazetted terms (such as city and country names) and implicit location-indicative words (such as local landmarks, sport teams and dialectal terms) are extracted and used in modelling taggers. Additionally, it also exploits user profile data such as user-declared locations and time zone information in a stacking framework to enhance the prediction accuracy (Han et al., 2014).

3.3 Twitter data set

We used all of the tweets from 2014[5] obtained from GNIP Decahose,[6] which provides 10% of tweets randomly selected from Twitter. In a pre-filtering step, we remove the 33.5% of posts marked as retweets (which are less interesting for our use cases) and the 70.5% that were marked as non-English (which our tagger is not designed for). The remaining tweets (23.3% of the tweets in the GNIP decahose overall) are processed using the pipeline in Figure 1 and stored if a medical entity was found.

4 Results

In this section, we explore the tweets that contain medical entities to understand what information it might be possible to extract from them. We first have a closer look at the medical entities extracted by `Micromed` and the extended coverage obtained from the geotagger. The coverage of `LIW-META` is further displayed showing statistics for several large cities.

4.1 Medical entities

The statistics for the number of tweets at each phase of the pipeline are summarised in Table 2. 27 million tweets had at least one medical entity, corresponding to 1.0 tweets per second (83k tweets per day) from the GNIP decahose, which would correspond to 10 tweets per second on the full live Twitter stream. Unsurprisingly, this proportion containing medical information is only a small fraction (around 0.2%) of the tweets in the Decahose stream.

1.5% drop in F-score

[4]`https://github.com/IBMMRL/medinfo2015`

[5]Apart from a gap from February 25 to March 22 in our dataset

[6]`https://gnip.com/sources/twitter`

Stage	Total	Per day	Kept
Decahose	$12{,}000\times10^6$	$36{,}254\times10^3$	–
Pre-filtered	$2{,}800\times10^6$	$8{,}459\times10^3$	23.3%
Medical	28×10^6	83×10^3	1%

Table 2: Statistics for tweet numbers initially, pre-filtered (removing non-En and retweets) and discarding tweets without medical entities

We have listed the most frequent annotated entities for each type in Table 3. Some entries are not particularly surprising: substances like *marijuana* or *caffeine*) and symptoms like *tired* or *hungry* are likely to be reflective of the frequency of people using or experiencing these. However diseases such as *heart attack* are less likely to indicate actually occurrences of that disease. Since the volume of tweets with medical entities makes it difficult to interpret the context of the entities mentioned, we have used the MALLET (McCallum, 2002) implementation of topic modelling (Blei et al., 2003) to group the tweets by topic.

Table 4 shows 5 topics for *heart attack*. Except for topic 3, related to the memory of people who suffered the disease, in most cases the use of the term seems to have a figurative connotation related to excitement, which indicates that additional work is required to identify tweets to discard figurative terms (and possibly historical events).

Table 5 shows the topics for *marijuana*. In most cases, the topics are related to legalisation of marijuana in the USA. Whether this has a correlation with actual usage rates, and thus potential impact in public policy for example, requires further investigation.

Topics for entity *tired* are shown in Table 6. In some topics, *tired* seems to be used figuratively to express being bored or impatient. Again, the ability to accurately identify figurative uses of terms could be valuable.

4.2 Geolocation

Location information for each tweet is needed, for instance, to identify the location of an outbreak. Overall, 4.8% of tweets come with GPS labels in our English GNIP collection. Not all tweets are equally predictable so we have calibrated LIW-META by selectively choosing reliable prediction indicators. We tested whether the overall prediction is more reliable when its sub-predictions agree with each other and we found that the overall prediction is more accurate when it agrees with predictions based on user declared locations. This calibrated setting achieves 0.938 precision and 0.214 recall using all geotagged tweet data for evaluation. Our Twitter set offers 0.6 million GPS-labelled tweets while Twitter + LIW-META generates 8.9 million tagging results.

4.3 Geotagged tweets with medical entities

The subset of tweets containing medical entities have been enhanced with location information from the geotagger. Figure 2 shows the number of tweets for three large cities (New York City, London and Chicago) during part of the first half of 2014. The geotagger used here significantly increases the number of health-related tweets that can be identified belonging to these large cities.

5 Discussion

From the large number of tweets being posted every second, just a small fraction of 0.2% (10 per second) contain medical terms. Despite this, a large number of tweets still provide relevant health information.

Twitter poses additional challenges compared to traditional NLP in medical literature and clinical text. Many tweets lack standard grammatical structure or possess abbreviations and misspellings (Baldwin et al., 2013). The use of figurative language in Twitter may be more frequent than other domains (it is clearly very common in our data for many of the frequent symptoms and diseases), although it is particularly important to disambiguate this here for most of the proposed used cases. However there are cases in which the context of the entity makes a medical entity seem legitimate to the tagger (e.g. *heart attack*), so additional filtering might be required.

6 Conclusions

This paper augments in-domain NLP tools to extract and analyse medical information in Twitter. We find the overall proportion of tweets with medical entities is small, nonetheless, we are able to harvest a respectable num-

Disease	Frequency	Pharm. Sub.	Frequency	Symptom	Frequency
heart attack	374810	marijuana	379838	tired	5075812
cancer	268988	caffeine	114526	hungry	2885491
diabetes	175992	cannabis	100233	pain	1724314
stroke	161549	heroin	93723	headache	980699
aids	131792	alcohol	64957	stress	947341

Table 3: Most frequent entities annotated by `Micromed` per entity type.

1	love, guy, put, feel, direction, knew, mtvhottest, heart, https, line
2	phone, mini, dropped, alarm, drop, screen, show, fire, case, find
3	dad, died, find, massive, ago, couldn, told, years, today, days
4	heart, attack, read, seconds, reading, part, summer, book, words, min
5	eat, food, eating, bacon, burger, plate, cheese, grill, pizza, ate

Table 4: Top 5 topics for entity *heart attack*

1	http, tv, legalization, live, job, reporter, fight, vending, machine, quit
2	arrested, possession, police, jail, texas, charges, arrest, son, man, officer
3	tax, million, weed, legalizeit, year, shouldbelegal, sales, revenue, taxes, billion
4	legalized, states, bowl, super, legal, legalize, seattle, united, teams, recreational
5	alcohol, marijuana, dangerous, california, worse, difference, safe, decide, tobacco, human

Table 5: Top 5 topics for entity *marijuana*

1	tired, haha, damn, xd, ah, la, tmr, meh, hmm, uh
2	tired, omg, damn, stand, understatement, joke, soooo, social, omfg, soooooo
3	tired, anymore, isn, point, word, part, fight, basically, helping, state
4	don, wanna, feel, sleep, understand, worry, honestly, numb, aware, bothered
5	tired, soo, sleep, damn, gosh, fucken, darn, crabby, frick, aswell

Table 6: Top 5 topics for entity *tired*

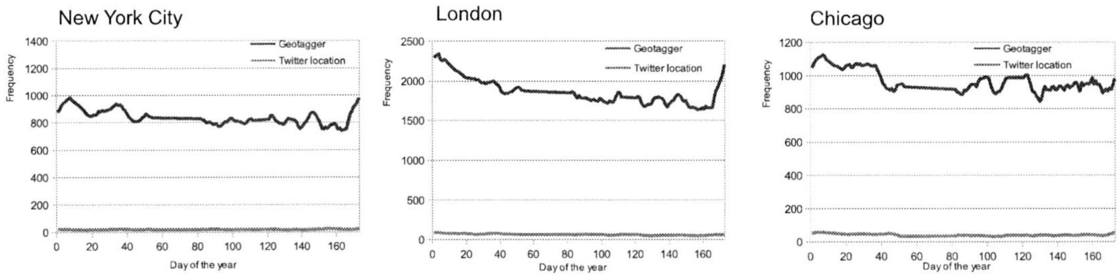

Figure 2: Seven day rolling average of tweets with medical entities count per day for New York city, London and Chicago for January–June 2014

ber of refined medical entities due to the sheer volumes of Twitter data. We extract frequent medical entities in three pre-defined categories, highlight the collocations with entities and investigate topics where an entity is mentioned. By further assigning entities with geographical locations, we can obtain better local medical trend signals which makes public surveillance more plausible. Overall, we have found evidence for the plausibility of public health surveillance using Twitter, although there is much scope to expand on our data analysis in the future.

References

Alan R Aronson and François-Michel Lang. 2010. An overview of metamap: historical perspective and recent advances. *Journal of the American Medical Informatics Association*, 17(3):229–236.

Timothy Baldwin, Paul Cook, Marco Lui, Andrew MacKinlay, and Li Wang. 2013. How noisy social media text, how diffrnt social media sources. In *Proceedings of the Sixth International Joint Conference on Natural Language Processing*, pages 356–364.

David M Blei, Andrew Y Ng, and Michael I Jordan. 2003. Latent dirichlet allocation. *the Journal of machine Learning research*, 3:993–1022.

Olivier Bodenreider. 2004. The unified medical language system (umls): integrating biomedical terminology. *Nucleic acids research*, 32(suppl 1):D267–D270.

Delroy Cameron, Gary A Smith, Raminta Daniulaityte, Amit P Sheth, Drashti Dave, Lu Chen, Gaurish Anand, Robert Carlson, Kera Z Watkins, and Russel Falck. 2013. Predose: A semantic web platform for drug abuse epidemiology using social media. *Journal of biomedical informatics*, 46(6):985–997.

Herman Anthony Carneiro and Eleftherios Mylonakis. 2009. Google trends: a web-based tool for real-time surveillance of disease outbreaks. *Clinical infectious diseases*, 49(10):1557–1564.

Zhiyuan Cheng, James Caverlee, and Kyumin Lee. 2010. You are where you tweet: a content-based approach to geo-locating Twitter users. In *Proceedings of the 19th ACM Conference on Information and Knowledge Management (CIKM 2010)*, pages 759–768, Toronto, Canada.

Nigel Collier, Son Doan, Ai Kawazoe, Reiko Matsuda Goodwin, Mike Conway, Yoshio Tateno, Quoc-Hung Ngo, Dinh Dien, Asanee Kawtrakul, Koichi Takeuchi, et al. 2008. Biocaster: detecting public health rumors with a web-based text mining system. *Bioinformatics*, 24(24):2940–2941.

Jeremy U Espino, Michael M Wagner, Fu-Chang Tsui, H Su, Robert T Olszewski, Z Lie, Wendy Chapman, Xiaoming Zeng, Lili Ma, Z Lu, et al. 2004. The rods open source project: removing a barrier to syndromic surveillance. *Medinfo*, 2004:1192–6.

Jenny Rose Finkel, Trond Grenager, and Christopher Manning. 2005. Incorporating non-local information into information extraction systems by gibbs sampling. In *Proceedings of the 43rd Annual Meeting on Association for Computational Linguistics*, pages 363–370. Association for Computational Linguistics.

Bo Han, Paul Cook, and Timothy Baldwin. 2013. A stacking-based approach to twitter user geolocation prediction. In *Proceedings of the 51st Annual Meeting of the Association for Computational Linguistics (ACL 2013)*, pages 7–12, Sofia, Bulgaria, August.

Bo Han, Paul Cook, and Timothy Baldwin. 2014. Text-based Twitter user geolocation prediction. *Journal Artificial Intelligence Research (JAIR)*, 49:451–500.

Antonio Jimeno Yepes, Andrew MacKinlay, Bo Han, and Qiang Chen. 2015. Identifying Diseases, Drugs and Symptoms in Twitter. *MEDINFO*.

Daniel Kershaw, Matthew Rowe, and Patrick Stacey. 2014. Towards tracking and analysing regional alcohol consumption patterns in the uk through the use of social media. In *Proceedings of the 2014 ACM conference on Web science*, pages 220–228. ACM.

John D. Lafferty, Andrew McCallum, and Fernando C. N. Pereira. 2001. Conditional random fields: Probabilistic models for segmenting and labeling sequence data. In *Proceedings of the Eighteenth International Conference on Machine Learning*, ICML '01, pages 282–289, San Francisco, CA, USA. Morgan Kaufmann Publishers Inc.

Vasileios Lampos and Nello Cristianini. 2010. Tracking the flu pandemic by monitoring the social web. In *Cognitive Information Processing (CIP), 2010 2nd International Workshop on*, pages 411–416. IEEE.

Andrew McCallum. 2002. Mallet: A machine learning for language toolkit.

Alejandro Metke-Jimenez, Sarvnaz Karimi, and Cecile Paris. 2014. Evaluation of text-processing algorithms for adverse drug event extraction from social media. In *Proceedings of the first international workshop on Social media retrieval and analysis*, pages 15–20. ACM.

Azadeh Nikfarjam, Abeed Sarker, Karen O'Connor, Rachel Ginn, and Graciela Gonzalez. 2015. Pharmacovigilance from social media: mining adverse drug reaction mentions using sequence labeling with word embedding cluster features. *Journal of the American Medical Informatics Association*, page ocu041.

Peter Nsubuga, Mark E. White, Stephen B. Thacker, Mark A. Anderson, Stephen B. Blount, Claire V. Broome, Tom M. Chiller, Victoria Espitia, Rubina Imtiaz, Dan Sosin, Donna F. Stroup, Robert V. Tauxe, Maya Vijayaraghavan, and Murray Trostle. 2006. Public health surveillance: a tool for targeting and monitoring interventions. In *Disease Control Priorities in Developing Countries. 2nd edition.* World Bank.

Michael J Paul and Mark Dredze. 2011. You are what you tweet: Analyzing twitter for public health. In *ICWSM*, pages 265–272.

Michael J Paul and Mark Dredze. 2012. A model for mining public health topics from twitter. *Health*, 11:16–6.

Kyle W Prier, Matthew S Smith, Christophe Giraud-Carrier, and Carl L Hanson. 2011. Identifying health-related topics on twitter. In *Social computing, behavioral-cultural modeling and prediction*, pages 18–25. Springer.

Abeed Sarker, Rachel Ginn, Azadeh Nikfarjam, Karen O'Connor, Karen Smith, Swetha Jayaraman, Tejaswi Upadhaya, and Graciela Gonzalez. 2015. Utilizing social media data for pharmacovigilance: A review. *Journal of biomedical informatics*, 54:202–212.

Isabel Segura-Bedmar, Santiago de la Pena, and Paloma Martınez. 2014. Extracting drug indications and adverse drug reactions from spanish health social media. *ACL 2014*, page 98.

Alessio Signorini, Alberto Maria Segre, and Philip M Polgreen. 2011. The use of twitter to track levels of disease activity and public concern in the us during the influenza a h1n1 pandemic. *PloS one*, 6(5):e19467.

WHO et al. 2002. The importance of pharmacovigilance. *Geneva: World Health Organization.*

Elad Yom-Tov and Evgeniy Gabrilovich. 2013. Postmarket drug surveillance without trial costs: discovery of adverse drug reactions through large-scale analysis of web search queries. *Journal of medical Internet research*, 15(6).

Clinical Abbreviation Disambiguation Using Neural Word Embeddings

Yonghui Wu, Jun Xu, Yaoyun Zhang, Hua Xu
School of Biomedical Informatics
The University of Texas Health Science Center at Houston
Houston TX, USA
{Yonghui.wu, Jun.Xu, Yaoyun.Zhang, Hua.Xu}@uth.tmc.edu

Abstract

This study examined the use of neural word embeddings for clinical abbreviation disambiguation, a special case of word sense disambiguation (WSD). We investigated three different methods for deriving word embeddings from a large unlabeled clinical corpus: one existing method called Surrounding based embedding feature (SBE), and two newly developed methods: Left-Right surrounding based embedding feature (LR_SBE) and MAX surrounding based embedding feature (MAX_SBE). We then added these word embeddings as additional features to a Support Vector Machines (SVM) based WSD system. Evaluation using the clinical abbreviation datasets from both the Vanderbilt University and the University of Minnesota showed that neural word embedding features improved the performance of the SVM-based clinical abbreviation disambiguation system. More specifically, the new MAX_SBE method outperformed the other two methods and achieved the state-of-the-art performance on both clinical abbreviation datasets.

1 Introduction

Abbreviations are frequently used in clinical notes and often represent important clinical concepts such as diseases and procedures. However, it is still challenging to handle clinical abbreviations. In a previous study (Wu et al., 2012), we examined three widely used clinical Natural Language Processing (NLP) systems and found that all of them have limited capability to accurately identify clinical abbreviations, especially for ambiguous abbreviations (abbreviations with multiple senses, e.g., "pt" can represent "patient" or "physical therapy"). The prevalence of ambiguous clinical abbreviations is very high. A study (Liu et al., 2001b) examining the abbreviations in the Unified Medical Language System (UMLS) reported that 33.1% of them have more than one sense. In reality, the ambiguity problem of clinical abbreviations could be even higher, as existing knowledge bases (e.g., the UMLS) have low coverage of abbreviations' senses (around 38% to 50%) (Xu, Stetson, et al., 2007).

Clinical abbreviation disambiguation is a particular case of the Word Sense Disambiguation (WSD), which is to "computationally determine which sense of a word is activated by its context" (Navigli, 2009). WSD has been extensively studied in the field of NLP (Lee and Ng, 2002). Researchers have developed different WSD methods including knowledge-based methods (Ponzetto and Navigli, 2010), supervised machine learning methods (Brown et al., 1991) and unsupervised machine learning based methods (Chasin et al., 2014; Yarowsky, 1995) for general English text. As the intrinsic linguistic essentials shared in between, researchers have applied similar methods to biomedical literature and clinical text (Schuemie et al., 2005). For example, researchers have conducted studies to disambiguate important entities in biomedical literature, such as gene names. (Xu, Fan, et al., 2007) Much work has been done for disambiguation of abbreviations in clinical text (Moon et al., 2013; S. Moon et al., 2012; Pakhomov et al., 2005; Wu, Denny, et al., 2013; Xu et al., 2012). Various types of WSD approaches have been proposed for clinical abbreviations, including traditional supervised machine learning based approaches with optimized features (Joshi et al., 2006; Moon et al., 2013; S. Moon et al., 2012), vector space model based methods (Pakhomov et al., 2005; Xu et al., 2012), algorithms based on

Proceedings of the 2015 Workshop on Biomedical Natural Language Processing (BioNLP 2015), pages 171–176,
Beijing, China, July 30, 2015. ©2015 Association for Computational Linguistics

hyper-dimensional computing (Moon et al., 2013), as well as recent unsupervised methods based on topic-modeling-based approaches (Chasin et al., 2014). Furthermore, there is also a study to recognize and disambiguate abbreviations in real-time when physicians are authorizing the notes (Wu, Denny, et al., 2013).

Among all these methods, supervised machine learning methods often show good performances, when annotated corpora are available (Liu et al., 2004). A few studies have proposed methods to automatically generate "pseudo" training corpus from biomedical/clinical text, by replacing the expanded long forms by their corresponding abbreviations (Liu et al., 2001a) (Pakhomov, 2002). In the recent 2013 Share/CLEF challenge on clinical abbreviation normalization (Suominen et al., 2013), a hybrid system developed by our group, which combines the supervised machine learning method, the profile-based method, as well as existing knowledge bases achieved the best performance (Wu, Tang, et al., 2013).

Over the last few years, there has been increasing interest in training word embeddings from large unlabeled corpora using deep neural networks. Word embedding is typically represented as a dense real-valued low dimensional matrix M of size V×D, where V is the vocabulary size and D is the predefined embedding dimension. Each row of the matrix is associated with a word in the vocabulary, and each column of the matrix represents a latent feature. Several neural network based training algorithms have been proposed. Bengio (Bengio et al., 2003) and Mikolov (Mikolov et al., 2013) proposed algorithms to train word embeddings by maximizing the probability of a word given by the previous word. Collobert (Collobert et al., 2011) proposed a neural network to train word embeddings using ranking lost criteria with negative sampling. The experimental results showed that the ranking based word embeddings derived from the entire English Wikipedia corpus greatly improved a number of NLP tasks in the general English text. Previous studies have found that the neural word embeddings could represent abundant semantic meanings in the real-valued matrix, which could be useful features for different NLP tasks including WSD. In 2014, Li et al. (Li et al., 2014) proposed two methods to derive word embedding features for WSD, including the "TF-IDF based Embedding" (TBE) feature, and the "Surrounding Based Embedding" (SBE) feature. The experimental results on the MSH collection data and the WISE collection data showed that the SBE method achieved better performance. In the biomedical domain, Tang et al. (Tang et al., 2013) used the popular word2vec package to generate word embeddings and showed that the word embedding features improved the F1-score of a baseline NER system by 0.49% (from 70.0% to 70.49%).

Nevertheless, there is no study that investigates the use of neural word embeddings for WSD in the medical domain, i.e., clinical abbreviation disambiguation. In this study, we developed two new word embeddings methods to generate WSD features from a large unlabeled clinical corpus. We compared them with the existing SBE method proposed by Li et al. for disambiguation of clinical abbreviations in two datasets from Vanderbilt University and the University of Minnesota. Our results showed that clinical abbreviation disambiguation could benefit from a much larger unlabeled corpus and our newly developed embedding features outperformed the SBE. To the best of our knowledge, this is the first study using the word embeddings trained from a large unlabeled clinical corpus to improve the performance of clinical abbreviation disambiguation methods.

2 Methods

2.1 Datasets

This study used the annotated abbreviation datasets from the Vanderbilt University Hospital's (VUH) admission notes, as well as the clinical notes from the University of Minnesota-affiliated (UMN) Fairview Health Services in the Twin Cities. The VUH dataset contains 25 abbreviations. For each abbreviation, up to 200 sentences containing the abbreviation were randomly selected and manually annotated by domain experts. The UMN dataset contains 75 abbreviations and 500 sentences were randomly selected and annotated for each abbreviation. Detailed information for the two datasets can be found in (Wu, Denny, et al., 2013) and (Sungrim Moon et al., 2012) respectively. In order to train the neural word embeddings, we utilized the unlabeled clinical notes from the Multiparameter Intelligent Monitoring in Intensive Care (MIMIC) II corpus (Saeed et al., 2011). The MIMIC II corpus is composed of 403,871 notes from four different note types, including discharge, radiology, ECG and ECHO. Table 1 shows the detailed information about the three datasets.

Dataset	#ABBR	#Sense	Size
VUH	25	103	4,721 sentences
UMN	75	352	37,500 sentences
MIMIC II	N/A	N/A	403,871 notes

Table 1. Statistics of the two abbreviation datasets and the unlabeled clinical corpus

2.2 Supervised machine learning-based WSD method

In this study, we used Support Vector Machines (SVMs), which is a supervised machine learning algorithm that has achieved state-of-the-art performances on a number of WSD datasets. (Cabezas et al., 2001; Hui et al., 2004; Lee and Ng, 2002) We used the implementation of SVMs in the libsvm package [a]. The details of the SVM-based WSD system can be found in our previous study (Wu, Denny, et al., 2013).

2.3 Conventional features

Previous research has identified a number of useful features for WSD. (Wu, Denny, et al., 2013) In this study, we constructed a baseline SVM-based WSD classifier by including the following proven features for clinical abbreviation disambiguation:

1). Word features - words within a window of the target abbreviation. We used the Snowball Stemmer from the python NLTK (Natural Language Toolkit) package to stem the words;

2). Word feature with direction - The relative direction (left side or right side) of stemmed words in feature set 1 towards the target abbreviation;

3). Position feature - The distance between the feature word and the target abbreviation;

4). Word formation features from the abbreviation itself - include: a) special characters such as "-" and "."; b) features derived from the different combination of numbers and letters; c) the number of uppercase letters.

2.4 Word embedding features

This study proposed two new strategies of deriving distributed WSD features from neural word embeddings, including the "MAX" surrounding based embedding features (MAX_SBE) and the Left-Right surrounding based embedding features (LR_SBE). In addition, we compared the two proposed embedding features with the best embedding features reported by Li et al. in 2014 – the surrounding based embedding (SBE) feature.

Surrounding based embedding feature (SBE)

Li et al. proposed the SBE feature, in 2014. The SBE feature for a target word was derived by aggregating the embedding row vectors of the surrounding words within a predefined window size (***k***), as shown in Equation 1.

$$SBE(w) = \sum_{i=j-k}^{j+k} Emb(S(i)) \quad (1)$$

Where ***w*** is the target word to disambiguate, ***j*** is the index of ***w***, ***S*** is the sentence containing ***w***, ***S(i)*** is the word indexed by position ***i*** in sentence ***S***, and ***k*** is the predefined window size. Previous study from Li et al. showed that the SBE feature achieved the best performance in general English domain.

Left-Right surrounding based embedding feature (LR_SBE)

The LR_SBE is a variation of SBE. Instead of summing up over all of the surrounding word, the LR-SBE composed of the left-side SBE – the SBE from the left-side surrounding words, and the right side SBE – the SBE from the right-side surround words. Previous research has shown that the performance of WSD can be improved by considering the relative word feature with directions (left side or right side). Thus, we assumed that the direction information could help the word embedding feature as well. Equation 2 and 3 show the calculation of LR-SBE embedding features.

$$SBE_{Right}(w) = \sum_{i=j+1}^{j+k} Emb(S(i)) \quad (2)$$

$$SBE_{Left}(w) = \sum_{i=j-k}^{j-1} Emb(S(i)) \quad (3)$$

MAX surrounding based embedding feature (MAX_SBE)

The MAX-SBE feature is generated by taking the MAX score of each embedding dimension over all the surrounding words. As each column of the embedding matrix represents a latent feature, the surrounding words that have a high association with a particular semantic meaning are more likely to have a higher score in a particular latent feature. The intuition of MAX_SBE is that the high-score latent features are more important to describe the word semantics. It is more likely that the WSD performance can be improved by

[a] http://www.csie.ntu.edu.tw/~cjlin/papers/ libsvm.pdf

keeping those high-score latent features over all the surrounding words. Equation 4 shows the calculation of MAX_SBE feature, where $\boldsymbol{Emb}_{\mathrm{j}}$ denotes the $\boldsymbol{j}_{\mathrm{th}}$ dimension of the embedding matrix.

$$MAX\_SBE(w)_j = MAX\{Emb_j(S(i))\}$$
$$w.r.t.\ j-k \le i \le j+k, S(i) \ne w \quad (4)$$

3 Experiments and evaluation

We implemented the neural network based word embedding algorithm from Collobert et al. (Collobert et al., 2011) and trained the word embedding matrix on the unlabeled MIMIC II corpus. We used the suggested parameters to train the neural network with a hidden layer size of 300, a fixed learning rate of 0.01, and an embedding dimension of 50.

For each abbreviation in a dataset, we trained an SVMs model using the conventional features as the baseline, where the model parameters and the window size were optimized by 10-fold cross validation. To reduce the parameter tuning effort, we select a set of unified model parameters for all the abbreviations. To assess the effect of word embedding features, we added each type of word embedding features (SBE, LR_SBE, or MAX_SBE) to the conventional features and then re-trained the SVM classifier using the optimized parameters. We then reported the (Macro) average accuracy across all abbreviations in either the VUH dataset or the UMN dataset based on the results from 10-fold cross validation.

4 Results

Dataset	Features	Average Accuracy (%)
VUH	Baseline (SVMs)	92.19
	+SBE	92.70
	+LR_SBE	92.86
	+MAX_SBE	**93.01**
UMN	Baseline (SVM)	94.97
	+SBE	95.36
	+LR_SBE	95.46
	+MAX_SBE	**95.79**

Table 2. Average accuracy of the WSD systems using different word embedding features on both VUH and UMN datasets

According to 10-fold cross validation, we set the optimized window size of 3 for both datasets. Table 2 shows the macro average accuracy of using different embedding features on the VUH and the UMN abbreviation datasets. The baseline system (SVMs classifier using conventional features) achieved an accuracy of 92.19% and an accuracy of 94.97% on the VUH and the UMN dataset, respectively. The baseline performance on the VUH dataset is lower than that in the UMN dataset. All three types of embedding features (SBE, LR_SBE, and MAX_SBE) improved the average accuracy when compared with the baseline system, with improvements of 0.51%, 0.67, 0.82% for the VUH dataset and 0.39%, 0.49% and 0.82% for the UMN dataset, for SBE, LR_SBE, and MAX_SBE, respectively. We used Wilcoxon test to compare the embedding features. The test results show that the best embedding features in this study (MAX_SBE) outperformed the SBE feature with a significant p-value of 0.004 on the VUH dataset and 7.05e-05 on the UMN dataset.

5 Discussion

This study demonstrates that the word embedding features derived from a large unlabeled corpus could remarkably improve the performance of the SVM-based clinical abbreviation disambiguation system. To the best of our knowledge, this is the first study that investigates the use of neural word embeddings for WSD in clinical text. The most relevant work is a study by Li et al. (Li et al., 2014), where they utilized the algorithm implemented in word2vec to derive embedding features for WSD on a biomedical literature dataset (MSH collection) and a general English dataset (Science WISE dataset). However, the unlabeled dataset used for training the word embedding was relatively small (7,741 abstracts in the MSH dataset and 2,943 abstracts in the WISE dataset), and the proposed WSD method was to directly calculate the cosine similarity. In this study, we proposed two new embedding features and explored a much larger unlabeled clinical corpus (403,871 notes). Our evaluation showed that the proposed LR_SBE feature and the MAX_SBE feature outperformed the SBE feature by Li et al. Among them, the MAX_SBE embedding feature achieved the best average accuracy on both the VUH and UMN datasets, indicating the potential of this new embedding algorithm in WSD tasks.

In fact, all word embedding features improved the performance of the baseline WSD system that uses conventional features only, indicating the usefulness of neural word embeddings in WSD tasks. The LR_SBE feature outperformed the SBE feature, denoting that it is helpful to consider the relative directions even for the real-valued word embedding features. This is consistent with the findings reported in the supervised machine learning based WSD methods using linguistic features. The MAX_SBE feature outperformed the other two types of embedding features, suggesting that the major dimension of the embedding matrix is more powerful for describing semantic meanings. The MAX_SBE word feature is related to the work from Collobert et al., where they designed a MAX convolutional layer in their deep neural network to weight and select the major dimensions among the context words. Our research shows that simply taking the major dimensions from the embedding matrix of context words works well for clinical abbreviation disambiguation.

The neural word embeddings could represent abundant semantic meanings and capture multi-aspect relations from unlabeled corpora, which may generate novel, useful features for various NLP tasks, as demonstrated in the open domain. (Collobert et al., 2011; Li et al., 2014) This study demonstrates its usefulness for clinical abbreviation disambiguation. In addition to WSD, we believe such word embedding features can benefit other NLP tasks in the medical domain.

This study has limitations. The evaluation datasets are composed of the frequently used abbreviations that have enough training samples. For example, the UMN dataset is a balanced dataset that there are exactly 500 samples for each of the abbreviations. We only used the embedding features from the surrounding words, where some semantically important words out off the window were missed. Similar to the study of capturing long distance conventional features, e.g., the syntactic feature, there are possible approaches that can capture long distance features from embedding matrix. Le et al. (Le and Mikolov, 2014) proposed a distributed representation of sentence and documents, which could be a potential solution. In the future, we plan to investigate different approaches that can capture the sentence level distributed representation feature and paragraph level distributed representation feature. We will also examine the word embedding features using deep neural network based classifiers.

6 Conclusion

This paper examined the neural word embedding features for the disambiguation of clinical abbreviations. We proposed two novel word embedding features and compared them with an existing word embedding feature in an SVM-based WSD classifier. Evaluation using two clinical abbreviation datasets showed that all word embedding features derived from a large unlabeled corpus could improve WSD performance, with MAX_SBE achieving the best performance.

7 Acknowledgement

This study was supported by grants from the NLM 2R01LM010681-05, NIGMS 1R01GM103859 and 1R01GM102282. We would like to thank the University of Minnesota and the 2014 SemEval challenge organizers for the development of corpora used in this study.

Reference

Bengio, Y., Ducharme, R., et al. 2003. A neural probabilistic language model. *The Journal of Machine Learning Research*, 3:1137-1155.

Brown, P. F., Pietra, S. A. D., et al. 1991. Word-sense disambiguation using statistical methods. In *Proceedings of the 29th annual meeting on Association for Computational Linguistics*. pp.264-270. Berkeley, California.

Cabezas, C., Resnik, P., et al. 2001. Supervised sense tagging using support vector machines. In *The Proceedings of the Second International Workshop on Evaluating Word Sense Disambiguation Systems*. pp.59-62. Toulouse, France.

Chasin, R., Rumshisky, A., et al. 2014. Word sense disambiguation in the clinical domain: a comparison of knowledge-rich and knowledge-poor unsupervised methods. *J Am Med Inform Assoc*, 21(5):842-849.

Collobert, R., Weston, J., et al. 2011. Natural Language Processing (Almost) from Scratch. *J. Mach. Learn. Res.*, 12:2493-2537.

Hui, H., Giles, L., et al. (2004). Two supervised learning approaches for name disambiguation in author citations. In *Digital Libraries, 2004. Proceedings of the 2004 Joint ACM/IEEE Conference on* (pp.296-305).

Joshi, M., Pakhomov, S., et al. 2006. A comparative study of supervised learning as applied to acronym expansion in clinical reports. *AMIA Annu Symp Proc*:399-403.

Le, Q., and Mikolov, T. (2014). Distributed Representations of Sentences and Documents. In *Proceedings of The 31st International Conference on Machine Learning* (pp.1188-1196).

Lee, Y. K., and Ng, H. T. 2002. An empirical evaluation of knowledge sources and learning algorithms for word sense disambiguation. In *Proceedings of the ACL-02 conference on Empirical methods in natural language processing - Volume 10*. pp.41-48.

Li, C., Ji, L., et al. (2014). Acronym Disambiguation Using Word Embedding. In *Proceedings of the Twenty-Ninth AAAI Conference on Artificial Intelligence*.

Liu, H., Lussier, Y. A., et al. 2001a. Disambiguating ambiguous biomedical terms in biomedical narrative text: an unsupervised method. *J Biomed Inform*, 34(4):249-261.

Liu, H., Lussier, Y. A., et al. 2001b. A study of abbreviations in the UMLS. *Proc AMIA Symp*:393-397.

Liu, H., Teller, V., et al. 2004. A multi-aspect comparison study of supervised word sense disambiguation. *J Am Med Inform Assoc*, 11(4):320-331.

Mikolov, T., Chen, K., et al. 2013. Efficient estimation of word representations in vector space. *arXiv preprint arXiv:1301.3781*.

Moon, S., Berster, B., et al. 2013. Word sense disambiguation of clinical abbreviations with hyperdimensional computing. In *AMIA Annu Symp Proc*.

Moon, S., Pakhomov, S., et al. 2012. *Clinical Abbreviation Sense Inventory*. http://purl.umn.edu/137703

Moon, S., Pakhomov, S., et al. 2012. Automated disambiguation of acronyms and abbreviations in clinical texts: window and training size considerations. *AMIA Annu Symp Proc*, 2012:1310-1319.

Navigli, R. 2009. Word sense disambiguation: A survey. *ACM Comput. Surv.*, 41(2):1-69.

Pakhomov, S. 2002. Semi-supervised Maximum Entropy based approach to acronym and abbreviation normalization in medical texts. In *Proceedings of the 40th Annual Meeting on Association for Computational Linguistics*. pp.160-167. Philadelphia, Pennsylvania.

Pakhomov, S., Pedersen, T., et al. 2005. Abbreviation and acronym disambiguation in clinical discourse. *AMIA Annu Symp Proc*:589-593.

Ponzetto, S. P., and Navigli, R. 2010. Knowledge-rich Word Sense Disambiguation rivaling supervised systems. In *Proceedings of the 48th Annual Meeting of the Association for Computational Linguistics*. pp.1522-1531. Uppsala, Sweden.

Saeed, M., Villarroel, M., et al. 2011. Multiparameter Intelligent Monitoring in Intensive Care II: a public-access intensive care unit database. *Crit Care Med*, 39(5):952-960.

Schuemie, M. J., Kors, J. A., et al. 2005. Word sense disambiguation in the biomedical domain: an overview. *J Comput Biol*, 12(5):554-565.

Suominen, H., Salanterä, S., et al. (2013). Overview of the ShARe/CLEF eHealth Evaluation Lab 2013. In P. Forner, H. Müller, et al. (Eds.), *Information Access Evaluation. Multilinguality, Multimodality, and Visualization* (Vol. 8138, pp.212-231): Springer Berlin Heidelberg.

Tang, B., Cao, H., et al. 2013. Recognizing clinical entities in hospital discharge summaries using Structural Support Vector Machines with word representation features. *BMC Medical Informatics and Decision Making*, 13(Suppl 1):S1.

Wu, Y., Denny, J. C., et al. (2013). A prototype application for real-time recognition and disambiguation of clinical abbreviations. In *Proceedings of the 7th international workshop on Data and text mining in biomedical informatics* (pp.7-8). San Francisco, California, USA: ACM.

Wu, Y., Denny, J. C., et al. 2012. A comparative study of current Clinical Natural Language Processing systems on handling abbreviations in discharge summaries. *AMIA Annu Symp Proc*, 2012:997-1003.

Wu, Y., Tang, B., et al. (2013). Clinical Acronym/Abbreviation Normalization using a Hybrid Approach. In *Proceedings of CLEF 2013*.

Xu, H., Fan, J. W., et al. 2007. Gene symbol disambiguation using knowledge-based profiles. *Bioinformatics*, 23(8):1015-1022.

Xu, H., Stetson, P. D., et al. 2007. A study of abbreviations in clinical notes. *AMIA Annu Symp Proc*:821-825.

Xu, H., Stetson, P. D., et al. 2012. Combining corpus-derived sense profiles with estimated frequency information to disambiguate clinical abbreviations. *AMIA Annu Symp Proc*, 2012:1004-1013.

Yarowsky, D. 1995. Unsupervised word sense disambiguation rivaling supervised methods. In *Proceedings of the 33rd annual meeting on Association for Computational Linguistics*. pp.189-196. Cambridge, Massachusetts.

Representing Clinical Diagnostic Criteria in Quality Data Model Using Natural Language Processing

Na Hong
Mayo Clinic/ Rochester, MN, USA
Institute of Medical Information, Chinese Academy of Medical Sciences/ Beijing, China
Hong.na@mayo.edu

Dingcheng Li
Mayo Clinic/ Rochester, MN, USA
Li.dingcheng@mayo.edu

Yue Yu
Mayo Clinic/ Rochester, MN, USA
School of Public Health, Jilin University/ Changchun, China
Yu.yue@mayo.edu

Hongfang Liu
Mayo Clinic/ Rochester, MN, USA
Liu.Hongfang@mayo.edu

Christopher G. Chute
Johns Hopkins University/ Baltimore, MD, USA
chute@jhu.edu

Guoqian Jiang
Mayo Clinic/ Rochester, MN, USA
Jiang.guoqian@mayo.edu

Abstract

Constructing standard and computable clinical diagnostic criteria is an important and challenging research area in clinical informatics community. In this study, we present our framework and methods for representing clinical diagnostic criteria in Quality Data Model (QDM) using natural language processing (NLP) technologies. We used a clinical NLP tool known as cTAKES for preprocessing of textual diagnostic criteria. We created mappings between cTAKES type system and QDM elements in both datatype and data levels. We evaluated the performance of our NLP-based approach by annotating 218 individual diagnostic criteria in the categories of Symptom and Laboratory Test. In conclusion, our NLP-based approach is a feasible solution in developing diagnostic criteria representation and computerization.

1 Introduction

The term *diagnostic criteria* designates the specific combination of signs, symptoms, and test results that the clinician uses to attempt to determine the correct diagnosis[1]. It is one kind of the most valuable sources of knowledge for supporting clinical decision-making and improving patient care (Yager and Mcintyre, 2014). Diagnostic criteria are a critical evidence resource of clinical decision support system; however, diagnostic criteria are usually described without uniform standard, scattered over different media such as medical textbooks, literatures and clinical practice guidelines, and mostly in free text formats. Several methods based on natural language processing (NLP) technology have been reported and used in structuring free-text-based clinical guidelines, clinical notes and electronic health records (EHRs), as (Rea, etc., 2012) and (Ohno-Machado, etc., 2013). However, there are not sufficient researches on using NLP-based approaches to support the formalization of free-text diagnostic criteria. To achieve computable diagnostic criteria, we consider that a computable model to represent diagnosis criteria and the use of clinical NLP applications to support the modeling are two essential research areas.

Current efforts on development of international recommendation standard models in clinical domains have laid the foundation for modeling and representing computable diagnostic criteria. National Quality Forum (NQF) Quality Data Model (QDM) (Quality Data Model, 2015) as an information model that describes clinical concepts in a standardized format. It allows quality measure developers and many clinical researchers or performers to describe clearly and unambiguously the data required to calculate the performance measure. QDM is designed with the purpose to

[1] https://en.wikipedia.org/wiki/Medical_diagnosis#Diagnostic_criteria

Proceedings of the 2015 Workshop on Biomedical Natural Language Processing (BioNLP 2015), pages 177–182,
Beijing, China, July 30, 2015.

allowing EHRs (Li, etc., 2012) and other clinical electronic systems to share a common understanding and interpretation of the clinical data. In a previous study, researcher Jiang (2015) evaluated the application feasibility of QDM through a data-driven approach and demonstrated that the use of QDM is feasible in building a standards-based information model for representing computable diagnostic criteria.

On clinical NLP studies, many NLP tools currently are applied in the clinical unstructured free text processing and also support terminology annotation, such as Health Information Text Extraction tool (HITex)[2], MetaMap (Aronson and Lang, 2010), OpenNLP[3] and Clinical Text Analysis and Knowledge Extraction System (cTAKES) (Savova, etc., 2010). Some studies compared the performance of the frequently used NLP tools, and the results showed that cTAKES scored best in both performance and usability. cTAKES is an open source Apache project and it is a NLP system for extraction of information from electronic medical record clinical free-text. cTAKES was built on the Unstructured Information Management Architecture (UIMA) framework which is an open source framework designed by IBM and a series of comprehensive NLP methods (Bruce, 2012). In this study, we use cTAKES as a NLP tool to support the formalization of diagnostic criteria.

The objective of our study is to describe our efforts in developing a semi-automatic approach using NLP to facilitate the representation of clinical diagnostic criteria in QDM.

2 Materials & Methods

2.1 Materials

cTAKES: The components of cTAKES are specifically trained for the clinical domain, and create rich linguistic and semantic annotations that can be utilized by clinical decision support systems and clinical research [4]. cTAKES discovers clinical named entities and clinical events using a dictionary lookup algorithm and a subset of the Unified Medical Language System (UMLS) [5], mainly including the following mentions: disease/disorders, sign/symptoms, medications, anatomical sites and procedures. Besides, cTAKES extract named entity attributes and assigns values for the attributes such as UMLS concept unique identifiers (CUIs) and SNOMED CT codes, polarity, uncertainty, conditional, etc. In this study, we used the cTAKES version 3.2.1.

NQF QDM: The QDM consists of criteria for data elements, relationships for relating data element criteria to each other, and functions for filtering criteria to the subset of data elements that are of interest[6]. The basic components of the QDM include: category (e.g., Symptom), datatype (e.g., Symptom, Active), attribute (e.g., information about severity, start Datetime, stop Datetime, and ordinality), and value set comprising concept codes from one or more code systems. In this study, we used the QDM version 4.1 (Quality Data Model, 2015).

2.2 Methods

Figure 1 shows a framework we designed for the NLP-supported QDM modeling of diagnostic criteria. The framework comprises three modules. The first module is an NLP annotation module. We use cTAKES as a NLP tool to support structured representation of diagnostic criteria. The second module is a data model transformation between cTAKES type system and QDM elements. The transformation is supported using both manual mapping strategies and machine learning algorithms. The third module is a unified web interface for human review. As the output, all collected data elements, value sets and logic expressions of diagnostic criteria are formalized by using QDM-based standard representation.

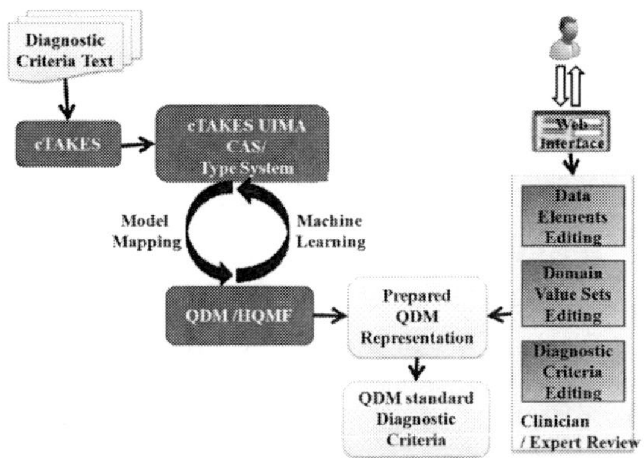

Figure 1. A framework for the NLP-supported QDM modeling of diagnostic criteria.

[2] https://www.i2b2.org/software/projects/hitex/hitex_manual.html.

[3] https://opennlp.apache.org/

[4] http://en.wikipedia.org/wiki/CTAKES

[5] http://www.nlm.nih.gov/research/umls/

[6] http://www.healthit.gov/quality-data-model.

2.2.1 NLP annotation

We first used the cTAKES to perform NLP annotation on textual diagnostic criteria. cTAKES is a modular system of pipelined components combining rule-based and machine learning techniques, introduced in (Savova, Masanz, etc., 2010). As an operable interface, UIMA provides the tooling for selecting which descriptors are used together and determining the order of the descriptors, see detail in (cTAKES 3.2 Component Use Guide, 2015). Dictionaries such as UMLS, SNOMED CT and RxNorm are integrated into cTAKES clinical pipeline.

2.2.2 Data Model Transformation

We implemented the model mapping and data transformation on two levels: the datatype-level mapping and the data-level mapping.

(1) Datatype-level Mapping

We created the datatype-level mappings between cTAKES UIMA Common Analysis System (CAS) type system and QDM datatypes, as well as corresponding attributes and features between these two heterogeneous schemas (Figure 2). We established the mapping relations through analyzing the textual definitions of datatypes in both models. Datatype-level mappings are mainly focused on 7 selected QDM datatypes and 8 cTAKES types that frequently appear in diagnostic criteria.

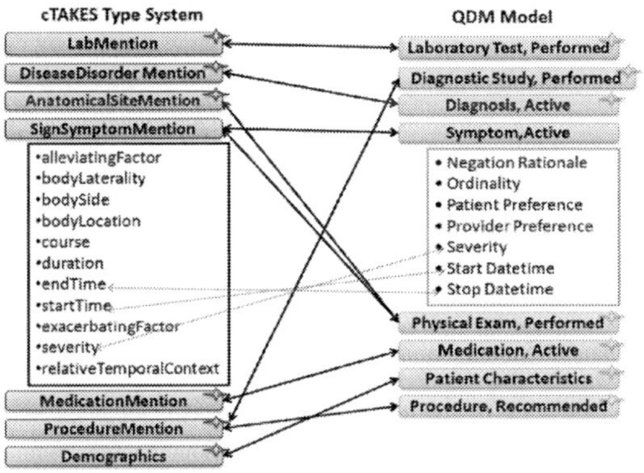

Figure 2. Datatype-level Mapping between cTAKES type system and QDM elements

(2) Data-level Mapping

The second level of mapping analysis is the data-level mappings which are created between the data structure of cTAKES CAS and QDM Health Quality Measure Format (HQMF) (HQMF Templates for QDM December 2013, 2015). The cTAKES processes text and stores the results in the UIMA-CAS structure, whereas the HQMF as a standard format is used to represent QDM-based eMeasure data. All cTAKES instance data output as CAS XML data and are converted into HQMF XML data using the data-level mapping rules.

Figure 3 illustrates the data-level mapping rules between CAS and HQMF elements we created for QDM datatype <u>Laboratory Test, Performed.</u>

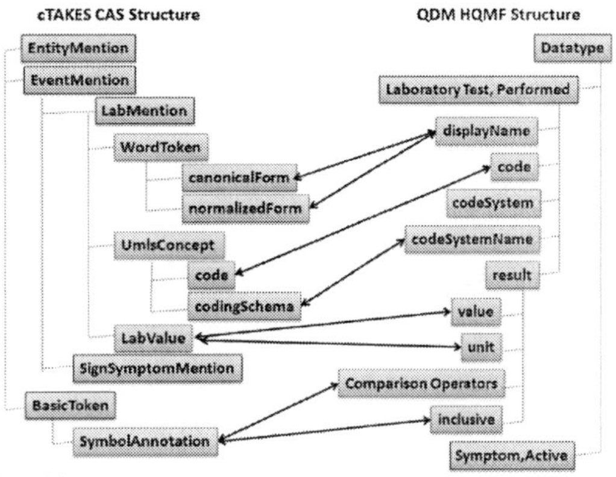

Figure 3. Data-level mappings between cTAKES CAS XML and QDM HQMF XML

2.2.3 Evaluation

For evaluation, we manually annotated a collection of individual criteria with QDM datatypes and attributes. We used the manual annotations as gold standard and evaluated the performance of NLP-based annotations. Two authors (HN, GJ) reviewed the annotations and the consensus was resolved through discussions. Three standard measures were used to describe the performance of the NLP module: precision, recall and F-measure.

3 Results

To implement experiment and evaluation, we first collected 218 individual criteria in the Symptom and Laboratory Test categories. The individual criteria were extracted manually from the text of 44 diagnostic criteria in 13 different clinical topics (an example of textual diagnostic criteria is shown in Appendix A). All the diagnostic criteria are collected from a number of sources including medical textbooks, journal papers, documents issued by professional organization (such as the World Health Organization - WHO) and Internet. Table 1 shows the number of individual criteria of the 13 clinical topics. We used a cTAKES (V.3.2.1) NLP analysis engine known as the <u>*AggregatePlaintxtUMLSProcessor*</u> and processed the test criteria. Using the datatype-level mapping rules we created and the cTAKES annotation results of two EventMentions (LabMention and SignSymptomMention), our algorithm automatically allocated a QDM datatype for each individual diagnostic criteria.

Table 1 Clinical Topics Distribution of 218 individual criteria.

Clinical Topic	Number of Symptom Criteria	Number of laboratory test Criteria
Allergy	6	0
Cardiology	45	15
Critical Care	37	26
Dermatology	18	5
Diabetes	2	9
Endocrinology and Metabolism	27	5
Gastroenterology	1	2
Hematology	0	4
Immunology	2	3
Infectious Disease	0	4
Nephrology	0	4
Neurology	1	1
orthopedics	1	0
Total	140	78

The allocation results could reflect the automatic mapping classification performance for QDM datatypes (Laboratory Test, Performed and Symptom, Active). For example, Figure 4 and Figure 5 show the text of two individual diagnostic criteria with CTAKES annotations in *LabMention and SignSymptomMention.*

Example1: Thrombocytopenia (platelets <100,000 cells/mm3)

Thrombocytopenia (platelets <100,000 cells/mm3)

Figure 4. "platelets" is highlighted as an cTAKES annotation in *LabMention*

Example2: Gastrointestinal-hepatic dysfunction: Moderate (diarrhea, nausea/vomiting, abdominal pain)

Gastrointestinal hepatic dysfunction:
Moderate (diarrhea, nausea/vomiting, abdominal pain)

Figure 5. "diarrhea, nausea, vomiting, abdominal pain" are highlighted as the cTAKES annotations in *SignSymptomMention.*

After automatically mapping based on our mapping rules between two systems, diagnostic criteria in free-text are transformed into QDM based HQML XML structure. One of the QDM data examples attached in the Appendix B. Table 2 shows the evaluation results in terms of whether mapping rules correctly allocate a QDM datatype for an individual criterion.

To evaluate the performance of data-level mappings, we tested the mapping results of 78 individual diagnostic criteria which were annotated manually using the *QDM datatype Laboratory Test, Performed*. The test was mainly focused on the mapping performance of four attribute elements, including code/code system, laboratory test value, measurement and unit, comparison operator. Table 3 shows the evaluation results of elements mapping in the QDM datatype *Laboratory Test, Performed*.

Table 2. Performance of the Datatype-level Mapping Results

QDM Datatype	Laboratory Test, Performed	Symptom, Active
Precision	94.0%	69.2%
Recall	80.8%	59.3%
F-score	86.9%	63.9%

Table 3. Performance of the Data-level Elements Mapping Results (QDM: Laboratory Test, Performed)

QDM Element	Code/ Code System	Value	Unit	Operator
Precision	94.0%	96.3%	100%	61.9%
Recall	80.8%	98.2%	53%	26.5%
F-score	86.9%	97.2%	69.3%	37.1%

4 Discussion

To bridge the semantic gap between cTAKES type system and QDM Model, we performed critical element analysis and created element mappings in both datatype and data levels. As cTAKES UIMA-CAS and QDM both are comprehensive models with independent structures, more semantic analysis need to be studied in order to extend our current mapping rules, e.g., the mapping analysis on QDM temporal representation and cTAKES temporal type. Furthermore, there exist elements that could not be directly mapped between two models under different contexts.

Previous studies investigated the eligibility criteria in clinical trial protocol and developed approaches (known as EliXR) for eligibility criteria extraction and semantic representation, and used hierarchical clustering for dynamic categorization of such criteria (Weng , etc., 2011) (Luo, etc.,2011). In future, we will develop machine learning-based methods leveraging the EliXR approach to enable the analysis for a large amount of clinical diagnostic criteria data.

The study demonstrated overall performance of cTAKES used for generating the QDM-based representation of diagnostic criteria. The evaluation results in Table 2 indicated that criteria in the Laboratory Test category could be automatically classified into the QDM datatype effectively; whereas the performance for classifying criteria in the Symptom category was sub-optimal. The reason is mainly because that cTAKES uses the *SignSymptomMention* that doesn't distinguish between a sign and a symptom. The evaluation results in Table 3 indicated that the code/code system and value mappings could acquire satisfactory performance whereas the performance for the unit annotation is good in precision but sub-optimal in recall. In addition, the operator recognition was insufficient, for examples, in criteria *'Sézary cells with a diameter > 14 um representing > 20% of the circulating lymphocytes'*, '%' is annotated as Symbol but '>' is not recognized in cTAKES that cause low precision. Above all, the mapping rules were able to generate validated QDM datatypes and related elements, covering most typical model elements used in diagnostic criteria. NLP-based technologies could provide a semi-automatic way to support the preliminary classification and enable a pattern-based QDM representation.

5 Conclusion

In this study, we demonstrated that clinical NLP tool (e.g., cTAKES) could support the QDM modeling of free-text diagnostic criteria in a semi-automatic way. We are actively working on developing machine learning algorithms to improve the performance of our NLP-based approaches for representing clinical diagnostic criteria in QDM.

Reference

Joel Yager, John S. Mcintyre. 2014. *DSM-5 Clinical and Public Health Committee: Challenges and Considerations*, American Journal of Psychiatry, 171: 142-44.

Susan Rea, Jyotishman Pathak, Guergana Savova, Thomas A. Oniki, Les Westberg, Calvin E. Beebe, Cui Tao,Craig G. Parker, Peter J. Haug, Stanley M. Huff, Christopher G. Chute. 2012. *Building a robust, scalable and standards-driven infrastructure for secondary use of EHR data: The SHARPn project*, Journal of Biomedical Informatics, 45: 763-71.

Lucila Ohno-Machado, Prakash Nadkarni, Kevin Johnson. 2013. *Natural language processing: algorithms and tools to extract computable information from EHRs and from the biomedical literature*, Journal of the American Medical Informatics Association, 20: 805-05.

Alan R Aronson , François-Michel Lang. 2010. *An overview of MetaMap: historical perspective and recent advances*, Journal of the American Medical Informatics Association, 17: 229-36.

Quality Data Model.[cited May 10, 2015],Available from: http://www.healthit.gov/quality-data-model.

Guoqian Jiang, Harold R. Solbrig, Jyotishman Pathak, Christopher G. Chute. 2015. *Developing a Standards-based Information Model for Representing Computable Diagnostic Criteria: A Feasibility Study of the NQF Quality Data Model*, MedInfo (in submission)

Guergana K. Savova, James J. Masanz, Philip V. Ogren , Jiaping Zheng , Sunghwan Sohn , Karin C. Kipper-Schuler ,Christopher G. Chute. 2010. *Mayo clinical Text Analysis and Knowledge Extraction System (cTAKES): architecture, component evaluation and applications*, J Am Med Inform Assoc, 17: 507-13.

Lars-Erik Bruce. 2012. *Apache UIMA and Mayo cTAKES UIMA and how it is used in the clinical domain.* [cited May 10, 2015]: Available from: http://www.uio.no/studier/emner/matnat/ifi/INF5880/v12/undervisningsmateriale/seminar.pdf

cTAKES 3.2 Component Use Guide.[cited May 10, 2015],Available from: https://cwiki.apache.org/confluence/display/CTAKES/cTAKES+3.2+Component+Use+Guide.

National Quality Forum, HQMF. 2013. *Templates for QDM December.* [cited May 10, 2015]: Available from: http://www.healthit.gov/sites/default/files/qdm_hqmf_templates_dec2013.pdf

Chunhua Weng , Xiaoying Wu , Zhihui Luo , Mary Regina Boland , Dimitri Theodoratos , Stephen B Johnson. 2011. *EliXR: an approach to eligibility criteria extraction and representation*, J Am Med Inform Assoc, 18 Suppl 1: i116-24.

Zhihui Luo, Meliha Yetisgen-Yildiz, Chunhua Weng. 2011. *Dynamic categorization of clinical research eligibility criteria by hierarchical clustering*, J Biomed Inform, 44: 927-35.

Dingcheng Li, Cory M Endle, Sahana Murthy, Craig Stancl, Dale Suesse, Davide Sottara, Stanley M. Huff, Christopher G. Chute, Jyotishman Pathak. 2012. *Modeling and executing electronic health records driven phenotyping algorithms using the NQF Quality Data Model and JBoss(R) Drools Engine*, AMIA Annu Symp Proc, 2012: 532-41.

Appendix

(A) An example of textual diagnostic criteria for diabetes mellitus

1) *A1C ≥ 6.5%. The test should be performed in a laboratory using a method that is NGSP certified and standardized to the DCCT assay.*

OR

2) *FPG ≥ 126 mg/dl (7.0 mmol/l). Fasting is defined as no caloric intake for at least 8 h.*

OR

3) *2-h plasma glucose ≥ 200 mg/dl (11.1mmol/l) during an OGTT. The test should be performed as described by the World Health Organization, using a glucose load containing the equivalent of 75 g anhydrous glucose dissolved in water.*

OR

4) *In a patient with classic symptoms of hyperglycemia or hyperglycemic crisis, a random plasma glucose ≥ 200 mg/dl (11.1 mmol/l).*

In the absence of unequivocal hyperglycemia, criteria 1–3 should be confirmed by repeat testing.

Reference: American Diabetes Association.2012. *Diagnosis and Classification of Diabetes Mellitus*, Diabetes Care. 35(suppl_1):S64-S71.

(B) An QDM representation of diagnostic criterion based on cTAKES annotation

Example text: *Thrombocytopenia (platelets <100,000 cells/mm3)*


```
<sourceOf typeCode="PRCN">
    <conjunctionCode code="AND"/>
      <act classCode="ACT" moodCode="EVN" isCriterionInd="true"><!-- Laboratory Test, Result pattern -->
          <templateId root="2.16.840.1.113883.3.560.1.12"/>
            <id root="c5244e91-3c2e-4863-ae87-a48556b9e3ae"/>
            <code code="30954-2" displayName="Results" codeSystem="2.16.840.1.113883.6.1"/>
            <sourceOf typeCode="COMP">
             <observation classCode="OBS" moodCode="EVN" isCriterionInd="true">
              <code code=" 2.16.840.1.113883.3.117.1.7.1.267 " displayName=" Platelets Count LOINC Value Set"
                 codeSystem="2.16.840.1.113883.3.560.101.1"/>
              <title>Laboratory Test, Result: platelets (result < 100,000  cells/mm3)</title>
              <statusCode code="completed"/>
              <sourceOf typeCode="REFR">
               <observation classCode="OBS" moodCode="EVN" isCriterionInd="true">
                <templateId root="2.16.840.1.113883.3.560.1.1019.3"/>
                <code code="385676005" codeSystem="2.16.840.1.113883.6.96" displayName="result"
                  codeSystemName="SNOMED-CT"/>
                <value xsi:type="IVL_PQ">
                  <high value="100,000" unit="cells/mm3" inclusive="false"/>
                </value>
               </observation>
              </sourceOf>
             </observation>
            </sourceOf>
          </act>
      </sourceOf>
```

Author Index